Mathematics for Engineering

Mathematics for Engineering

W. Bolton

Newnes
An imprint of Butterworth-Heinemann
Linacre House, Jordan Hill, Oxford OX2 8DP
A division of Reed Educational and Professional Publishing Ltd

A member of the Reed Elsevier plc group

OXFORD BOSTON JOHANNESBURG
MELBOURNE NEW DELHI SINGAPORE

First published 1995
Reprinted 1996

British Library Cataloguing in Publication Data
Bolton, W.
1 Mathematics for Engineering
I. Title
510.2462

ISBN 0 7506 2268 7

Library of Congress Cataloging in Publication Data
Bolton, W.
Mathematics for Engineering/W. Bolton
p. cm
Includes index.
ISBN 0 7506 2268 7
1. Engineering mathematics. I Title.
TA 330.B6S 199S
515'-1'02462–dc20

Printed by Martins The Printers, Berwick upon Tweed.

Contents

22 Work

Preface

This book has been written to provide a comprehensive coverage of the mandatory unit in Mathematics for Engineering (Unit 8) for the Advanced GNVQ in Engineering. It aims to develop the mathematical skills essential to engineering at this level. In this text the development of mathematical skills is closely linked to engineering, with each of the four sections of Algebra, Trigonometry, Graphs and Calculus starting with chapters primarily concerned with the development of mathematical skills and followed by chapters in which the mathematical skills are used to develop engineering principles. While the text has been specifically written for the GNVQ unit, it is likely to be of relevance in many other engineering courses where mathematical skills have to be developed from a base which assumes only basic algebra.

The text is divided into four sections:

Section A: Algebra
This covers element 8.1 in the GNVQ unit.

Section B: Trigonometry
This covers elements 8.2 and 8.5 in the GNVQ unit.

Section C: Graphs
This covers element 8.3 in the GNVQ unit.

Section D: Calculus
This covers element 8.4 in the GNVQ unit.

At the beginning of each section the chapters in the section, the aims of the section and the skills required are listed. Each chapter includes many worked examples, revision problems and further problems, answers to all problems being given.

W. Bolton

Section A Algebra

1 Algebraic equations
2 Linear equations
3 Quadratic equations
4 Exponentials
5 Logarithms
6 Straight-line motion
7 Electrical circuits

The aims of this section are to enable the reader to:

- Manipulate expressions involving fractions and powers.
- Manipulate algebraic equations in order to solve engineering problems.
- Substitute values into algebraic equations and obtain solutions.
- Solve simultaneous linear equations.
- Solve quadratic equations.
- Manipulate and solve equations involving logarithmic and exponential expressions.
- Manipulate and solve equations involved in electrical circuit analysis and mechanics problems.

This section is about developing the skills to use algebra in the solution of engineering problems. It is assumed that the reader is not completely unfamiliar with algebra and can handle fractions and powers, though revision is included in chapter 1. Chapter 12 on graphs might well be studied together with section A since it gives a graphical interpretation of what is given an algebraic interpretation in section A.

Chapter 1 is the basic chapter for the section, revising basic concepts of algebra and applying them to the solution of equations. Chapters 2, 3, 4 and 5 all assume the concepts given in chapter 1. Chapter 4 assumes a basic knowledge of graphs. Chapter 5 assumes that chapter 4 has been completed. Chapter 6 shows the application of chapters 1, 2 and 3 in a study of straight-line motion. Chapter 7 shows the application of chapters 1, 2 and 3 in a study of basic electrical circuit analysis.

1 Algebraic equations

1.1 Introduction

In the study of motion in a straight line with a constant acceleration we have the equation

$$v = u + at$$

Given v, u and t can you find a value for a? In the analysis of an electrical circuit we have the equation for resistors in parallel

$$\frac{1}{R} = \frac{1}{R_1} + \frac{1}{R_2}$$

Given values for R_1 and R_2 can you find a value for R? A study of the oscillation of a simple pendulum gives the equation

$$T = 2\pi\sqrt{\frac{L}{g}}$$

Given values of T, π and L can you find a value for g?

This chapter is about the basic techniques that can be used in manipulating algebraic expressions and equations. Chapters 2, 3, 4 and 5 are concerned with more complex equations. Chapters 6 and 7 illustrate the use of these techniques in a consideration of straight-line motion and electrical circuit analysis.

1.1.1 The use of letters

In algebra *letters* are used to represent quantities. Thus in $V = IR$, the V represents the value of the voltage across a resistor with a resistance value of R and I the value of the current through it. The quantities represented by the letters might vary or be constant. Thus in the equation $V = IR$ we might have a constant value for R but V and I can vary.

Letters are also used to represent units. Thus we might have a voltage of two volts and write this as 2 V. The V here does not represent a quantity but is just a shorthand way of writing the unit of voltage. In textbooks, units are written as V but when the letter is used to represent a quantity it is written in italics as V.

1.2 Manipulating fractions

The term on the top of a fraction is called the *numerator* and the term on the bottom the *denominator*. If the top and bottom of a fraction are multiplied by the same number then the value of the fraction is unchanged. This is because we are effectively multiplying the fraction by 1. Thus

$$\frac{1}{2} \text{ has the same value as } \frac{1}{2} \times \frac{4}{4}$$

$$\frac{a}{b} \text{ has the same value as } \frac{a}{b} \times \frac{c}{c}$$

1.2.1 Adding or subtracting fractions

Suppose we want to add two fractions, e.g. $\frac{1}{2} + \frac{1}{4}$. Note that the answer is *not* $\frac{1}{6}$. To carry out such an addition both the fractions must have the same denominator. This we can achieve by multiplying the denominator of the $\frac{1}{2}$ by 4 and the denominator of the $\frac{1}{4}$ by 2, both then becoming 8. In order not to alter the value of the fraction we need to multiply the top and bottom of each by the same number, hence

$$\frac{1}{2} + \frac{1}{4} \text{ has the same value as } \frac{1}{2} \times \frac{4}{4} + \frac{1}{4} \times \frac{2}{2} \text{ or } \frac{1 \times 4 + 1 \times 2}{2 \times 4} = \frac{6}{8}$$

$$\frac{a}{b} + \frac{c}{d} \text{ has the same value as } \frac{a}{b} \times \frac{d}{d} + \frac{c}{d} \times \frac{b}{b} \text{ or } \frac{ad + bc}{bd}$$

and for subtraction

$$\frac{a}{b} - \frac{c}{d} \text{ has the same value as } \frac{a}{b} \times \frac{d}{d} - \frac{c}{d} \times \frac{b}{b} \text{ or } \frac{ad - bc}{bd}$$

Example

Simplify the fractions (a) $\frac{1}{2} + \frac{2}{3}$, (b) $\frac{5}{7} - \frac{1}{3}$, (c) $\frac{1}{x} + \frac{2}{y}$.

(a) Simplification, in this case, involves bringing both the fractions to the same denominator of 2×3. We thus have

$$\frac{1}{2} \times \frac{3}{3} + \frac{2}{3} \times \frac{2}{2} \text{ and so } \frac{1 \times 3 + 2 \times 2}{2 \times 3} \text{ or } \frac{7}{6}$$

(b) Bringing both fractions to the denominator of 7×3, we have

$$\frac{5}{7} \times \frac{3}{3} - \frac{1}{3} \times \frac{7}{7} \text{ and so } \frac{5 \times 3 - 1 \times 7}{7 \times 3} \text{ or } \frac{8}{21}$$

(c) Bringing both fractions to the denominator of $x \times y$, we have

$$\frac{1}{x} \times \frac{y}{y} + \frac{2}{y} \times \frac{x}{x} \text{ and so } \frac{y + 2x}{xy}$$

Revision

1 Simplify the following fractions:

(a) $\frac{2}{3} + \frac{1}{6}$, (b) $\frac{1}{5} + \frac{1}{3}$, (c) $\frac{4}{5} - \frac{1}{2}$, (d) $\frac{3}{2} - \frac{1}{3}$, (e) $\frac{1}{4} + \frac{2}{5}$

2 Simplify the following fractions:

$$\text{(a) } \frac{2}{a}+\frac{3}{b}, \text{ (b) } \frac{1}{R_1}+\frac{1}{R_2}, \text{ (c) } \frac{3}{x}-\frac{5}{y}, \text{ (d) } \frac{a}{2}+\frac{b}{3}$$

1.2.2 Multiplication and division of fractions

Suppose we multiply two fractions, e.g. $\frac{1}{2}$ and $\frac{1}{4}$. What we are asking is what is half of a quarter. This can be obtained by multiplying the two numerators to obtain the new numerator and the two denominators to obtain the new denominator, i.e.

$$\frac{1}{2}\times\frac{1}{4} \text{ has the same value as } \frac{1\times 1}{2\times 4}=\frac{1}{8}$$

$$\frac{a}{b}\times\frac{c}{d} \text{ has the same value as } \frac{a\times c}{b\times d}=\frac{ac}{bd}$$

Suppose we divide $\frac{1}{2}$ by $\frac{1}{4}$. What we are asking is how many quarters are in a half, the answer being 2. This is obtained as follows:

$$\frac{1}{2}\div\frac{1}{4} \text{ is } \frac{\frac{1}{2}}{\frac{1}{4}}$$

We can simplify this by multiplying the top and bottom of the fraction by 4 to give

$$\frac{\frac{1}{2}\times 4}{\frac{1}{4}\times 4}=\frac{2}{1}=2$$

In general we can write this operation as the rule

$$\frac{\frac{a}{b}}{\frac{c}{d}}=\frac{a}{b}\times\frac{d}{c}$$

so that effectively we take the fraction that is the denominator, invert it and then multiply the numerator fraction by it.

Example

Simplify the following fractions:

$$\text{(a) } \frac{1}{2}\times\frac{4}{3}, \text{ (b) } \frac{1}{4}\times\frac{1}{3}, \text{ (c) } \frac{1}{x}\times\frac{2}{y}, \text{ (d) } \frac{2}{3}\div\frac{1}{2}, \text{ (e) } \frac{2}{a}\div\frac{1}{b}$$

(a) For this multiplication of two fractions we have

$$\frac{1 \times 4}{2 \times 3} = \frac{4}{6} \text{ or } \frac{2}{3}$$

(b) For this multiplication of two fractions we have

$$\frac{1 \times 1}{4 \times 3} = \frac{1}{12}$$

(c) For this multiplication of two fractions we have

$$\frac{1 \times 2}{x \times y} = \frac{2}{xy}$$

(d) Multiplying the numerator fraction by the inverted denominator fraction gives

$$\frac{\frac{2}{3}}{\frac{1}{2}} = \frac{2}{3} \times \frac{2}{1} = \frac{4}{3}$$

(e) Multiplying the numerator fraction by the inverted denominator fraction gives

$$\frac{\frac{2}{a}}{\frac{1}{b}} = \frac{2}{a} \times \frac{b}{1} = \frac{2b}{a}$$

Revision

3 Simplify the following fractions:

(a) $\frac{1}{5} \times \frac{1}{3}$, (b) $\frac{2}{3} \times \frac{4}{7}$, (c) $\frac{x}{3} \times \frac{1}{5}$, (d) $\frac{a}{4} \times \frac{b}{3}$, (e) $\frac{1}{5} \div \frac{3}{2}$, (f) $\frac{a}{3} \div \frac{b}{4}$,

(g) $\frac{x}{4} \div \frac{x}{7}$

1.3 Manipulating powers

If we have 2×2 then we can represent this as 2^2; likewise $2 \times 2 \times 2$ is 2^3. In general we have, where a represents some number,

a^1 has the value a,
a^2 has the value $a \times a$,
a^3 has the value $a \times a \times a$,
and so on.

Since $2 \times 2 \times 2$ is the same as $2 \times (2 \times 2)$ then 2^3 has the same value as $2^1 \times 2^2$. In general we have as the basic rule for the multiplication of two terms involving powers

a^{p+q} has the same value as $a^p \times a^q$

You may wonder what the value a^0 is. Consider $a^0 \times a^5$. According to the above rule, this must have the value a^{0+5} and so a^5. The only way this can occur is if a^0 has the value 1.

Suppose we have $(ab)^2$. We can write this as $ab \times ab$ and hence it can be written as $a \times a \times b \times b$ and so $(ab)^2 = a^2b^2$. In general

$$(ab)^n = a^nb^n$$

Example

Simplify (a) $3^2 \times 3^4 \times 3^{-1}$, (b) $x^2 \times x^5$, (c) $(x^2y)^3$

(a) Because the terms are multiplied we can add the powers to give 3^{2+4-1} or 3^5.

(b) Because the terms are multiplied we can add the powers to give x^{2+5} or x^7.

(c) Each of the terms in the brackets is cubed and so we can write this as $(x^2)^3 \times (y)^3$ and hence x^6y^3.

Revision

4 Simplify the following:

(a) $2^3 \times 2^4$, (b) $a^2 \times a^5$, (c) $x^2 \times x^4 \times x^3$, (d) $(ab^2)^5$, (e) $(2x)^2 \times (2x)^4$

1.3.1 Negative powers and division

Powers are not restricted to just positive numbers, we can also have powers with negative numbers. Suppose we now have $\frac{1}{2} \times 2 \times 2 \times 2$. This has the value 4. What we have is $\frac{1}{2} \times 2^3 = 2^2$. Thus if we write 2^{-1} for $\frac{1}{2}$ then we can use the above rule for the multiplication of powers and have $2^{3-1} = 2^2$. In general we have

$\frac{1}{a}$ has the value a^{-1},

$\frac{1}{a \times a}$ has the value a^{-2},

$\frac{1}{a \times a \times a}$ has the value a^{-3},

and so on.

Thus, in general,

$$\frac{1}{a^n} = a^{-n}$$

where n is some number. As a consequence we can write

$$\frac{a^p}{a^q} = a^p a^{-q} = a^{p-q}$$

Note that

$$\left(\frac{a}{b}\right)^n = \frac{a^n}{b^n}$$

Example

Simplify (a) $\frac{a^4}{a^2}$, (b) $\left(\frac{x^2}{y}\right)^3$, (c) $\frac{a^4 \times a^2}{a^3}$

(a) The terms are divided so we subtract the powers. Hence a^{4-2} or a^2.
(b) Both the numerator and denominator are cubed. Hence we have

$$\frac{(x^2)^3}{y^3} \text{ or } \frac{x^6}{y^3} \text{ or } x^6y^{-3}$$

(c) Multiplying the terms in the numerator gives a^6. Hence we have a^{6-3} or a^3.

Revision

5 Simplify the following:

(a) $a^4 \times a^{-2}$, (b) $\frac{2^4}{2^2}$, (c) $\frac{a^5}{a^2 \times a^4}$, (d) $\frac{(2x)^2 \times (3x)^3}{2x^4}$, (e) $2^3 \times 2^{-4}$

1.3.2 Fractional powers

Consider what powers we need to multiply to obtain a^1. We could use the multiplication rule to write $a^1 = a^{1/2}a^{1/2}$. But a can also be written as the product of square roots, i.e. $\sqrt{a} \times \sqrt{a}$. Thus $a^{1/2}$ is the same as the square root of a. We could have written $a^1 = a^{1/3} \times a^{1/3} \times a^{1/3}$ and a as the product of cube roots, i.e. $\sqrt[3]{a} \times \sqrt[3]{a} \times \sqrt[3]{a}$. Thus $a^{1/3}$ is the cube root of a. If you are not sure of these relationships, try some number in place of a. For example, $4^1 = \sqrt{4} \times \sqrt{4}$, i.e. 2×2 and $8^1 = \sqrt[3]{8} \times \sqrt[3]{8} \times \sqrt[3]{8}$ or $2 \times 2 \times 2$.

Example

Simplify the following:

(a) $2^{1/2} \times 2^2$, (b) $a^{-1/2} \times a^3$, (c) $\frac{a^2 \times \sqrt{a}}{a^3}$

(a) $2^{1/2} \times 2^2 = 2^{1/2+2} = 2^{5/2}$

(b) $a^{-1/2} \times a^3 = a^{-1/2+3} = a^{5/2}$.

(c) We can write the square root as $a^{1/2}$ and so we have

$$\frac{a^2 \times a^{1/2}}{a^3} = a^{5/2} \times a^{-3} = a^{-1/2} = \frac{1}{\sqrt{a}}$$

Revision

6 Simplify the following:

(a) $a^3 \times a^{1/2}$, (b) $x\sqrt{x}$, (c) $\frac{x^2}{\sqrt{x^3}}$, (d) $2^2 \times \frac{\sqrt{2}}{2^3}$, (e) $a^{5/2}a^2a^{1/3}$, (f) $x^{0.1}x^{1.5}$

1.3.3 Using a calculator

You may have a calculator which can be easily used for the working out of powers. It will have a key labelled x^y. Thus, for example, to use this to obtain 2^4: the procedure could be: press the key for 2, then press the x^y key, then press 4 and finally the key for = to give the answer. In this case 16. To obtain 2^{-4}, the procedure could be: press the key for 2, then press the x^y key, then the key for 4 followed by the +/– key, and finally = to give the answer of 0.0625.

Example

Use a calculator to evaluate (a) 5^3, (b) $3^{5/2}$.

(a) The key sequences are: press 5, press x^y, press 3, press =. The result is 125

(b) The key sequences are: press 3, press x^y, press 2.5, press =. The result is 15.6 (to one decimal place).

Revision

7 Use a calculator to evaluate:

(a) 3^5, (b) 2^{-5}, (c) $5^{0.3}$, (d) $3^{1/3}$, (e) $2^{1.5}$, (f) $3.5^{1.2}$, (g) $1.2^{-0.67}$, (h) $2.7^{-0.5}$

1.4 Manipulating simple equations

The term *equation* is used when there is an exact balance between what is on one side of the equals sign (=) and what is on the other side. Thus with the equation $V = IR$, the numerical value on the left-hand side of the equals sign must be the same as the numerical value on the right-hand side. Thus, if V has the value 2 then the value of IR must have the value 2.

Thus if we have the equation $x + 12 = 20$, then the numerical value of the left-hand side of the equation, i.e. $x + 12$, must be the same as the numerical value on the right-hand side of the equation, i.e. 20. It does not matter what x represents, the expressions on the left and right of the equals sign have the same numerical value. We can then consider what value x

must have if this equality is to be true. This is termed solving an equation. *Solving an equation* means finding the particular numerical value of the unknown which makes the equation balance. Thus, for the above equation, x must have the value 8 for the value of the expression on the left-hand side of the equals sign to equal the value on the right. A basic rule which would have enabled us to determine that result is: *adding the same quantity to, or subtracting the same quantity from, both sides of an equation does not change the equality*. Thus if we subtract 12 from both sides of the above equation, then $x + 12 - 12 = 20 - 12$ and so $x = 20 - 12 = 8$. As a check of the answer obtained, we can always replace x by the value found. The equality must still hold. Thus $8 + 12 = 20$. The equality holds.

Another rule which can be used is: *multiplying, or dividing, both sides of an equation by the same non-zero quantity does not change the equality*. Thus if we have the equation $3x = 18$, then if we divide both sides of the equation by 3 we obtain

$$\frac{3x}{3} = \frac{18}{3}$$

and thus we have $x = 6$. As a check of the answer obtained, we can always replace x by the value found. Thus $3 \times 6 = 18$, the equality still holds.

If we have the equation

$$\frac{2x}{3} = \frac{1}{4}$$

then we can multiply both sides of the equation by 3 to give

$$3 \times \frac{2x}{3} = 3 \times \frac{1}{4}$$

and so

$$2x = \frac{3}{4}$$

If we now divide both sides of the equation by 2 we then obtain

$$\frac{2x}{2} = \frac{\frac{3}{4}}{2} = \frac{3}{8}$$

and so $x = \frac{3}{8}$.

In general, whatever mathematical operation we do to one side of an equation, provided we do the same to the other side of the equation then the balance is not affected. Thus we can manipulate equations without affecting their balance by, for example:

1 Adding the same quantity to both sides of the equation.
2 Subtracting the same quantity from both sides of the equation.
3 Multiplying both sides of the equation by the same quantity.

4 Dividing both sides of the equation by the same quantity.
5 Squaring both sides of the equation.
6 Taking the square roots of both sides of the equation.
7 Taking logarithms of both sides of the equation.

The following examples illustrate some of these manipulations.

Example

Solve the equation $2x - 4 = x + 1$.

If we add 4 to each side of the equation then

$$2x - 4 + 4 = x + 1 + 4$$

and so $2x = x + 5$. If we now subtract x from both sides of the equation

$$2x - x = x + 5 - x$$

and so we have $x = 5$ as the solution. We can check this by replacing x in the original equation by this value. Then we have $2 \times 5 - 4 = 5 + 1$.

Example

Solve the equation $\frac{2}{3}x = 8$.

Multiplying both sides of the equation by 3 gives

$$3 \times \frac{2}{3}x = 3 \times 8$$

and so $2x = 24$. Dividing both sides of the equation by 2 gives

$$\frac{2x}{2} = \frac{24}{2}$$

and so the solution is $x = 12$. We can check this by replacing x in the original equation by this value. Then we have $\frac{2}{3} \times 12 = 8$.

Example

The voltage V across a resistance R is related to the current through it by the equation $V = IR$. When $V = 2$ V then $I = 0.1$ A. What is the value of the resistance?

Writing the equation with the numbers substituted gives $2 = 0.1R$. Multiplying both sides of the equation by 10 gives $2 \times 10 = 10 \times 0.1R$. Hence $20 = 1R$ and so the resistance R is 20 Ω.

Example

Solve the equation $\frac{2x+3}{3} = \frac{x+1}{4}$.

Multiplying both sides of the equation by 3 gives

$$3\left(\frac{2x+3}{3}\right) = 3\left(\frac{x+1}{4}\right)$$

Hence

$$2x + 3 = 3\left(\frac{x+1}{4}\right)$$

Multiplying both sides of the equation by 4 gives

$$4(2x + 3) = 4 \times 3\left(\frac{x+1}{4}\right) = 3(x+1)$$

and so $8x + 12 = 3x + 3$. Subtracting 12 from both sides of the equation gives

$$8x + 12 - 12 = 3x + 3 - 12$$

and so $8x = 3x - 9$. Subtracting $3x$ from both sides of the equation gives

$$8x - 3x = 3x - 3x - 9$$

Hence $5x = -9$ and so $x = -\frac{9}{5}$.

Example

Solve the equation $2 = \sqrt{\frac{L}{10}}$.

Squaring both sides of the equation gives

$$4 = \frac{L}{10}$$

Hence, when both sides are multiplied by 10, we have $L = 40$.

Example

Solve the equation $100 = R^2 + 75$.

Subtracting 75 from each side of the equation gives $25 = R^2$. Taking the square root of both sides of the equation then gives $R = 5$.

Revision

8 Solve the following equations:

(a) $x + 7 = 12$, (b) $2x + 3 = x + 5$, (c) $x + 3 = 4x$, (d) $\frac{1}{2}x = 12$,

(e) $\frac{2}{3}x = 4$, (f) $\frac{1}{2}x + 2 = x + 1$, (g) $\frac{2x}{5} = \frac{12}{15}$, (h) $\frac{x+1}{2} = 4$,

(i) $x + 4 = \frac{1}{2}$, (j) $2x = \frac{5}{3}$, (k) $20 = 5\sqrt{x}$, (l) $4 + x^2 = 20$

9 (a) The voltage V across a resistance R is related to the current through it by the equation $V = IR$. When $V = 10$ V then $I = 2$ A. What is the value of the resistance? The resistance unit will be Ω.
(b) The extension x of a spring is related to the stretching force F by the equation $F = kx$, where k is a constant. When $F = 10$ N then $x = 20$ mm. What is the value of k? The unit of k will be N/mm.
(c) The pressure p exerted by a force F applied over an area A is given by the equation $p = F/A$. When $p = 1000$ N/m^2 and $A = 0.01$ m^2, what is the value of F? The unit of F will be N.
(d) The impedance Z of an electrical circuit is given by $Z^2 = R^2 + X^2$. With $Z = 10$ Ω and $R = 5$ Ω, what is the value of X?

1.4.1 Transposition

There is another way of considering the manipulation operations discussed in section 1.4. Consider again the examples discussed there.

For the equation $x + 12 = 20$, subtracting 12 from each side gives

$$x + 12 - 12 = 20 - 12$$

and so $x = 20 - 12$. Effectively what we have done with this operation is to move the +12 from the left-hand side of the equation to the right-hand side when it becomes −12, as illustrated below This is termed *transposition*.

$$x + 12 = 20$$

$$x \qquad = 20 - 12$$

With the equation

$$\frac{2x}{3} = \frac{1}{4}$$

We can solve this by multiplying both sides of the equation by 3 and dividing both sides by 2 to give

$$x = \frac{3 \times 1}{2 \times 4}$$

We can consider that the operations we have carried out are equivalent to having the 2 in the numerator of the fraction on the left moved across the equals sign and downwards to become the denominator on the right and the 3 moved from the denominator on the left moved across the equals sign to become the numerator on the right. Thus the number that is being used to multiply a quantity on the left-hand side of the equation becomes when transposed to the right-hand side a division. A number that is being used to divide a quantity on the left-hand side of the equation becomes when transposed to the right-hand side a multiplication.

The term *transposition* is used when a quantity is moved from one side of an equation to the other side. The following are basic rules for use with transposition:

1 A quantity which is added on the left hand side of an equation side becomes subtracted on the right hand side.
2 A quantity which is subtracted on the left hand side of an equation becomes added on the right hand side.
3 A quantity which is multiplying on the right-hand side of an equation becomes a dividing quantity on the left-hand side.
4 A quantity which is dividing on the left-hand side of an equation becomes a multiplying quantity on the right-hand side.

Example

Solve the equation $2x - 4 = x + 1$.

This is a repeat of the example in the previous section. Now consider the same example with transposition. We have $2x - 4 = x + 1$. If we move the -4 from the left-hand side to the right-hand side of the equation it becomes $+4$ and we obtain $2x = x + 1 + 4$. If we move the $+x$ from the right-hand side to the left-hand side of the equation it becomes $-x$ and so we have $2x - x = 1 + 4$. Hence, as before, the solution is $x = 5$.

Example

Solve the equation $\frac{2}{3}x = 8$.

This is a repeat of the example in the previous section. Now consider the same example with transposition. We have

$$\tfrac{2}{3}x = 8$$

The multiplication by 2 on the left hand side becomes a division by 2 on the right-hand side. Hence we can write

$$\frac{x}{3} = \frac{8}{2}$$

The division by 3 on the left-hand side of the equation becomes a multiplication by 3 on the right-hand side. Thus $x = \frac{8}{2} \times 3 = 12$.

Example

The voltage V across a resistance R is related to the current through it by the equation $V = IR$. When $V = 2$ V then $I = 0.1$ A. What is the value of the resistance?

This is the example considered in the previous section. Now by transposition, the equation $2 = 0.1R$ has the multiplication by 0.1 on the right-hand side of the equation becoming a division by 0.1 on the left-hand side. Hence

$$\frac{2}{0.1} = R$$

Thus, as before, $R = 20\ \Omega$.

Example

Solve the equation $\dfrac{2x+3}{3} = \dfrac{x+1}{4}$.

This is the example considered in the previous section. Now by transposition, dividing by 3 on the left-hand side of the equation becomes multiplication by 3 on the right-hand side. Hence we have

$$2x + 3 = 3\left(\frac{x+1}{4}\right)$$

Dividing by 4 on the right-hand side of the equation becomes multiplication by 4 on the left-hand side. Hence

$$4(2x + 3) = 3(x + 1)$$

Thus $8x + 12 = 3x + 3$ and so $5x = -9$ and $x = -\frac{9}{5}$.

Revision

10 Solve the equations given in revision problem 8 by the use of transposition.

1.4.2 Manipulation of formulae

Suppose we have the equation $F = kx$ and we want to solve the equation for x in terms of the other quantities. Writing the equation as

$$kx = F$$

then transposing the k from the left-hand side to the right-hand side (or dividing both sides by k) gives

$$x = \frac{F}{k}$$

As another example, consider the following equation for the variation of resistance R of a conductor with temperature t:

$$R_t = R_0(1 + \alpha t)$$

where R_t is the resistance at temperature t, R_0 the resistance at 0°C and α a constant called the temperature coefficient of resistance. We might need to rearrange the equation so that we express α in terms of the other variables. As a first step we can multiply out the brackets to give

$$R_t = R_0 + R_0\alpha t$$

Transposing the R_0 from the left-hand to the right-hand side of the equation gives

$$R_t - R_0 = R_0\alpha t$$

Reversing the sides of the equation so that we have the term involving α on the left-hand side, we then have

$$R_0\alpha t = R_t - R_0$$

Transposing the R_0t from the left-hand to right-hand sides gives

$$\alpha = \frac{R_t - R_0}{R_0 t}$$

Example

Rearrange the following equation for resistances in parallel to obtain R_1 in terms of the other variables:

$$\frac{1}{R_1} + \frac{1}{R_2} = \frac{1}{R}$$

Transposing $1/R_2$ from the left-hand side to the right-hand side of the equation gives

$$\frac{1}{R_1} = \frac{1}{R} - \frac{1}{R_2}$$

Arranging the fractions on the right-hand side of the equation with a common denominator, then

$$\frac{1}{R_1} = \frac{R_2 - R}{RR_2}$$

We can invert the fractions provided we do the same to both sides of the equation. Effectively what we are doing here is transposing the numerators and denominators on both sides of the equation. Thus

$$R_1 = \frac{RR_2}{R_2 - R}$$

Example

The surface area A of a closed cylinder is given by the equation

$$A = 2\pi r^2 + 2\pi rh$$

where r is the radius of the cylinder and h its height. Rearrange the equation to give h in terms of the other variables.

Transposing the $2\pi r^2$ term gives

$$A - 2\pi r^2 = 2\pi rh$$

Reversing the sides the equation gives

$$2\pi rh = A - 2\pi r^2$$

Transposing the $2\pi r$ from the left-hand to right-hand sides gives

$$h = \frac{A - 2\pi r^2}{2\pi r}$$

Example

The power P dissipated by passing a current I through a resistance R is given by the equation $P = I^2R$. Rearrange the equation to give I in terms of the other variables.

Writing the equation as $I^2R = P$ and then transposing the R gives

$$I^2 = \frac{P}{R}$$

Taking the square roots of both sides of the equation then gives

$$I = \sqrt{\frac{P}{R}}$$

Revision

11 Rearrange the following equations to give the indicated variable in terms of the remaining variables:

(a) $v = u + at$, for a, (b) $F_1r_1 = F_2r_2$, for r_1, (c) $\frac{F}{A} = E \times \frac{x}{L}$, for x,

(d) $T^2 = \frac{4\pi^2 L}{g}$, for L, (e) $Z^2 = R^2 + X^2$, for X, (f) $pV = mRT$, for m,

(g) $E = \frac{1}{2}mv^2$, for v, (h) $s = ut + \frac{1}{2}at^2$, for a, (i) $Z = \frac{bd^2}{6}$, for d.

12 The length L of a bar of metal at t°C is related to the length L_0 at 0°C by the equation $L = L_0(1 + \alpha t)$. If the length at 100°C is 1.004 m for a bar of length 1 m at 0°C, what is the value of α?

13 The stress σ acting on a material is related to the force F acting over a cross-sectional area A of the material by the equation

$$\sigma = \frac{F}{A}$$

If the stress is 1000 N/mm^2 and the area 4 mm^2, what is the force?

14 The relationship between the pressure p, the volume V and the temperature T on the kelvin scale can be written as

$$\frac{pV}{T} = mR$$

where m is the mass of the gas and R the characteristic gas constant. What will be the volume of a mass of 2.0 kg of air at a temperature of 293 K and a pressure of 1 100 000 N/m^2 (Pa) if the characteristic gas constant is 287 J/kg K?

1.5 Units

It is not only numerical values that must balance when there is an equation. The units of the quantities must also balance. For example, the area A of a rectangle is the product of the lengths b and w of the sides, i.e. $A = bw$. The units of the area must therefore be the product of the units of b and w if the units on both sides of the equation are to balance. Thus if b and w are both in metres then the unit of area is metre × metre, or square metres (m^2). The equation for the stress σ acting on a body when it is subject to a force F acting on a length of the material with a cross-sectional area A is

$$\sigma = \frac{F}{A}$$

Thus if the units of F are newtons (N) and the area square metres (m^2) then, for the units to be the same on both sides of the equation:

$$\text{unit of stress} = \frac{\text{newtons}}{\text{square metres}} = \text{newtons/square metre}$$

or, using the unit symbols,

$$\text{unit of stress} = \frac{\text{N}}{\text{m}^2} = \text{N/m}^2$$

This unit of N/m^2 is given a special name of the pascal (Pa).

Velocity is distance/time and so the unit of velocity can be written as the unit of distance divided by the unit of time. For the distance in metres and the time in seconds, then the unit of velocity is metres per second, i.e. m/s or m s^{-1}. For the equation describing straight line motion of $v = at$, if the unit of v is metres/second then the unit of the at must be metres per second. Thus, if the unit of t is seconds, we must have

$$\frac{\text{metre}}{\text{second}} = \frac{\text{metre}}{\text{second}^2} \times \text{second}$$

or, in symbols,

$$\frac{\text{m}}{\text{s}} = \frac{\text{m}}{\text{s}^2} \times \text{s}$$

Hence the unit of the acceleration a must be m/s^2 for the units on both sides of the equation to balance.

Example

When a force F acts on a body of mass m it accelerates with an acceleration a given by the equation $F = ma$. If m has the unit of kg and a the unit m/s^2, what is the unit of F?

For equality of units to occur on both sides of the equation

$$\text{unit of } F = \text{kg} \times \frac{\text{m}}{\text{s}^2} = \text{kg m/s}^2$$

This unit is given a special name of the newton (N).

Example

The pressure p due to a column of liquid of height h and density ρ is given by $p = h\rho g$, where g is the acceleration due to gravity. If h has the unit of m, ρ the unit kg/m^3 and g the unit m/s^2, what is the unit of the pressure?

For equality of units to occur on both sides of the equation

$$\text{unit of pressure} = \text{m} \times \frac{\text{kg}}{\text{m}^3} \times \frac{\text{m}}{\text{s}^2} = \frac{\text{kg}}{\text{m} \times \text{s}^2}$$

Note that in the previous example we obtained the unit of newton (N) as being the special name for kg m/s^2. Thus we can write the unit of pressure as

$$\text{unit of pressure} = \frac{\text{kg} \times \text{m}}{\text{s}^2} \times \frac{1}{\text{m}^2} = \text{N/m}^2$$

This unit of N/m^2 is given a special name, the pascal (Pa).

Example

For uniformly accelerated motion in a straight line, the velocity v after a time t is given by $v = u + at$, where u is the initial velocity and a the acceleration. If v has the unit m/s, u the unit m/s and t the unit s, what must be the unit of a?

For the units to balance we must have the unit of each term on the right-hand side of the equation the same as the unit on the left-hand side. Thus the unit of at must be m/s. Hence, since t has the unit of s,

$$\text{unit of } a = \frac{\text{unit of } at}{\text{unit of } t} = \frac{\text{m/s}}{\text{s}} = \frac{\text{m}}{\text{s}^2} = \text{m/s}^2$$

Revision

15 The pressure p acting on a surface of area A when it is subject to a force F is given by $p = F/A$. If A has the unit of m^2 and F the unit of N, what will be the unit of the pressure?

16 The torque T resulting from a force F being applied at the end of a lever arm of radius r is given by $T = Fr$. What is the unit of the torque when F has the unit of N and r the unit of m?

17 The specific heat capacity c of a material is defined by the equation

$$c = \frac{Q}{mt}$$

where Q is the heat transfer needed for a mass m of a material to change its temperature by t. If Q has the unit of joule (J), m the unit of kg and t the unit of kelvin (K), what is the unit of c?

18 The electrical resistivity ρ of a conductor is given by the equation

$$\rho = \frac{RA}{L}$$

where R is the resistance of a strip of length L and cross-sectional area A. What is the unit of resistivity if R has the unit Ω, L the unit m and A the unit m^2?

19 Kinetic energy E is given by the equation $E = \frac{1}{2}mv^2$, where m is the mass in kg, v the velocity in m/s. What will be the unit of E?

Problems

1 Simplify the following fractions:

(a) $\frac{3}{5}+\frac{1}{3}$, (b) $\frac{1}{5}-\frac{1}{6}$, (c) $\frac{4}{7}+\frac{1}{3}$, (d) $\frac{1}{a}+\frac{b}{c}$, (e) $\frac{2}{3}\times\frac{4}{7}$, (f) $\frac{1}{5}\times\frac{2}{3}$,

(g) $\frac{x}{3}\times\frac{1}{y}$, (h) $\frac{3}{4}\div\frac{1}{3}$, (i) $\frac{4}{5}\div\frac{1}{7}$, (j) $\frac{a}{b}\div\frac{5}{4}$, (k) $\dfrac{\frac{1}{a}}{\frac{1}{a}+\frac{1}{b}}$

2 Simplify the following:

(a) $3^2 \times 3^6$, (b) $a^3 \times a^2$, (c) $a^5 \times a^{-2}$, (d) $\frac{a^4}{a^2}$, (e) $(2 \times 3)^2$, (f) $\left(\frac{1}{x^2}\right)^3$,

(g) $a^{1/2}a^2a^3$, (h) $\frac{x\sqrt{x}}{x^3}$, (i) $\frac{a^3}{\sqrt{a}}$, (j) $\sqrt{x^5}$, (k) $(-4)^{-2}$, (l) $2^5 2^{-2} 2^{1/2}$,

(m) $(a^{-1}b^{-2})^{-1}$, (n) $a^{-1} + a^{-2}$, (o) $\frac{3a^2b^{-1}}{12b^3a}$

3 Use a calculator to evaluate (a) 2^7, (b) 2^{-4}, (c) $3^{4/5}$, (d) $5^{0.7}$.

4 When a resistance R_1 is in series with a pair of parallel resistances R_2 and R_3, the total resistance R is given by

$$R_1 + \frac{R_2R_3}{R_2+R_3}$$

Does this have the same value as:

(a) $\frac{R_1R_2+R_1R_3+R_2R_3}{R_2+R_3}$, (b) $R_1+\frac{R_2}{R_2+R_3}+\frac{R_3}{R_2+R_3}$?

5 Solve the following equations:

(a) $2x + 4 = 5 - x$, (b) $x - 4 = 12$, (c) $2x = \frac{3}{4}$, (d) $\frac{3}{2}x = \frac{2}{5}$, (e) $\frac{x+1}{4} = 3$,

(f) $x - 3 = \frac{x+1}{2}$, (g) $\frac{x-1}{2} = \frac{x+2}{3}$, (h) $\frac{1}{2}(x+2) = 5$, (i) $4(x+2) = 10$,

(j) $\frac{x+1}{6} = \frac{2x-3}{2}$, (k) $\frac{x}{2}+\frac{x}{4} = 12$, (l) $\frac{2x-1}{3}+1 = 2$

6 The area A of a circle is given by the equation

$$A = \frac{\pi d^2}{4}$$

Solve the equation for d given that the area is 5000 mm^2.

7 The velocity v of an object is given by the equation

$$v = 10 + 2t$$

Solve the equation for the time t when $v = 20$ m/s.

8 The length L of a bar of metal at t°C is related to the length L_0 at 0°C by the equation

$$L = L_0(1 + 0.0002t)$$

What is the length at 100°C for a bar of length 1 m at 0°C?

9 The specific latent heat L of a material is defined by the equation

$$L = \frac{Q}{m}$$

where Q is the heat transfer needed to change the state of a mass m of the material. If Q has the unit of joule (J) and m the unit of kg, what is the unit of L?

10 The density ρ of a material is defined by the equation

$$\rho = \frac{m}{V}$$

What is the unit of density if m has the unit kg and V the unit m^3?

11 For the bending of a beam, the bending moment M is given by the equation

$$M = \frac{EI}{R}$$

If E, the tensile modulus, has the unit N/m^2, I, the second moment of area, the unit m^4, and R the radius of a bent beam the unit of m, what must be the unit of M?

12 Rearrange the following equations to give the indicated variable in terms of the remaining variables:

(a) $p_1 - p_2 = h\rho g$, for h, (b) $\rho = \dfrac{RA}{L}$, for R, (c) $E = V - Ir$, for r,

(d) $I\alpha = T - mr(a + g)$, for T, (e) $s = ut + \frac{Ft^2}{2m}$, for F,

(f) $E = \frac{mgL}{Ax}$, for x, (g) $\frac{M}{I} = \frac{E}{R}$, for R, (h) $E = \frac{1}{2}mv^2$, for m,

(i) $V = \frac{1}{3}\pi r^2 h$, for r, (j) $P = I^2R$, for R, (k) $m_1a_1 = -m_2a_2$, for a_2,

(l) $F = \frac{mv^2}{r}$, for v, (m) $C = \varepsilon_0\varepsilon_r\frac{A}{d}$, for d

13 For a gas that obeys Charles' law

$$\frac{V_1}{T_1} = \frac{V_2}{T_2}$$

where V_1 is the initial volume and T_1 the initial temperature, V_2 the final volume and T_2 the final temperature. Rearrange the equation to express T_2 in terms of the other variables. Hence determine the final temperature when the volume changes from 1 m^3 at 300 K to 2 m^3.

14 For a beam simply supported over a span of L and carrying a central load of F plus a uniformly distributed load of w per metre of beam, then the reaction R at the beam supports is given by the equation

$$R = \tfrac{1}{2}(W + wL)$$

Rearrange the equation so that w is expressed in terms of the other variables. Hence determine the value of w when L = 3 m, W = 10 kN and R = 11 kN.

2 Linear equations

2.1 Introduction

The term *linear equation* is used for an equation in which the unknown quantity is only raised to the power 1. For example

$$2x + 5 = 0$$

is a linear equation since the unknown, the x, is only to the power 1. An equation which is *not* linear is

$$2x^2 + 5 = 0$$

In this equation the x is raised to the power 2. Such an equation is termed a *quadratic equation*. Quadratic equations are discussed in chapter 3.

Many equations in engineering are linear equations. For example, for motion in a straight line with a constant acceleration, we have

$$v = u + at$$

with perhaps the velocity v after a time t being the unknown, u a given initial velocity and a a given acceleration. Putting given values of, say, $u = 10$ m/s, $a = 4$ m/s^2 and $t = 2$ s in the equation we obtain as the solution for v

$$v = 10 + 4 \times 2 = 18 \text{ m/s}$$

There is just one value of v which fits the equation. This is always the case with a linear equation (but *not* with a quadratic equation, see chapter 3).

The above is a linear equation with just one unknown. In analysing situations in engineering we often end up with linear equations which have two unknowns. For example, in analysing an electrical circuit which has two meshes with a current I_1 in one and I_2 in the other we might obtain the equations for each mesh as

$$2I_1 - I_2 = 3 \text{ and } I_1 + 3I_2 = 4$$

Both these are linear equations since the unknowns are only to the power 1. We have thus two unknowns and two equations. Because both equations must give the same values for both the unknowns, they are referred to as *simultaneous equations*. Because they are linear equations we will still only obtain just one value of each unknown which fits the equations.

This chapter is about solving linear equations when they contain just one unknown and simultaneous linear equations when they contain two unknowns. Chapters 6 and 7 give more details of such equations in relation to motion in a straight line and electrical circuit analysis.

2.2 Solving linear equations in one unknown

The basic rules for solving linear equations in one unknown can be summarised as:

1 Bring all the terms in the unknown to one side and all the constant terms to the other side.
2 Transpose any numbers which are multiplying or dividing the unknown.

Thus, for example, if we have

$$3x + 4 = x + 8$$

then moving all the x terms to the left-hand side of the equation and all the constants to the right-hand side gives

$$3x - x = 8 - 4$$

and so $2x = 4$. Transposing the 2 then gives

$$x = \frac{4}{2} = 2$$

We can check this value by substituting it into the original equation, i.e. the equation $3x + 4 = x + 8$. We then obtain $3 \times 2 + 4 = 2 + 8$ and so $10 = 10$.

Example

Solve the linear equation $3x + 4 = 8 - x$.

Moving all the x terms to the left-hand side of the equation and the constants to the right-hand side, gives

$$3x + x = 8 - 4$$

and so $4x = 4$. Transposing the 4 in front of the x from the left-hand side to the right-hand side of the equation then gives

$$x = \frac{4}{4} = 1$$

We can check this value by substituting it into the original equation, i.e. $3x + 4 = 8 - x$, when we have $3 \times 1 + 4 = 8 - 1$ and so $7 = 7$.

Example

Solve the linear equation $\frac{2}{x+1} = \frac{3}{2x-2}$.

This is a linear equation because when we transpose it to remove the denominators on both sides of the equation we end up with an equation with the unknown to just the power 1.

$$2(2x - 2) = 3(x + 1)$$

and so

$$4x - 4 = 3x + 3$$

Hence with all the terms involving the unknown on the left-hand side and all the constants on the right, $4x - 3x = 4 + 3$ and we obtain the solution $x = 7$.

We can check this value by substituting it in the original equation.

$$\frac{2}{7+1} = \frac{3}{14-2}$$

Thus $\frac{2}{8} = \frac{3}{12}$ and $\frac{1}{4} = \frac{1}{4}$.

Example

In a calculation of the stress σ acting on the reinforcing rod in a reinforced concrete column the following equation was obtained:

$$400 \times 10^{-6}\sigma + 2.73 \times 10^5 = 3.00 \times 10^5$$

Determine the value of the stress. Its unit will be Pa.

Rearranging the equation so that the unknown is on the left-hand side and the constants on the right-hand side gives

$$400 \times 10^{-6}\sigma = 3.00 \times 10^5 - 2.73 \times 10^5 = 0.27 \times 10^5$$

Transposing the 400×10^{-6} gives

$$\sigma = \frac{0.27 \times 10^5}{400 \times 10^{-6}} = 6.75 \times 10^7 \text{ Pa}$$

We can check this value by substituting it in the original equation. Thus

$$400 \times 10^{-6} \times 6.75 \times 10^7 + 2.73 \times 10^5 = 3.00 \times 10^5$$

and $0.27 \times 10^5 + 2.73 \times 10^5 = 3.00 \times 10^5$, i.e. $3.00 \times 10^5 = 3.00 \times 10^5$.

Revision

1 Solve the following linear equations:

(a) $2x + 7 = x + 8$, (b) $2(x + 4) = 12$, (c) $3(x - 2) = 2(x + 1) - 5$,

(d) $2.5x + 1.2 = 4.3 - 1.1x$, (e) $2x + \frac{1}{2} = 4x - \frac{5}{2} + x$, (f) $\frac{2}{3}x - \frac{1}{2} = \frac{7}{6}x + \frac{3}{4}$,

(g) $1.6x + 0.5x + 12 = 10 - 0.3x$, (h) $\frac{2}{x-1} = \frac{5}{x+2}$, (i) $\frac{x}{x+2} - 2 = \frac{2}{x+2}$,

(j) $\frac{1}{3}(x-2) = \frac{4}{5}$, (k) $\frac{1}{4}x = \frac{16}{3}(2-x)$, (l) $0.5(x-2) + 1.2(x+1) = 12$,

(m) $5.0 \times 10^9 x - 6.1 \times 10^{10} = 0$.

2 (a) Kirchhoff's voltage law when applied to an electrical circuit gave the equation $2I + 4I + 10I = 12$. What is the current I? The current is in amps.
(b) The resistance R_t of a conductor at a temperature t is given by the equation $R_t = R_0(1 + \alpha t)$. Determine α if $R_t = 20\ \Omega$, $t = 20°\text{C}$ and $R_0 = 18.4\ \Omega$.
(c) The resistivity ρ of a conductor is given by $\rho = RA/L$. Determine the resistance R if $\rho = 1.72 \times 10^{-8}\ \Omega$ m, $L = 100$ m and $A = 0.8 \times 10^{-6}$ m^2.

2.3 Simultaneous equations

Consider the situation where we have an equation with two unknowns, e.g.

$$x + 2y = 5$$

If $y = 1$ then $x = 3$, if $y = 2$ then $x = 1$, if $y = 3$ then $x = -1$, and so on. Thus we have an infinite set of solutions. However, if we have a second equation involving the same two variables, e.g.

$$2x + y = 4$$

then there is only a single solution that will simultaneously satisfy both equations. Thus for this second equation, if $y = 1$ then $x = 3/2$, if $y = 2$ then $x = 1$, if $y = 3$ then $x = 1/2$, and so. There is only solution that satisfies both equations and that is that $y = 2$ and $x = 1$. The two equations are said to be *simultaneous equations* because they both simultaneously apply for the two variables.

There are three methods we can use to solve simultaneous equations:

1 By substitution.
2 By elimination.
3 By using graphs.

In this chapter we will only consider the first and second methods, the third method being discussed in chapter 12.

2.3.1 Solution by substitution

Consider the two equations

$$x + 2y = 5 \text{ and } 2x + y = 4$$

We can rearrange the first equation to obtain the unknown x in terms of the other variable and the constants, i.e.

$$x = 5 - 2y$$

If we now substitute this expression for x into the second equation we obtain

$$2(5 - 2y) + y = 4$$

$$10 - 4y + y = 4$$

Thus, by transposition, we can obtain $4y - y = 10 - 4$. Hence $3y = 6$ and so $y = 2$. We can substitute this value of y in one of the original equations, say $x + 2y = 5$, to give $x + 2 \times 2 = 5$. Hence $x = 1$. Thus the solutions are $y = 2$ and $x = 1$.

We can check that these values are correct by substituting them both in the other equation, $2x + y = 4$. Thus $2 \times 1 + 2 = 4$ and thus $4 = 4$.

The procedure can thus be summarised as:

1 From one of the equations, obtain one of the variables, say x, in terms of constants and the other variable.
2 Substitute this variable x into the second equation.
3 Solve the second equation for x.
4 The other variable y can then be obtained by substituting the determined value of y in one of the original equations.

Example

Solve the following simultaneous equations by the substitution method:

$$x + 2y = 8 \text{ and } 2x - y = 1$$

Using the first equation to obtain x in terms of the other variable, then

$$x = 8 - 2y$$

Now, substituting this value of x into the second equation gives

$$2(8 - 2y) - y = 1$$

and so

$$16 - 4y - y = 1$$

Hence we have $5y = 15$ and so $y = 3$. Substituting this value into the first equation, $x + 2y = 8$, gives $x + 2 \times 3 = 8$ and so $x = 2$. The solutions are thus $y = 3$ and $x = 2$.

We can check this by substituting these values into the second equation, $2x - y = 1$, to give $2 \times 2 - 3 = 1$.

Example

Solve the following simultaneous equations by the substitution method:

$$\frac{x}{2}+\frac{y}{3}=4 \text{ and } \frac{x}{2}-\frac{y}{3}=0$$

If we multiply both sides of the first equation by 6 then we can eliminate the fractions and so give

$$3x + 2y = 24$$

These terms can be rearranged, by transposing the $2y$ from the left-hand to right-hand side of the equation, to give

$$3x = 24 - 2y$$

and so

$$x = \frac{24-2y}{3}$$

To simplify matters, before substituting this value of x, we eliminate the fraction in the second equation by multiplying both sides of the equation by 6 to give

$$3x - 2y = 0$$

Hence

$$3\left(\frac{24-2y}{3}\right) - 2y = 0$$

$$24 - 2y - 2y = 0$$

Thus $4y = 24$ and so $y = 6$. Substituting this value in the first equation, for simplicity in the form for which the fractions have been eliminated, gives $3x + 2 \times 6 = 24$ and so $3x = 24 - 12$. Hence $3x = 12$ and so $x = 4$. The solutions are thus $y = 6$ and $x = 4$.

We can check these values by substituting them into the second equation to give $\frac{4}{2}-\frac{6}{3}=0$.

Example

When Kirchhoff's laws are applied to an electrical circuit the following two equations are produced, the currents being in amps:

$$2I_1 + 3(I_1 - I_2) = 7 \text{ and } 4I_2 + 3(I_2 - I_1) = 1$$

Determine the values of the two currents I_1 and I_2.

Rearranging the first equation to collect together all the terms involving each of the variables gives

$$2I_1 + 3I_1 - 3I_2 = 7$$

$$5I_1 - 3I_2 = 7$$

Likewise for the second equation,

$$4I_2 + 3I_2 - 3I_1 = 1$$

and so

$$-3I_1 + 7I_2 = 1$$

Using the first equation we can obtain

$$I_1 = \frac{7 + 3I_2}{5}$$

Substituting this into the second equation gives

$$-3\left(\frac{7 + 3I_2}{5}\right) + 7I_2 = 1$$

Multiplying throughout by 5 gives

$$-3(7 + 3I_2) + 35I_2 = 5$$

and so

$$-21 - 9I_2 + 35I_2 = 5$$

Thus $26I_2 = 26$ and hence $I_2 = 1$ A.

Substituting this value of I_2 into the first equation, $5I_1 - 3I_2 = 7$, gives $5I_1 - 3 = 7$ and so $I_1 = 2$ A. Thus the solution is $I_1 = 2$ A and $I_2 = 1$ A.

We can check these values by substituting then into the second equation, $-3I_1 + 7I_2 = 1$, to give $-3 \times 2 + 7 \times 1 = 1$ and $1 = 1$.

Revision

3 Solve, by substitution, the following sets of simultaneous equations:

(a) $2x + y = 3$ and $4x - y = 3$, (b) $3x + y = 7$, $2x + 3y = 7$,

(c) $x + 4y = 13$, $2x + y = 5$, (d) $x + 2y = 2$, $3x + y = 11$,

(e) $3x + 4y = 5$, $x + y = 1$, (f) $x - y = 7$, $2x + y = 8$,

(g) $4x - y = -1$, $2x + y = 7$, (h) $2x + 3y = 4$, $6x - y = 2$,

(i) $\frac{1}{x+y} = \frac{1}{3}$, $\frac{2}{3x-y} = \frac{4}{10}$, (j) $x + y = 6$, $\frac{1}{2}x + \frac{3}{4}y = 2$.

4 For a balanced beam we have when considering the equilibrium of the forces

$$R_1 + R_2 = 5.5$$

and as a result of taking moments

$$0.4R_1 + 3.4R_2 - 0.85 = 6.3$$

R_1 and R_2 are the reaction forces at the supports and are in units of kN. Hence determine the reaction forces.

5 When applying Kirchhoff's laws to the analysis of an electrical circuit, the following equations were obtained:

$$25I_1 + 20I_2 = 5 \text{ and } 5I_1 - 10I_2 = -15$$

Determine the currents I_1 and I_2.

6 The forces acting on a beam are such that

$$5F_1 + 2F_2 = 7 \text{ and } 4F_1 - F_2 = 3$$

Determine the values of F_1 and F_2.

7 The velocity v of a moving object is given by the equation $v = u + at$. If $v = 5$ m/s at $t = 1$ s and $v = 11$ m/s at $t = 3$ s, what are the values of u and a?

2.3.2 Solution by elimination

Solving simultaneous equations by elimination relies on the rule that: *one equation can be added to another or subtracted from one another without changing the simultaneous solutions.* This is because if one equation balances and a second equation balances then the sum of the two equations also balances.

Consider, as in the previous section, the two equations

$$x + 2y = 5 \text{ and } 2x + y = 4$$

We want to add, or subtract, them in such a way that we eliminate one of the variables. Suppose we double the second equation and then subtract it from the first equation.

$$
\begin{array}{rll}
 & x + 2y & = 5 \\
\text{minus} & 4x + 2y & = 8 \\
\hline
 & -3x & = -3
\end{array}
$$

Thus $x = 1$. We can substitute this value in the first equation to give

$$1 + 2y = 5$$

Hence $2y = 5 - 1$ and $y = 2$. The solution is thus $x = 1$ and $y = 2$.

We can check these values by substituting them into the second (the second because we already substituted into the first equation) equation, $2x + y = 4$ to give $2 \times 1 + 2 = 4$ and hence $4 = 4$.

The procedure for solving simultaneous equations by elimination can thus be summarised as:

1 Multiply or divide by a constant both sides of one of the equations so that one of the terms has the same form in both equations.
2 Add or subtract the two equations to eliminate one of the variables and give the value of the other.
3 Substitute the value of the variable into one of the simultaneous equations to obtain the other variable.

Example

Solve the following simultaneous equations by the elimination method:

$$x + 2y = 8 \text{ and } 2x - y = 1$$

This is the example considered in the previous section and solved by the substitution method.

We can eliminate the y variable term by doubling the second equation to give $-2y$ in that equation and adding it to the first equation which has $+2y$.

$$
\begin{array}{rll}
 & x + 2y & = 8 \\
\text{plus} & 4x - 2y & = 2 \\
\hline
 & 5x & = 10
\end{array}
$$

Thus $x = 2$. We can substitute this in the first equation, $x + 2y = 8$, to give $2 + 2y = 8$ and hence $y = 3$. The solution is thus $x = 2$ and $y = 3$.

We can check these values by substituting them in the second equation, $2x - y = 1$, to give $2 \times 2 - 3 = 1$ and hence $1 = 1$.

Example

Solve the following simultaneous equations by the elimination method:

$$\frac{x}{2}+\frac{y}{3}=4 \text{ and } \frac{x}{2}-\frac{y}{3}=0$$

This is the example considered in the previous section and solved by the substitution method.

It generally simplifies matters if we multiply both sides of the first equation by 6 to eliminate the fractions. Thus

$$3x + 2y = 24$$

Likewise, we can eliminate the fractions in the second equation by multiplying by 6 to give

$$3x - 2y = 0$$

We can eliminate the y term by adding these two equations. Thus

$$\begin{array}{rll} & 3x + 2y & = 24 \\ \text{plus} & 3x - 2y & = 0 \\ \hline & 6x & = 24 \end{array}$$

Thus $x = 4$. We can substitute this into the first equation, $3x + 2y = 24$, to give $3 \times 4 + 2y = 24$. Hence $2y = 12$ and so $y = 6$. Thus the solution is $x = 4$ and $y = 6$.

We can check these values by substituting them into the second equation, $3x - 2y = 0$, to give $3 \times 4 - 2 \times 6 = 0$ and hence $0 = 0$.

Example

When Kirchhoff's laws are applied to an electrical circuit the following two equations are produced, the current being in amps:

$$2I_1 + 3(I_1 - I_2) = 7 \text{ and } 4I_2 + 3(I_2 - I_1) = 1$$

Determine, by elimination, the values of the two currents I_1 and I_2.

This is the example considered in the previous section and solved by substitution.

Rearranging the first equation to collect together all the terms involving each of the variables gives

$$2I_1 + 3I_1 - 3I_2 = 7$$

$$5I_1 - 3I_2 = 7$$

Likewise for the second equation,

$$4I_2 + 3I_2 - 3I_1 = 1$$

$$-3I_1 + 7I_2 = 1$$

If we multiply the first equation by 7 and the second equation by 3 then we obtain:

$$\begin{array}{lrl} & 35I_1 - 21I_2 & = 49 \\ \text{plus} & -9I_1 + 21I_2 & = \ \ 3 \\ \hline & 26I_1 & = 52 \end{array}$$

Thus $26I_1 = 52$ and so we have $I_1 = 2$ A. Substituting this value into the first equation, $5I_1 - 3I_2 = 7$, gives $10 - 3I_2 = 7$. Hence $I_2 = 1$ A. Thus the solution is $I_1 = 2$ A and $I_2 = 1$ A.

We can check these values by substituting them into the second equation, $-3I_1 + 7I_2 = 1$. Thus $-3 \times 2 + 7 \times 1 = 1$ and so $1 = 1$.

Revision

8 Solve, by elimination, revision problems 3, 4 ,5, 6 and 7.

Problems

1 Solve the following linear equations:

(a) $x + 5 - 2x = x + 3$, (b) $0.4x + 1.2x = 6 - 0.1x$,

(c) $3(x + 5) = 7(x + 1)$, (d) $23.2x + 10.9 = 1.6x - 4.3$,

(e) $2.5 \times 10^6 x - 2.1 \times 10^9 = 0$, (f) $\frac{1}{2}(x + 6) = \frac{1}{3}(x - 12)$,

(g) $\frac{1}{5}x - \frac{5}{8} = \frac{9}{10}x - \frac{9}{20}$, (h) $\frac{x+5}{2x-2} = 12$, (i) $\frac{1}{x} + \frac{1}{4} = \frac{1}{12}$.

2 For resistances in parallel the following equation was obtained:

$$\frac{1}{R} + \frac{1}{10} = \frac{1}{5}$$

What is the value of R? The resistance is in ohms.

3 When applying the principle of moments to a beam the following equation was obtained:

$$4F + 2 \times 10 - 12 \times 2 = 0$$

What is the value of F?

4 Solve the following simultaneous equation by (i) substitution, and (ii) elimination:

(a) $4x + 3y = 5$ and $x - 3y = 5$, (b) $2x - y = 5$ and $x + 4y = 7$,

(c) $2x + 3y = 10$ and $x + y = 3$, (d) $4x + y = 10$ and $x + 5y = 12$,

(e) $3x + 2y = 7$ and $x - 4y = 7$, (f) $x - 2y = 6$ and $2x + 3y = 5$,

(g) $x + 2y = 9$ and $5x - y = 1$, (h) $x + y = 8$ and $4x - 2y = 2$,

(i) $\frac{1}{2}x + \frac{2}{3}y = 5$ and $\frac{1}{4}x + y = \frac{13}{2}$, (j) $\frac{1}{x+2y} = \frac{2}{14}$, $\frac{5}{3x-2y} = -\frac{10}{6}$.

5 With a machine the effort E that is required to overcome a load L is given by the equation $E = a + bL$, where a and b are constants. If an effort of 6 N is required to overcome a load of 8 N and an effort of 8 N to overcome a load of 12 N, determine the values of a and b.

6 Kirchhoff's laws applied to a circuit gave the following equations for the currents in two parts of the circuit:

$$2I_1 + 5I_2 = 4.5 \text{ and } 5I_1 + 3I_2 = 6.5$$

Determine the values of the currents I_1 and I_2, both currents being in amps.

7 The forces acting on a beam are such that

$$2F_1 + 3F_2 = 13 \text{ and } 4F_1 - 2F_2 = 2$$

Determine the values of F_1 and F_2.

8 The velocity v of a moving object is given by the equation $v = u + at$. If $v = 3$ m/s at $t = 2$ s and $v = 4$ m/s at $t = 4$ s, what are the values of u and a?

9 The velocity v of a moving object is given by the equation $v = u + at$. If $v = 5$ m/s at $t = 2$ s and $v = 7$ m/s at $t = 3$ s, what are the values of u and a?

10 For a beam in equilibrium we have

$$R_1 + R_2 = 80 \text{ and } 1.5R_1 + 120 = 4.5R_2$$

Determine the reactive forces R_1 and R_2.

3 Quadratic equations

3.1 Introduction

A *linear equation* is one where the highest power of a variable is 1 (see chapter 2). Such equations have the general form

$$ax + b = 0$$

where x is the variable and a and b are constants. There is just one solution, $x = -b/a$.

The term *quadratic equation* is used for an equation where the highest power of a variable is 2. Such equations have the general form

$$ax^2 + bx + c = 0$$

where x is the variable and a, b and c are constants. Such equations have two solutions and the obtaining of such solutions is discussed in this chapter.

Quadratic equations occur often in engineering. An example of such an equation in engineering occurs with uniformly accelerated motion in a straight line, namely

$$s = ut + \tfrac{1}{2}at^2$$

where s is the distance travelled in time t, u the initial velocity and a the acceleration. This equation can be rewritten in the general form indicated above, of $ax^2 + bx + c = 0$, as

$$\tfrac{1}{2}at^2 + ut - s = 0$$

See chapter 6 for further discussion of this straight-line motion equation.

The linear equation and the quadratic equation are just two examples of what are termed *polynomials*. A polynomial is the term used for any equation involving powers of the variable which are positive integers. Such powers can be 1, 2, 3, 4, 5, etc. For example, $x^4 + 4x^3 + 2x^2 + 5x + 2 = 0$ is a polynomial with the highest power being 4. In this chapter the discussion is restricted to just the quadratic equation. The constants a, b, c, etc. that are used to multiply the x terms are termed *coefficients*. Thus we might have with a polynomial having the highest power 3

$$ax^3 + bx^2 + cx^1 + dx^0 = 0$$

Example

Which of the following are quadratic equations?

(a) $2x + 6 = 0$, (b) $x^2 + 3x + 2 = 0$, (c) $3x = 5$, (d) $x^2 = 4 + 2x$,

(e) $2x^3 + 5x^2 + 3x + 1 = 0$, (f) $x^2 = 20$, (g) $\frac{1}{x} + 2x = 4$

Equations (a) and (c) are linear equations since they can be written in the form $ax + b = 0$. Thus (c) can be written as $3x - 5 = 0$. Equations (b), (d), (f) and (g) are quadratic equations since they can be written in the form $ax^2 + bx + c = 0$, even if b or c are zero (a must not be zero if it is to be a quadratic). Equation (d) can be written as $x^2 - 2x - 4 = 0$ and equation (f) as $x^2 + 0x - 20 = 0$. When both sides of equation (g) are multiplied by x we obtain $1 + 2x^2 = 4x$ which rearranges to give the equation $2x^2 - 4x + 1 = 0$. Equation (e) has the highest power of the variable as 3 and so is not a linear or quadratic equation. All the equations are polynomials.

Revision

1 Which of the following equations are (i) linear equations, (ii) quadratic equations?

(a) $x = 3$, (b) $2x + 5 = 0$, (c) $x^2 = 4$, (d) $x^2 + 5x + 2 = 0$, (e) $x^3 + x^2 = 4$,

(f) $4x^2 + 3x = 5$, (g) $\frac{1}{x+1} = 3$, (h) $\frac{1}{x} - 3x = 5$

3.2 Factors

To factor a number means to write it as the product of smaller numbers. Thus, for example, we can factor 12 to give $12 = 3 \times 4$. The 3 and 4 are factors of 12. If the 3 and the 4 are multiplied together then 12 is obtained. To factor a polynomial means to write it as the product of simpler polynomials. Thus for the quadratic expression $x^2 + 5x + 6$ we can write

$$x^2 + 5x + 6 = (x + 2)(x + 3)$$

$(x + 2)$ and $(x + 3)$ are factors. If the two factors are multiplied together then the $x^2 + 5x + 6$ is obtained. Note that, in general,

$$(a + b)(c + d) = a(c + d) + b(c + d) = ac + ad + bc + bd$$

Hence

$$(x + 2)(x + 3) = x(x + 3) + 2(x + 3) = x^2 + 3x + 2x + 6 = x^2 + 5x + 6$$

If we have $u \times v = 0$ then we must have either $u = 0$ or $v = 0$ or both u and v are 0. This is because 0 times any number is 0. Thus if we have the quadratic equation $x^2 + 5x + 6 = 0$ and rewrite it as $(x + 2)(x + 3) = 0$, then we must have either $x + 2 = 0$, or $x + 3 = 0$ or both equal to 0. This means that the solutions to the quadratic equation are the solutions of these two linear equations, i.e. $x = -2$ and $x = -3$. These values are called the *roots* of the equation. We can check these values by substituting them into the quadratic equation. Thus for $x = -2$ we have $4 - 10 + 6 = 0$ and thus $0 = 0$. For $x = -3$ we have $9 - 15 + 6 = 0$ and thus $0 = 0$.

Thus we can solve a quadratic equation by the following:

1 Factorise the quadratic.
2 Set each factor equal to 0.
3 Solve the resulting linear equations.

Consider a quadratic equation which gives the factors of $(x + a)$ and $(x + b)$, i.e. we have

$$(x + a)(x + b) = 0$$

Multiplying this out gives

$$x^2 + (a + b)x + ab = 0$$

When the coefficient of the x^2 term is 1, then the coefficient of the x term is the sum of the roots and the constant, the coefficient of the x^0 term, is the product of the roots.

The following are some commonly encountered forms of equations and their factors:

$x^2 + (a + b)x + ab$	$(x + a)(x + b)$	[1]
$x^2 + 2ax + a^2$	$(x + a)(x + a)$ or $(x + a)^2$	[2]
$x^2 - 2ax + a^2$	$(x - a)(x - a)$ or $(x - a)^2$	[3]
$x^2 - a^2$	$(x + a)(x - a)$	[4]
$x^2 + ax$	$x(x + a)$	[5]

Example

Factorise and hence solve the quadratic equation $x^2 - 3x + 2 = 0$.

This equation is of the form given by equation [1] above. To factorise this equation we need to find the two numbers which when multiplied together will give 2 and which when added together will give -3. If we multiply -1 and -2 we obtain 2 and the addition of -1 and -2 gives -3. Thus we can write

$$(x - 1)(x - 2) = 0$$

The solutions are thus given by $x - 1 = 0$, i.e. $x = 1$, and $x - 2 = 0$, i.e. $x = 2$.

We can check these values by substituting them into the orginal equation, $x^2 - 3x + 2$. Thus, for $x = 1$ we have $1 - 3 + 2 = 0$ and so $0 = 0$. For $x = 2$ we have $4 - 6 + 2 = 0$ and so $0 = 0$.

Example

Factorise and hence solve the quadratic equation $x^2 - 4 = 0$.

This equation is of the form given by equation [4] above. We can write this equation in the form $x^2 + 0x - 4 = 0$. We thus need to find a product of two numbers which gives –4 and a sum which gives 0. Hence the factors are:

$$(x - 2)(x + 2) = 0$$

Thus the solutions are $x - 2 = 0$, i.e. $x = 2$, and $x + 2 = 0$, i.e. $x = -2$. These values can be checked by substituting them in the original equation, $x^2 - 4 = 0$. With $x = 2$ we have $4 - 4 = 0$ and with $x = -2$ we have $4 - 4 = 0$.

Example

Factorise and hence solve the quadratic equation $x^2 + 6x + 9 = 0$.

This is of the form given by equation [2] above. The product of two numbers is 9 and the sum is 6. Hence

$$(x + 3)(x + 3) = 0$$

The solutions are thus given by $x + 3 = 0$, i.e. $x = -3$, and $x + 3 = 0$, i.e. $x = -3$. There are two roots with the same value. We can check these values by substituting them in the original equation, $x^2 + 6x + 9 = 0$. Thus $9 - 18 + 9 = 0$.

Example

Factorise and hence solve the quadratic equation $x^2 + 5x = 0$.

This is of the form of equation [5] given above. We can rewrite this equation in the form $x(x + 5) = 0$. Hence the solutions are given by $x = 0$ or $x + 5 = 0$, i.e. $x = -5$. we can check these values by substituting them in the original equation, $x^2 + 5x = 0$. Thus with $x = 0$ we have $0 + 0 = 0$ and with $x = -5$ we have $25 - 25 = 0$.

Example

If the roots of a quadratic equation are +2 and +1, what is the equation?

We have $x = +2$ and so $x - 2 = 0$, and $x = +1$ and so $x - 1 = 0$. Thus we can write the equation as

$$(x - 2)(x - 1) = 0$$

Multiplying this out gives $x^2 - 3x + 2 = 0$

Revision

2 Factorise and hence solve the following quadratic equations:

(a) $x^2 + x - 2 = 0$, (b) $x^2 + 2x - 8 = 0$, (c) $x^2 - 5x + 4 = 0$,
(d) $x^2 - 4 = 0$, (e) $x^2 + 4x + 4 = 0$, (f) $x^2 - 2x + 1 = 0$, (g) $x^2 + 7x = 0$,

(h) $x^2 - 1 = 0$, (i) $x^2 + 2x - 3 = 0$, (j) $x^2 - 5x + 6 = 0$

3 Determine the quadratic equations giving the following roots:

(a) $x = 1$, $x = 4$, (b) $x = -1$, $x = 2$, (c) $x = 0$, $x = 6$, (d) $x = -2$, $x = -3$,

(e) $x = 2$, $x = 0$, (f) $x = -1$, $x = -2$, (g) $x = 3$, $x = 1$

3.3 Completing the square

Consider the equation $x^2 + 6x + 9 = 0$. This equation can be factorised to give $(x + 3)(x + 3) = 0$, i.e. $(x + 3)^2 = 0$. It is a perfect square, both the roots being the same.

Now consider the equation $x^2 + 6x + 2 = 0$. What are the factors? We can rewrite the equation as

$$x^2 + 6x = -2$$

If we add 9 to both sides of the equation then we obtain

$$x^2 + 6x + 9 = -2 + 9 = 7$$

The left-hand side of the equation has been made into a perfect square by the adding of the 9. Thus we can write

$$(x + 3)^2 = 7$$

This means that $x + 3$ must be one of the square roots of 7, i.e.

$$x + 3 = \pm\sqrt{7}$$

The plus or minus is because every positive number has two square roots, one positive and one negative. Thus we have $(+\sqrt{7}) \times (+\sqrt{7}) = 7$ and $(-\sqrt{7}) \times (-\sqrt{7}) = 7$. Hence

$$x = -3 \pm \sqrt{7}$$

The two solutions are thus $x = -3 + \sqrt{7}$ and $x = -3 - \sqrt{7}$.

This method of determining the roots of a quadratic equation is known as *completing the square*. In the above discussion the left-hand side of the equation was made into a perfect square by the adding of 9. How do we determine what number to add in order to make a perfect square? Any expression of the form $x^2 + ax$ becomes a perfect square when we add $(a/2)^2$, since

$$x^2 + ax + \left(\frac{a}{2}\right)^2 = \left(x + \frac{a}{2}\right)^2$$

Thus for $x^2 + 6x$ we have $a = 6$ and so $(a/2) = 3$. Hence we add $3^2 = 9$.

The procedure for determining the roots of a quadratic equation by completing the square can thus be summarised as:

1 Put the equation into the form $x^2 + ax = b$.
2 Determine the value of $(a/2)$.
3 Then add $(a/2)^2$ to both sides of the equation to give

$$x^2 + ax + (a/2)^2 = b + (a/2)^2$$

4 Hence obtain the equation

$$(x + a/2)^2 = b + (a/2)^2$$

5 Determine the two roots by taking the square root of both sides of the equation, i.e.

$$x + \frac{a}{2} = \pm\sqrt{b + \left(\frac{a}{2}\right)^2}$$

The above rule for completing the square only works if the coefficient of x^2, i.e. the number in front of x^2, is 1. However, if this is not the case we can simply divide throughout by that coefficient in order to make it 1.

Example

Use the method of completing the square to solve the quadratic equation $x^2 + 10x - 4 = 0$.

The quadratic equation can be written as

$$x^2 + 10x = 4$$

Adding $(10/2)^2$, i.e. 25, to both sides of the equation gives

$$x^2 + 10x + 25 = 4 + 25$$

Thus

$$(x + 5)^2 = 29$$

Hence, $x + 5 = \pm\sqrt{29} = \pm 5.39$ and the solutions of the quadratic equation are $x = +5.39 - 5 = 0.39$ and $x = -5.39 - 5 = -10.39$.

We can check these values by substituting them in the equation $x^2 + 10x - 4 = 0$. Thus, for $x = 0.39$ we have $0.39^2 + 3.9 - 4 = 0.05$, which, because of the rounding used to limit the number of decimal places in determining the root, is effectively zero. For the other solution of $x = -10.39$ we have $(-10.39)^2 + 103.9 - 4 = 0.05$, which, because of the rounding used to limit the number of decimal places in determining the root, is effectively zero.

Example

Use the method of completing the square to solve the quadratic equation $2x^2 + 5x - 4 = 0$.

The coefficient of the x^2 is 2. Thus to make the coefficient 1 we divide throughout by 2.

$$x^2 + 2.5x - 2 = 0$$

Thus we can write

$$x^2 + 2.5x = 2$$

Completing the square gives

$$x^2 + 2.5x + \left(\frac{2.5}{2}\right)^2 = 2 + \left(\frac{2.5}{2}\right)^2$$

and so

$$(x + 1.25)^2 = 2 + 1.25^2 = 3.56$$

This then gives $x + 1.25 = \pm\sqrt{3.56} = \pm 1.89$ and so the solutions of the equation are $x = +1.89 - 1.25 = 0.64$ and $x = -1.89 - 1.25 = -3.14$.

We can check these values by substituting them in the equation $2x^2 + 5x - 4 = 0$. Thus, with $x = 0.64$ we have $2(0.64)^2 + 3.2 - 4 = 0.02$ and so effectively zero. With $x = -3.14$ we have $19.7 - 15.7 - 4 = 0$.

Revision

4 Use the method of completing the square to solve the following quadratic equations:

(a) $x^2 + 8x + 12 = 0$, (b) $x^2 + 2x + 8 = 0$, (c) $3x^2 - x - 5 = 0$,

(d) $3x^2 - 6x - 11 = 0$, (e) $3x^2 + 4x - 1 = 0$, (f) $2x^2 + 5x - 4 = 0$,

(g) $x^2 + 4x - 8 = 0$, (h) $x^2 - 4x + 2 = 0$

3.4 Solving a quadratic equation by formula

Consider the quadratic equation $ax^2 + bx + c = 0$. To obtain the roots by the completing the square method, we divide throughout by a to give:

$$x^2 + \frac{b}{a}x + \frac{c}{a} = 0$$

This can be written as

$$x^2 + \frac{b}{a}x = -\frac{c}{a}$$

We complete the square by adding $(b/2a)^2$ to both sides of the equation. Hence

$$x^2 + \frac{b}{a}x + \left(\frac{b}{2a}\right)^2 = -\frac{c}{a} + \left(\frac{b}{2a}\right)^2$$

and so

$$\left(x + \frac{b}{2a}\right)^2 = -\frac{c}{a} + \left(\frac{b}{2a}\right)^2 = \frac{-4ac + b^2}{4a^2}$$

Taking the square root of both sides of the equation gives

$$x + \frac{b}{2a} = \pm\frac{\sqrt{b^2 - 4ac}}{2a}$$

Thus

$$x = -\frac{b}{2a} \pm \frac{\sqrt{b^2 - 4ac}}{2a}$$

and so we have the general formula for the solution of a quadratic equation

$$x = \frac{-b \pm \sqrt{b^2 - 4ac}}{2a}$$

Example

Use the formula to solve the quadratic equation $x^2 - 6x + 2 = 0$.

Comparing the equation with $ax^2 + bx + c = 0$, then $a = 1$, $b = -6$ and $c = 2$. Thus the formula gives

$$x = \frac{-b \pm \sqrt{b^2 - 4ac}}{2a} = \frac{-(-6) \pm \sqrt{(-6)^2 - 4 \times 1 \times 2}}{2 \times 1}$$

$$= \frac{6 \pm \sqrt{36 - 8}}{2} = \frac{6 \pm \sqrt{28}}{2} = \frac{6 \pm 5.29}{2} = 3 \pm 2.65$$

Hence the solutions are $x = 5.65$ and $x = 0.35$. We can check these values by substituting them in the equation $x^2 - 6x + 2 = 0$. Thus for the root $x = 5.65$ we have $31.9 - 33.9 + 2 = 0$ and for $x = 0.35$ we have $0.1 - 2.1 + 2 = 0$.

Example

Use the formula to solve the quadratic equation $3x^2 - 5x - 7 = 0$.

Comparing the equation with $ax^2 + bx + c = 0$, then $a = 3$, $b = -5$ and $c = -7$. Thus the formula gives

$$x = \frac{-b \pm \sqrt{b^2 - 4ac}}{2a} = \frac{-(-5) \pm \sqrt{(-5)^2 - 4 \times 3 \times (-7)}}{2 \times 3}$$

$$= \frac{5 \pm \sqrt{25 + 84}}{6} = \frac{5 \pm \sqrt{109}}{6} = \frac{5 \pm 10.44}{6} = 0.83 \pm 1.74$$

Hence the solutions are $x = 2.57$ or -0.91.

We can check these values by substitution in the original equation, $3x^2 - 5x - 7 = 0$. Thus, for $x = 2.57$ we have $19.81 - 12.85 - 7 = -0.04$. This is effectively zero, bearing in mind the rounding to two decimal places of the solution. For $x = -0.91$ we have $2.48 + 4.55 - 7 = 0.03$. This is effectively zero, bearing in mind the rounding to two decimal places of the solution.

Example

When an object is thrown vertically upwards, its height h varies with the time t according to the equation $h = 20t - 5t^2$. At what times will the height be 10 m?

This requires a solution of the equation $10 = 20t - 5t^2$. This, when rearranged, is $5t^2 - 20t + 10 = 0$. Comparing this with the quadratic equation $ax^2 + bx + c = 0$, then $a = 5$, $b = -20$ and $c = 10$. Using the formula we have

$$t = \frac{20 \pm \sqrt{20^2 - 4 \times 5 \times 10}}{2 \times 5} = \frac{20 \pm \sqrt{400 - 200}}{10} = \frac{20 \pm 14.1}{10}$$

$$= 2 \pm 1.41$$

Hence the solutions are $t = 3.41$ s and 0.59 s. we can check these values by substitution in the equation $10 = 20t - 5t^2$. Thus for $t = 3.41$ we have $68.2 - 58.1 = 10.1$, i.e. effectively 10 bearing in mind the rounding of the values of the roots in working to just one decimal place. For $t = 0.59$ we have $11.8 - 1.74 = 10.06$, i.e. effectively 10 bearing in mind the rounding of the values of the roots in working to just one decimal place.

Example

The total surface area A of a cylinder of radius r and height h is given by the equation $A = 2\pi r^2 + 2\pi rh$. If $h = 6$ cm, what will be the radius required to give a surface area of 88/7 cm^2? Take π as 22/7.

Putting the numbers in the equation gives

$$\frac{88}{7} = 2 \times \frac{22}{7}r^2 + 2 \times \frac{22}{7} \times 6r$$

Multiplying throughout by 7 and dividing by 44 gives

$$2 = r^2 + 6r$$

Hence we can write

$$r^2 + 6r - 2 = 0$$

Comparing this with the quadratic equation $ax^2 + bx + c = 0$, then $a = 1$, $b = 6$ and $c = -2$. Using the formula we have

$$r = \frac{-6 \pm \sqrt{36 - 4 \times 1 \times (-2)}}{2 \times 1} = \frac{-6 \pm \sqrt{44}}{2} = \frac{-6 \pm 6.63}{2}$$

Hence the solutions are $r = -6.32$ cm and $r = 0.32$ cm. The negative solution has no physical significance. Hence the solution is a radius of 0.32 cm. Note that often in physical problems involving quadratics, the negative solution may have no physical significance.

We can check this value of 0.32 cm by substitution in the equation $2 = r^2 + 6r$. Hence $0.10 + 1.92 = 2.02$, which is effectively 2 bearing in mind the rounding of the root value to two decimal places that has occurred.

Revision

5 Use the formula to solve the quadratic equations given in revision problem 4.

6 Use the formula to solve the following quadratic equations:

(a) $x^2 - 3x - 2 = 0$, (b) $x^2 + 8x + 12 = 0$, (c) $3x^2 + 12x - 1 = 0$,

(d) $3x^2 - 6x - 11 = 0$, (e) $x^2 + 3x - 7 = 0$

7 The height h risen by an object, after a time t, when thrown vertically upwards with an initial velocity u is given by the equation $h = ut - \frac{1}{2}gt^2$, where g is the acceleration due to gravity. Solve the quadratic equation for t if $u = 100$ m/s, $h = 150$ m and $g = 9.81$ m/s^2.

8 A rectangle has one side 3 cm longer than the other. What will be the dimensions of the rectangle if the diagonals have to have lengths of 10 cm? Hint: let one of the sides have a length x, then the other side has a length of $3 + x$. The Pythagoras theorem can then be used.

9 In an electrical circuit the reactance X is given by the equation

$$X = \omega L - \frac{1}{\omega C}$$

Solve the equation for ω, the unit being rad/s with the following given units, when $L = 0.5$ H, $C = 20$ μF (1 μF = 10^{-6} F) and $X = 100$ Ω.

10 The bending moment M a distance x along a beam is given by the equation

$$M = \frac{wx(L - x)}{2}$$

Determine the values of x when $M =$ 50 N m, $w = 3$ N/m and $L = 20$ m.

11 The motion of an object is described by the equation $s = ut + \frac{1}{2}at^2$. Determine t when $s = 96$ m, if $u = 10$ m/s and $a = 2$ m/s^2.

12 A rectangular sheet of metal 8 cm by 10 cm has a rectangle cut from its centre to leave a border all round the hole of width x. Determine the value of x if the area of the border is to be 20 cm^2.

13 A square sheet of metal of side 80 cm has a rectangle cut from one corner to leave an L-shaped piece with arms of width x. Determine the value of x if the area of the L-shaped piece is to be 1776 cm^2.

3.4.1 The nature of the roots

The formula for the roots of the quadratic equation is

$$x = \frac{-b \pm \sqrt{b^2 - 4ac}}{2a}$$

Consider the following three situations:

1 If we have $(b^2 - 4ac) > 0$, then the square root is of a positive number. There are then *two distinct roots*.
2 If we have $(b^2 - 4ac) = 0$, then the square root is zero and the formula gives just one value for x. Since a quadratic equation must have two roots, we say that the equation has *two coincident roots*.
3 If we have $(b^2 - 4ac) < 0$, then the square root is of a negative number. A new type of number has to be invented to enable such expressions to be solved. The number is referred to as a *complex number* and the roots

are said to be *imaginary* (the roots in 1 and 2 above are said to be *real*). Such numbers are not discussed further in this book.

Example

Have the following quadratic equations real roots?

(a) $x^2 - 5x - 2 = 0$, (b) $x^2 + 3x + 5 = 0$, (c) $x^2 - 4x + 4 = 0$

(a) Comparing the equation with $ax^2 + bx + c = 0$, then $a = 1$, $b = -5$ and $c = -2$. Thus $(b^2 - 4ac)$ is $25 + 8 = +33$. It is thus greater than 0 and so there are two distinct real roots.
(b) Comparing the equation with $ax^2 + bx + c = 0$, then $a = 1$, $b = 3$ and $c = 5$. Thus $(b^2 - 4ac)$ is $9 - 20 = -11$. It is thus less than 0 and so the roots are imaginary.
(c) Comparing the equation with $ax^2 + bx + c = 0$, then $a = 1$, $b = -4$ and $c = 4$. Thus $(b^2 - 4ac)$ is $16 - 16 = 0$. Because it is zero there are two coincident real roots.

Revision

14 Have the following quadratic equations real roots?

(a) $2x^2 + 3x + 4 = 0$, (b) $x^2 - 2x - 5 = 0$, (c) $x^2 - 3x + 8 = 0$,

(d) $x^2 + 2x + 1 = 0$, (e) $2x^2 - 5x + 1 = 0$

Problems

1 Factorise and hence solve the following quadratic equations:

(a) $x^2 + x - 2 = 0$, (b) $x^2 + 4x + 3 = 0$, (c) $x^2 - 4x + 3 = 0$,

(d) $x^2 + 2x + 1 = 0$, (e) $x^2 - 9 = 0$, (f) $x^2 - 25 = 0$, (g) $x^2 + 3x = 0$

2 Determine the quadratic equations giving the following roots:

(a) $x = 1, x = -1$, (b) $x = 2, x = 2$, (c) $x = -3, x = -1$, (d) $x = 0, x = 1$

3 Solve the following quadratic equations using (a) the method of completing the square, (b) the formula:

(a) $x^2 - 2x - 8 = 0$, (b) $x^2 + x - 3 = 0$, (c) $2x^2 + x - 3 = 0$,

(d) $3x^2 - 4x + 1 = 0$, (e) $x^2 + 2x - 5 = 0$, (f) $5x^2 + x - 3 = 0$,

(g) $4x^2 + 4x - 15 = 0$, (h) $5x^2 - 8x + 2 = 0$, (i) $4x^2 + 5x - 3 = 0$

4 The area of a rectangle is 100 cm². If it has a width which is 5 cm less than its length, what are the dimensions of the rectangle? Hint: let the length be x and the width is then $(x - 5)$.

5 A rectangle has a perimeter of 26 cm and an area of 30 cm^2. What are its dimensions? Hint: let x equal the length of one side, then the length of the other is $13 - x$.

6 An object is thrown vertically upwards with a velocity of 50 m/s. The height h of the object after a time t is given by $h = 50t - 4.9t^2$. Hence determine the times when the object is at a height of 100 m.

7 A thermocouple gives an e.m.f. E in μV which depends on the temperature θ in degrees C according to the equation $E = 6.88\theta - 0.0192\theta^2$. Solve the equation for a thermocouple for which the e.m.f. is 200 μV.

8 The bending moment M at a point a distance x along a beam is given by

$$M = \frac{wx(L - x)}{2}$$

Determine the values of x when $M =$ 50 N m, $w = 5$ N/m and $L = 20$ m.

9 The surface area A of a cylinder is given by $A = 2\pi r^2 h + 2\pi r^2$, where r is the radius and h the height. What height cylinder is required if, with $r = 5$ cm, the area is to be 125 cm^2?

10 The current i, in amperes, through a circuit element is related to the voltage v, in volts, across it by the equation $i = 0.001v^2 + 0.010v$. Determine the value of v when i is 0.004 A.

11 A square sheet of metal has a square hole cut from it so that the border all round the hole has a width of 4 mm. If the hole has an area of one quarter of the area of the original square, determine the length of the side of the square.

12 A right-angled triangle has a hypotenuse of length 18 mm and a total perimeter of length 40 mm. Determine the lengths of the other two sides.

13 A right-angled triangle is to have a hypotenuse of length 10 cm. If, for the other two sides one of them is to be longer than the other by 3 cm, what are the lengths of the sides?

14 Determine the two numbers which have a difference of 5 and a product of 266.

15 A rectangle has a length which is 3 cm greater than its width. If the area of the rectangle is 28 cm^2, what are its length and width?

16 A piece of wire of length 40 cm is to be bent into a rectangular shape to enclose an area of 100 cm^2. What will be the lengths of the sides of the rectangle?

17 Determine the number which added to its square gives 90.

18 A rectangular sheet of metal is 14 cm by 17 cm. Strips of the same width are cut off one side and one end. If the area remaining is 108 cm^2, what is the width of the pieces removed?

19 What are the two consecutive numbers which have a product of 156?

20 The length of a factory floor is to be 50 m more than its width and the total area 266 m^2. What will be the width and length?

21 Have the following quadratic equations real roots?

(a) $x^2 + 3x + 1 = 0$, (b) $x^2 - 5x - 2 = 0$, (c) $x^2 + 2x + 4 = 0$,

(d) $x^2 + 3x + 4 = 0$, (e) $2x^2 + 4x + 2 = 0$

4 Exponentials

4.1 Introduction

There are many situations in engineering science where we are concerned with the growth or decay of some process. Thus, for example, we might be concerned with the way a hot object cools, i.e. the way the temperature decreases with time. We might have the following data:

Temperature °C	60	49.6	42.0	36.3	32.1	28.9
Time in min.	0	1	2	3	4	5

Eventually the temperature settles down to 20°C. What equation can we use to describe the relationships between the temperature and time?

Perhaps we might be considering an electrical capacitor and the way the charge on its plates changes with time, growing when it is charged and decreasing when it is discharged. What equations can we use to describe these relationships?

This chapter is about the equations we can use to describe the growth or decay of processes.

4.2 Exponentials

We can describe growth and decay processes by an equation of the form

$$y = a^t$$

where a is some constant, and y the value of the quantity at a time t.

To illustrate this for a growing quantity, consider how the value of y changes with time when we have $a = 2$ and so the equation $y = 2^t$. At $t = 0$ we have $y = 2^0 = 1$, at $t = 1$ we have $y = 2^1 = 2$, at $t = 2$ we have $y = 2^2 = 4$, at $t = 3$ we have $y = 2^3 = 8$, and so on. Thus we have the data:

$y = 2^t$	1	2	4	8	16	32
t	0	1	2	3	4	5

This data describes a quantity that increases with time, starting at $t = 0$ with the value 1 and rapidly increasing as t increases. Figure 4.1(a) shows this data plotted as a graph (see chapter 12 for a discussion of graphs).

For a quantity that decreases with time, consider how the value of y changes with time when we have $a = 2$ and the equation $y = 2^{-t}$. At $t = 0$ we have $y = 2^0 = 1$, at $t = 1$ we have $y = 2^{-1} = 0.5$, at $t = 2$ we have $y = 2^{-2} = 0.25$, at $y = 3$ we have $y = 2^{-3} = 0.125$, and so on. Thus we have the data:

$y = 2^{-t}$	1	0.5	0.25	0.125	0.0625	0.03125
t	0	1	2	3	4	5

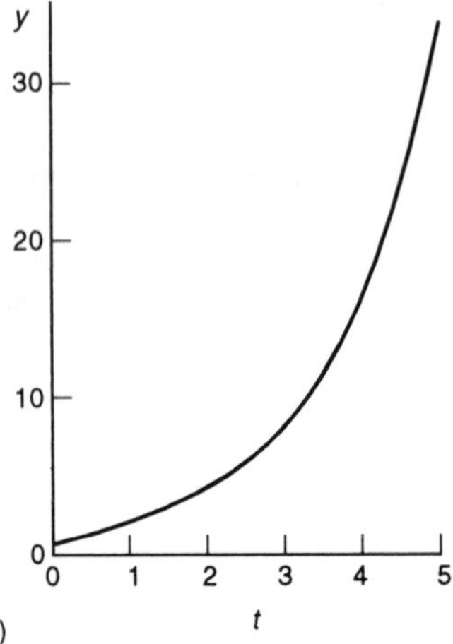

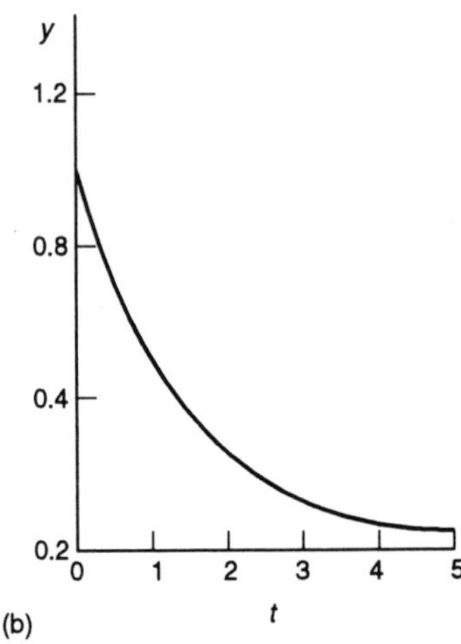

Figure 4.1 *(a)* $y = 2^t$, *(b)* $y = 2^{-t}$

This data describes a quantity that decreases with time, starting with a value 1 at $t = 0$ and rapidly decreasing in size as t increases. Figure 4.1(b) shows this data plotted as a graph. The equation $y = a^t$ describes a growth curve, the equation $y = a^{-t}$ a decay curve.

4.2.1 The exponential

This chapter is about quantities that can be described by the general equation $y = a^x$, where the varying quantity x in engineering is often time. Such equations are called ***exponentials*** with ***a*** being called the ***base***. An exponential is *not* a polynomial. The powers of a polynomial are constants, e.g. the 2 in x^2, whereas the power of an exponential is the variable. An exponential could be 2^x, 3^x, 4^x, etc. However, we usually standardise the base to one particular value. The most widely used exponential is e^x, where e is a constant with the value 2.728 281 828 ... and is often referred to as *the exponential*. Whenever an engineer refers to an exponential change he or she is almost invariably referring to an equation written in terms of e^x.

The following shows the values of e^x and e^{-x} for various values of x and figure 4.2 the resulting graphs. The e^x graph describes a growth curve, the e^{-x} a decay curve.

$y = e^x$	0.14	0.37	1	2.72	7.39	20.09	... infinite
x	−2	−1	0	1	2	3	... infinite

$y = e^{-x}$	7.39	2.72	1	0.37	0.14	0.05	... 0
x	−2	−1	0	1	2	3	... infinite

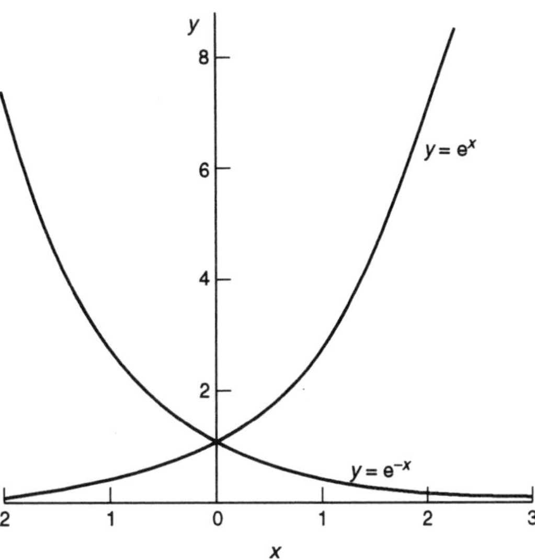

Figure 4.2 $y = \mathrm{e}^{x}$ *and* $y = \mathrm{e}^{-x}$

In a more general form we can write the exponential equation in the form $y = \mathrm{e}^{kx}$, or $y = \mathrm{e}^{-kx}$, where k is some constant. This constant k determines how fast y changes with x. The following data illustrates this:

x	0	1	2	3	... infinite
$y = \mathrm{e}^{-1x}$	1	0.368	0.135	0.050	... 0
$y = \mathrm{e}^{-2x}$	1	0.135	0.018	0.003	... 0

x	0	1	2	3	... infinite
$y = \mathrm{e}^{1x}$	1	2.718	7.389	20.086	... infinite
$y = \mathrm{e}^{2x}$	1	7.389	54.598	403.429	... infinite

The bigger k is the faster y decreases, or increases, with x.

When $x = 0$ then for $y = \mathrm{e}^{kx}$, or $y = \mathrm{e}^{-kx}$, $y = \mathrm{e}^{0}$ and so $y = 1$. This is thus the value of y that occurs when x is zero. Since we may often have an initial value other than 1, we write the equation in the form

$$y = A\ \mathrm{e}^{kx}$$

where A is the initial value of y at $x = 0$. For example, for the discharging of a capacitor in an electrical circuit we have, for the charge q on the capacitor at a time t, the equation

$$q = Q_0\ \mathrm{e}^{-t/CR}$$

When $t = 0$ then $q = Q_0$. The constant k is $1/CR$. The bigger the value of CR the smaller the value of $1/CR$ and so the slower the rate at which the capacitor becomes discharged.

One form of equation involving exponentials that is quite common is of the form

$$y = A - A\,e^{-kx}$$

When $x = 0$ then $e^0 = 1$ and so $y = A - A = 0$. The initial value is thus 0. As x increases then e^{-kx} decreases from 1 towards 0, eventually becoming zero when x is infinite. Thus the value of y increases as x increases. When x is very large then e^{-kx} becomes virtually 0 and so y becomes equal to A. Figure 4.3 shows the graph. It shows a quantity y which increases rapidly at first and then slows down to become eventually A.

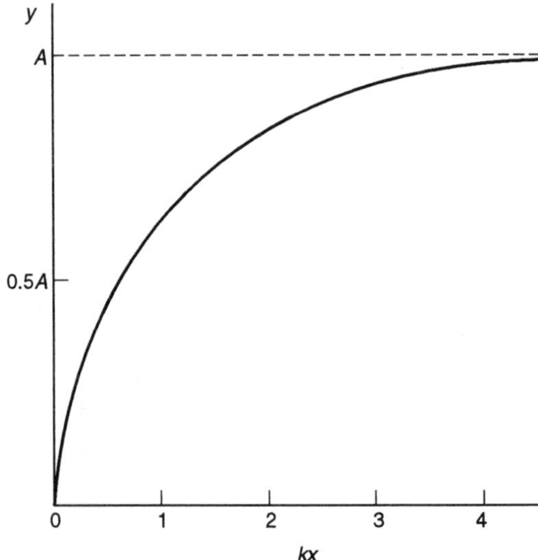

Figure 4.3 $y = A - A\,e^{-kx}$

For example, for a capacitor which starts with zero charge on its plates and is then charged we have the equation

$$q = Q_0 - Q_0\,e^{-t/CR}$$

When $t = 0$ then $e^0 = 1$ and so $q = Q_0 - Q_0 = 0$. As t increases, so the value of $e^{-t/CR}$ decreases and so q becomes more and more equal to Q_0.

Example

For an object cooling according to Newton's law, the temperature θ of the object varies with time t according to the equation $\theta = \theta_0\,e^{-kt}$, where θ_0 and k are constants. (a) Explain why this equation represents a quantity which is decreasing with time. (b) What is the value of the temperature at $t = 0$? (c) How will the rates at which the object cools change if in one instance $k = 0.01$ and in another $k = 0.02$ (the units of k are per °C)?

(a) If we assume that t and k will only have positive values, then the $-kt$ means that the power is negative and so the temperature decreases with time.
(b) When $t = 0$ then $e^{-kt} = 1$ and so $\theta = \theta_0$. Thus θ_0 is the initial value at the time $t = 0$.
(c) Doubling the value of k means that the object will cool faster, in fact it will cool twice as fast.

Example

The current i in amperes in an electrical circuit varies with time t according to the equation $i = 10(1 - e^{-t/0.4})$. What will be (a) the initial value of the current when $t = 0$, (b) the final value of the current at infinite time?

(a) When $t = 0$ then $e^{-t/0.4} = e^0 = 1$. Thus $i = 10(1 - 1) = 0$.
(b) When t becomes very large then $e^{-t/0.4}$ becomes 0. Thus we have $i = 10(1 - 0)$ and so the current becomes 10 A.

Revision

1 The number N of radioactive atoms in a sample of radioactive material decreases with time t and is described by the equation $N = N_0 e^{-\lambda t}$, where N_0 and λ are constants. (a) Explain why this equation represents a quantity which is decreasing with time. (b) What will be the number of radioactive atoms at time $t = 0$? (c) For a radioactive material that decreases only very slowly with time, will λ have a large or smaller value than with a radioactive material which decreases quickly with time?

2 The length L of a rod of material increases from some initial length with the temperature θ above that at which the initial length is measured and is described by the equation $L = L_0 e^{\alpha\theta}$, where L_0 and α are constants. (a) Explain why the equation represents a quantity which increases with time. (b) What will be the length of the rod when $\theta = 0$? (c) What will be the effect of a material having a higher value of α than some other material?

3 For an electrical circuit involving inductance, the current in amperes is related to the time t by the equation $i = 3(1 - e^{-10t})$. What is the value of the current when (a) $t = 0$, and (b) t is very large?

4 What are the values of y in the following equations when (i) $x = 0$, (ii) x is very large, i.e. infinite?

(a) $y = 2\ e^{3x}$, (b) $y = 10\ e^{-5x}$, (c) $y = 2(1 - e^{-2x})$, (d) $y = 2\ e^{-0.2x}$,

(e) $y = -4\ e^{-x/3}$, (f) $y = 0.5(1 - e^{-x/5})$, (g) $y = 4(1 - e^{-x/2})$, (h) $y = 10\ e^{4x}$,

(i) $y = 0.2 - 0.2\ e^{-3x}$

4.2.2 Using a calculator with exponentials

To find the value of a^x the x^y key can be used. For example, to find 2^x with $x = 4$, the key sequences would be:

press the 2 key,
press the x^y key,
press the 4 key,
press the = key.

The result is 16.

To find the value of e^x the e^x key can be used. For example, to find e^4, the key sequences would be:

press the 4 key,
press the e^x key (there is usually no need to press the = key).

The result is 54.598... To find e^{-4}, the key sequences would be:

press the 4 key,
press the +/– key,
press the e^x key.

The result is 0.0183...

Example

Use a calculator to determine the value of $2\ e^{t/3}$, for when $t = 5$.

The key sequence might be:

press the 5 key,
press the ÷ key,
press the 3 key,
press the = key,
press the e^x key,
press the × key,
press the 2 key,
press the = key.

The result is 10.59.

Example

Newton's law of cooling for an object can be written as $\theta = \theta_0 \, e^{-kt}$, where k is a constant, θ is the temperature at a time t and θ_0 the original temperature at $t = 0$. For $\theta_0 = 50$°C and $k = 0.002$ (unit s^{-1}), what will be the value of the temperature after a time of 100 s?

The equation to be solved is thus $q = 50 \, e^{-0.002t}$, with $t = 100$. The key sequences with a calculator might be:

press the keys for 0.002,
press the × key,
press the keys for 100,
press the = key,
press the +/– key,
press the e^x key,
press the × key,
press the keys for 50,
press the = key.

The result is 40.9°C

Revision

5 Use a calculator to determine the values of:

(a) e^6, (b) e^{-6}, (c) $e^{0.4}$, (d) $e^{-0.3}$, (e) $e^{12.2}$, (f) $e^{-5.6}$

6 Use a calculator to complete, to two decimal places, the following table of values:

x	0	+1	+2	+3	infinite
$y = 5 \, e^{1x}$					
$y = 5 \, e^{-1x}$					
$y = 5(1 - e^{-1x})$					

7 Use a calculator to complete, to two decimal places, the following table of values:

x	0	+1	+2	+3	infinite
$y = 10 \, e^{2x}$					
$y = 10 \, e^{-2x}$					
$y = 10(1 - e^{-2x})$					

8 Use a calculator to complete, to two decimal places, the following table of values:

x	0	+1	+2	+3	infinite
$y = 2\ e^{x/2}$					
$y = 2\ e^{-x/2}$					
$y = 2(1 - e^{-x/2})$					

9 Use a calculator to complete, to two decimal places, the following table of values:

x	0	+1	+2	+3	infinite
$y = 10\ e^{0.1x}$					
$y = 10\ e^{-0.1x}$					
$y = 10(1 - e^{-0.1x})$					

10 The voltage, in volts, across a capacitor is given by $20\ e^{-0.1t}$, where t is the time in seconds. Determine the voltage when t is (a) 1 s, (b) 10 s.

11 The atmospheric pressure p is related to the height h above the ground at which it is measured by the equation $p = p_0\ e^{-h/c}$, where c is a constant and p_0 the pressure at ground level where $h = 0$. Determine the pressure at a height of 1000 m if p_0 is 1.01×10^5 Pa and $c = 70\ 000$ (unit m^{-1}).

12 The current i, in amperes, in a circuit involving an inductor in series with a resistor when a voltage is E is applied to the circuit at time $t = 0$ is given by the equation

$$i = \frac{E}{R}(1 - e^{-Rt/L})$$

If R/L has the value 2 Ω/H (actually the same unit as seconds), what is the current when (a) $t = 0$, (b) $t = 1$ s?

13 The voltage v across a resistor in series with an inductor when a voltage E is applied to the circuit at time $t = 0$ is given by the equation

$$v = E(1 - e^{-t/T})$$

where T is the so-called time constant of the circuit. If $T = 0.5$ s, what is the voltage when (a) $t = 0$, (b) $t = 1$ s?

4.2.3 Why e?

Why choose this strange number 2.718... for the base? The reason is linked to the properties of expressions written in this way. For $y = e^x$, the rate of change of y with x, i.e. the slope of a graph of y against x, is equal to e^x. Thus we have for $y = e^x$

slope of graph of y against $x = y = e^x$

This topic is discussed in more detail in chapters 12 and 17.

4.3 Properties of exponentials

The properties of exponentials are the same as those described in chapter 1 for powers. Thus

$$a^x a^y = a^{x+y} \quad [1]$$

$$\frac{a^x}{a^y} = a^{x-y} \quad [2]$$

$$(a^x)^y = a^{xy} \quad [3]$$

$$(ab)^x = a^x b^x \quad [4]$$

$$\left(\frac{a}{b}\right)^x = \frac{a^x}{b^x} \quad [5]$$

where a and b are bases.

Example

Simplify the following:

(a) $2^x 2^{2x}$, (b) $e^{2t}e^{4t}$, (c) $(e^{2t})^{-3}$, (d) $\frac{e^{5x}}{e^{2x}}$, (e) $\frac{10\,e^{t/2}}{2\,e^{t/3}}$, (f) $\frac{1}{e^{2t}} + \frac{2}{e^{3t}}$

(a) Using equation [1] from the above list, then $2^x 2^{2x} = 2^{x+2x} = 2^{3x}$.
(b) Using equation [1] from the above list, then $e^{2t}e^{4t} = e^{2t+4t} = e^{6t}$.
(c) Using equation [3] from the above list, then $(e^{2t})^{-3} = e^{-6t}$.
(d) Using equation [2] from the above list, then

$$\frac{e^{5x}}{e^{2x}} = e^{5x-2x} = e^{3x}$$

(e) Using equation [2] from the above list, then

$$\frac{10\,e^{t/2}}{2\,e^{t/3}} = 5\,e^{\frac{t}{2}-\frac{t}{3}} = 5\,e^{t/6}$$

(f) Bringing the fraction to a common denominator

$$\frac{1}{e^{2t}} + \frac{2}{e^{3t}} = \frac{e^{3t} + 2\,e^{3t}}{e^{2t}e^{3t}} = \frac{e^{3t} + 2\,e^{3t}}{e^{5t}}$$

Alternatively we could just write the equation as

$$\frac{1}{e^{2t}} + \frac{2}{e^{3t}} = e^{-2t} + 2\,e^{-3t}$$

Revision

14 Simplify the following:

(a) $2^{3x}2^{5x}$, (b) $3^{2x}3^{-4x}$, (c) e^3e^5, (d) $e^{3t}e^{5t}$, (e) $e^{-5t}e^{3t}$, (f) $(e^{-4t})^3$, (g) $(1 + e^{2t})^2$,

(h) $\frac{1}{e^{3t}}$, (i) $\frac{e^{3t}}{e^{5t}}$, (j) $\left(\frac{1}{e^{4t}}\right)^2$, (k) $\frac{10\,e^{4t}}{2\,e^t}$, (l) $\frac{2\,e^{3t}e^{5t}}{5\,e^{4t}}$

Problems

1 If the voltage v, in volts, in an electrical circuit is given by the equation $v = 100\ e^{-2t}$, what will be the voltage when the time t is 0? Does the equation describe a voltage which is increasing or decreasing with time?

2 In a chemical reaction the amount C of the starting material left after a time t is given by the equation $C = C_0\ e^{-kt}$, where C_0 and k are constants. (a) Explain why this equation describes a decrease with time of C. (b) What was the amount of the material at $t = 0$? (c) If one reaction has a higher value of k than another, how will this affect the rate at which the amount of material C changes with time?

3 Determine the values of y in the following equations when (i) $x = 0$, (ii) x is infinite:

(a) $y = 2\ e^x$, (b) $y = 12\ e^{x/2}$, (c) $y = 3\ e^{-x}$, (d) $y = 10\ e^{-x/10}$, (e) $y = 2\ e^{-5x}$,

(f) $y = 2(1 - e^{-x/4})$, (g) $y = 12(1 - e^{-4x})$, (h) $y = 2(1 - e^{-x/1000})$

4 Use a calculator to determine, to two decimal places, the values of the following:

(a) e^4, (b) e^{-4}, (c) $e^{0.4}$, (d) $e^{-0.4}$, (e) $10\ e^{2/3}$, (f) $5\ e^{-5/9}$, (g) $-12\ e^{0.5}$,

(h) $10(1 - e^{-3})$, (i) $2(1 - e^{-0.1})$, (j) $4(1 - e^{-1/2})$

5 Use a calculator to complete, to two decimal places, the following table of values of y:

x	0	+1	+2	+3	infinite
$y = 10\ e^{0.2x}$					
$y = 10\ e^{-0.2x}$					
$y = 10(1 - e^{-0.2x})$					
$y = 10\ e^{x/10}$					
$y = 10\ e^{-x/10}$					
$y = 10(1 - e^{-x/10})$					

6 The charge q on a discharging capacitor is related to the time t by the equation $q = Q_0\, e^{-t/CR}$, where Q_0 is the charge at $t = 0$, R is the resistance in the circuit and C the capacitance. Determine the charge on a capacitor after a time of 0.2 s if initially the charge was 1 μC (1 μC = 10^{-6} C), R is 1 MΩ (1 MΩ = 10^6 Ω) and C is 4 μF (1 μF = 10^{-6} F). Note that with the units in seconds (s), coulombs (C), ohms (Ω) and farads (F), the resulting charge will be in coulombs.

7 The current i, in amperes, in a circuit with an inductor in series with a resistor is given by the equation $i = 4(1 - e^{-10t})$, where the time t is in seconds. Determine the current when (a) $t = 0$, (b) $t = 0.05$ s, (c) $t = 0.10$ s, (d) $t = 0.15$ s, (e) t = infinity.

8 The voltage v, in volts, across a capacitor after a time t, in seconds, is given by the equation $v = 10\, e^{-t/3}$. Determine the value of the voltage v after 2 s.

9 The resistance R, in ohms, of an electrical conductor at a temperature of θ°C is given by the equation $R = R_0\, e^{\alpha\theta}$. Determine the resistance at a temperature of 1000°C if R_0 is 5000 Ω and α is1.2 × 10^{-4} (unit per °C).

10 The current i, in amperes, in an electrical circuit varies with time t and is given by the equation $i = 2(1 - e^{-10t})$. Determine the current after times of (a) 0.1 s, (b) 0.2 s, (c) 0.3 s.

11 The amount N of a radioactive material decays with time t and is given by the equation $N = N_0\, e^{-0.7t}$, where t is in years. If at time $t = 0$ the amount of radioactive material is 1 g, what will be the amount after five years?

12 The atmospheric pressure p, in pascals, varies with the height h, in kilometres, above sea level according to the equation $p = p_0\, e^{-0.15h}$. If the pressure at sea level is 10^5 Pa, what will be the pressure at heights of (a) 1 km, (b) 2 km?

13 The voltage v, in volts, across an inductor in an electrical circuit varies with time t, in milliseconds, according to the equation $v = 200\, e^{-t/10}$. Determine the voltage after times of (a) 0.1 ms, (b) 0.5 ms.

14 When the voltage E to a circuit consisting of an inductor in series with a resistor is switched off, the voltage across the inductor varies with time t according to the equation $v = -E\, e^{-t/T}$, where T is the time constant of the circuit. If $T = 2$ s, determine the voltage when (a) $t = 0$, (b) $t = 1$ s.

15 When a voltage E is applied to a circuit consisting of a capacitor in series with a resistor at time $t = 0$, the voltage v across the capacitor varies with time according to the equation $v = E(1 - e^{-t/T})$, where T is the time constant of the circuit. If $T = 0.1$ s, determine the voltage when (a) $t = 0$, (b) $t = 0.1$ s.

16 The temperature θ, in °C, of a cooling object varies with time t, in minutes, according to the equation $\theta = 200\ e^{-0.04t}$. Determine the temperature when (a) $t = 0$, (b) $t = 10$ minutes, (c) t is infinite.

17 The concentration C, as a percentage, of the starting material in a chemical reaction varies with time t, in seconds, according to the equation $C = 100\ e^{-0.005t}$. Determine the concentration when (a) $t = 0$, (b) $t = 10$ s, (c) $t = 100$ s, (d) t is infinite.

18 Under one set of conditions the amplitude A of the oscillations of a system varies with time t according to the equation $A = A_0\ e^{kt}$. Under other conditions the amplitude varies according to the equation $y = A_0\ e^{-kt}$. If k is a positive number, how do the oscillations differ?

19 Simplify the following:

(a) $4^x 4^{2x}$, (b) $5^{-x} 5^{3x}$, (c) $e^{3x} e^{6x}$, (d) $e^{-x} e^{-3x}$, (e) $(e^{2t})^3$, (f) $(1 + e^{2t})^2$,

(g) $(2 + 3\ e^{4t})^2$, (h) $\dfrac{4}{e^{4t}}$, (i) $\dfrac{6\ e^{3t}}{2\ e^{5t}}$, (j) $\left(\dfrac{4}{2\ e^{3t}}\right)^2$

5 Logarithms

5.1 Introduction

Consider the equation $y = 2^x$. If we are given a value for x then we can calculate y. Thus, if we have $x = 3$ then $y = 2^3 = 8$. But suppose we are given a value for y and have to calculate x. How do we solve such an equation? For example, if we have $2.5 = 2^x$, what is x? How do we find the number to which we have to raise a power in order to get a value of 2.5? This chapter introduces a technique for dealing with such problems. It is called *logarithms*.

5.2 Logarithms

If $y = a^x$, then x is called the *logarithm to base a of y*. This can be written as

$$\log_a y = x$$

A logarithm of a number is thus the power to which a base has to be raised to be equal to the number. Logarithms to base 10 are called *common logarithms* and are written as $\log_{10}$ or just log or more commonly lg. In the absence of any special mention of base, lg should be taken to be to base 10. Logarithms to base e are called *natural logarithms* or *Napierian logarithms* and are written as $\log_e$ or more commonly ln.

Consider some logarithms to base 10. When $x = 0$ then $10^0 = 1$ and so $\lg_{10} 1 = 0$. When $x = 1$ then $10^1 = 10$ and so $\lg_{10} 10 = 1$. When $x = 100$ then $10^2 = 100$ and so $\lg_{10} 100 = 2$. Now consider negative values for x. When $x = -1$ then $10^{-1} = 0.1$ and so $\lg_{10} 0.1 = -1$. When $x = -2$ then $10^{-2} = 0.01$ and so $\lg_{10} 0.01 = -2$. Thus we have:

y	0.01	0.1	1	10	100
$\lg_{10} y$	−2	−1	0	1	2

Now consider natural logarithms. When $x = 0$ then $e^0 = 1$ and so we have $\ln 1 = 0$. When $x = 1$ then $e^1 = 2.73$ and so $\ln 2.73 = 1$. When $x = 2$ then $e^2 = 7.39$ and so $\ln 7.39 = 2$. Now consider negative values for x. When we have $x = -1$ then $e^{-1} = 0.37$ and so $\ln 0.37 = -1$. When $x = -2$ then $e^{-2} = 0.14$ and so $\ln 0.14 = -2$. Hence we have:

y	0.14	0.37	1	2.73	7.39
$\ln y$	−2	−1	0	1	2

In general, since $a^1 = a$ then $\lg_a a = 1$; since $a^0 = 1$ then $\lg_a 1 = 0$; and since $a^y = a^y$ then $\lg_a a^y = y$. For values of y greater than 1 we have positive values for the logarithm of y; for values of y less than 1 we have negative values for the logarithm of y. Note that we cannot have a value of $\lg_{10} y$, or $\ln y$, or indeed logarithms to any base, which is 0 or negative.

Example

Write the following equations in logarithmic form:

(a) $3.0 = 10^x$, (b) $0.28 = 10^x$, (c) $4.5 = e^x$, (d) $0.54 = e^x$, (e) $\frac{1}{2} = e^x$.

(a) We can write this as a logarithm to base 10 and so as $x = \lg_{10} 3.0$.
(b) We can write this as a logarithm to base 10 and so as $x = \lg_{10} 0.28$.
(c) We can write this as a logarithm to base e and so as $x = \ln 4.5$.
(d) We can write this as a logarithm to base e and so as $x = \ln 0.54$.
(e) We can write this as a logarithm to base e and so as $x = \ln \frac{1}{2}$.

Example

Write the following logarithms in exponential form:

(a) $x = \lg_{10} 5.0$, (b) $x = \lg_{10} 0.68$, (c) $x = \ln 3.0$, (d) $x = \ln 0.23$.

(a) We can write this as $5.0 = 10^x$.
(b) We can write this as $0.68 = 10^x$.
(c) We can write this as $3.0 = e^x$.
(d) We can write this as $0.23 = e^x$.

Example

Determine x in the equations (a) $\lg_{10} x = 2$, (b) $\lg_3 x = 3$.

(a) We can write this as $x = 10^2 = 100$.
(b) We can write this as $x = 3^3 = 27$.

Revision

1 Write the following equations in logarithmic form:

(a) $7.8 = 10^x$, (b) $0.97 = 10^x$, (c) $1.54 = e^x$, (d) $0.99 = e^x$, (e) $4.5 = 2^x$,

(f) $1.3 = 5^x$, (g) $4.5 = 10^x$, (h) $1.8 = e^x$

2 Write the following equations in exponential form:

(a) $x = \lg_{10} 7.5$, (b) $x = \lg_{10} 0.45$, (c) $x = \ln 9.9$, (d) $x = \ln 0.88$,

(e) $x = \lg_2 20$, (f) $x = \lg_5 12$, (g) $x = \ln 3$

3 Determine x in the following equations:

(a) $x = \lg_2 16$, (b) $x = \lg_{10} 100$, (c) $x = \lg_{10} 0.01$, (d) $\lg_{10} x = 4$,

(e) $\lg_3 x = 2$, (f) $\lg_2 x = 4$

5.2.1 Using a calculator for logarithms

Logarithms can be obtained using calculators. Thus for determining $\lg_{10} 5$ the procedure would be: press key 5, press key log (there is no need then to press =), to give the result 0.699. This means that

$$\lg_{10} 5 = 0.699 \text{ and so } 5 = 10^{0.699}$$

For determining ln 5 the procedure would be: press key 5, press ln (there is no need then to press =), to give the result 1.609. This means that

$$\ln 5 = 1.609 \text{ and so } 5 = e^{1.609}$$

Example

Determine, using logarithms, the value of x in the following equations:

(a) $3.0 = 10^x$, (b) $0.28 = 10^x$, (c) $4.5 = e^x$, (d) $0.54 = e^x$

(a) We can write this as a logarithm to base 10 and so as $x = \lg_{10} 3.0$. Hence, using a calculator, we key in 3.0, then press key log. The result is 0.477.
(b) We can write this as a logarithm to base 10 and so as $x = \lg_{10} 0.28$. Hence, using a calculator, we key in 0.28, then press key log. The result is −0.553.
(c) We can write this as a logarithm to base e and so as $x = \ln 4.5$. Hence, using a calculator, we key in 4.5, then press key ln. The result is 1.504.
(d) We can write this as a logarithm to base e and so as $x = \ln 0.54$. Hence, using a calculator, we key in 0.54, then press key ln. The result is −0.616.

Example

Determine x in the equation $1.54 = \ln (3x)$.

From the definition of the logarithm, then e to the power 1.54 must be $3x$.

$$3x = e^{1.54} = 4.665$$

Hence $x = 1.555$.

Example

Determine x in the equation $1.563 = \ln\left(\frac{2.645}{x}\right)$.

From the definition of the logarithm, then e to the power 1.563 must be $2.645/x$.

$$e^{1.563} = \frac{2.645}{x}$$

Hence

$$x = \frac{2.645}{e^{1.563}} = 2.645\,e^{-1.563} = 2.645 \times 0.210 = 0.555$$

Revision

4 Determine, using logarithms, the values of x to three decimal places in the following equations:

(a) $7.8 = 10^x$, (b) $0.97 = 10^x$, (c) $1.54 = e^x$, (d) $0.99 = e^x$, (e) $x = \lg_{10} 7.5$,

(f) $x = \lg_{10} 0.45$, (g) $x = \ln 9.9$, (h) $x = \ln 0.88$, (i) $\ln x = 5$,

(j) $\ln x = 0.1$, (k) $1.5 = \ln\left(\frac{x}{1.8}\right)$, (l) $2.76 = \ln\left(\frac{53.4}{x}\right)$,

(m) $1.89 = \lg_{10}\left(\frac{45}{x}\right)$, (n) $0.310 = \ln(5x)$, (o) $10 = \ln(2x)$

5.3 Properties of logarithms

The properties of logarithms follow directly from the properties of powers discussed in chapters 1 and 4. Consider the multiplication of two numbers. If we have $A = a^x$ and $B = a^y$, then $A \times B = a^{x+y}$. Using logarithms we have $\lg_a A = x$ and $\lg_a B = y$. The logarithm of the product means that

$$\lg_a (A \times B) = x + y$$

Thus

$$\lg_a (A \times B) = \lg_a A + \lg_a B$$

For example, $\lg (5 \times 2) = \lg 5 + \lg 2$.

Consider the division of two numbers. If we have $A = a^x$ and $B = a^y$, then $A \div B = a^{x-y}$. Thus

$$\lg_a\left(\frac{A}{B}\right) = x - y$$

and so

$$\lg_a\left(\frac{A}{B}\right) = \lg_a A - \lg_a B$$

For example, $\lg (5/2) = \lg 5 - \lg 2$.

Consider raising a number to a power. If we have $A = a^x$, then $A^n = a^{nx}$. Thus

$$\lg_a A^n = nx$$

Since $\lg_a A = x$, then

$$\lg_a A^n = n \lg_a A$$

For example, $\lg 5^2 = 2 \lg 5$. You can think of this in terms of the multiplication of two numbers, i.e. $\lg 5^2 = \lg (5 \times 5) = \lg 5 + \lg 5 = 2 \lg 5$.

Example

Evaluate the following logarithms:

(a) $\lg_{10} (2 \times 30)$, (b) $\ln (12 \times 0.02)$, (c) $\lg_{10} (20/7)$, (d) $\ln (2/0.03)$,

(e) $\lg_{10} 4^3$, (f) $\ln 3^7$, (g) $\ln \sqrt{6e}$.

(a) Using the rule for multiplication given above,

$$\lg_{10} (2 \times 30) = \lg_{10} 2 + \lg_{10} 30 = 0.301 + 1.477 = 1.778$$

(b) Using the rule for multiplication given above:

$$\ln (12 \times 0.02) = \ln 12 + \ln 0.02 = 2.485 - 3.912 = -1.427$$

(c) Using the rule for division given above,

$$\lg_{10}\left(\frac{20}{7}\right) = \lg_{10} 20 - \lg_{10} 7 = 1.301 - 0.845 = 0.456$$

(d) Using the rule for division given above,

$$\ln\left(\frac{2}{0.03}\right) = \ln 2 - \ln 0.03 = 0.693 - (-3.507) = 4.200$$

(e) Using the rule for powers given above,

$$\lg_{10} 4^3 = 3 \lg_{10} 4 = 3 \times 0.602 = 1.806$$

(f) Using the rule for powers given above,

$$\ln 3^7 = 7 \ln 3 = 7 \times 1.099 = 7.693$$

(g) We can write $\ln \sqrt{6e}$ as $\ln (6e)^{1/2}$ and so, using the rule for powers, as $\frac{1}{2} \ln 6e$. Using the rule for multiplication this can be written as

$$\tfrac{1}{2} \ln 6e = \tfrac{1}{2}(\ln 6 + \ln e) = \tfrac{1}{2}(1.796 + 1) = 1.396$$

Revision

5 Evaluate the following:

(a) $\lg_{10}(123 \times 513)$, (b) $\ln(0.003 \times 125)$, (c) $\lg_{10}(0.015/0.013)$,

(d) $\ln(3.18/578)$, (e) $\lg_{10} 0.012^6$, (f) $\ln 1.56^4$, (g) $\ln e^2$, (h) $\lg 10^{-6}$

5.4 Solving exponential equations

The following examples illustrate how logarithms, in particular natural logarithms, can be used to solve equations.

Example

Determine t in the equation $0.5 = 1 - e^{-t/3}$.

The equation can be rearranged to give $e^{-t/3} = 0.5$ and hence, since we have $e^{-t/3} = 1/e^{t/3}$, then

$$e^{t/3} = \frac{1}{0.5} = 2$$

Taking natural logarithms of both sides of the equation gives

$$\frac{t}{3} = \ln 2$$

Hence $t = 3 \ln 2 = 3 \times 0.693 = 2.079$.

Example

Determine x in the equation $5^x = 1.8$.

Taking natural logarithms of both sides of the equation gives

$$\ln(5^x) = \ln 1.8$$

Hence $x \ln 5 = \ln 1.8$ and so

$$x = \frac{\ln 1.8}{\ln 5} = \frac{0.588}{1.609} = 0.365$$

Alternatively, we could have used logarithms to base 10. Then we obtain $x \lg_{10} 5 = \lg_{10} 1.8$ and so $x = \lg_{10} 1.8/\lg_{10} 5 = 0.255/0.699 = 0.365$.

Example

Determine x in the equation $4^{x+5} = 0.7$.

Taking natural logarithms of both sides of the equation gives

$$(x+5)\ln 4 = \ln 0.7$$

Hence

$$x+5 = \frac{\ln 0.7}{\ln 4} = \frac{-0.357}{1.386} = -0.258$$

Thus $x = -0.258 - 5 = -5.258$.

Revision

6 Determine t in the following equations:

(a) $12 = 5\,e^{-2t}$, (b) $4 = 9\,e^{-3t}$, (c) $12 = 100(1 - e^{-t/2})$, (d) $2 = 15(1 - e^{3t})$

7 Determine x in the following equations:

(a) $2^x = 1.9$, (b) $4^x = 9.7$, (c) $4^{3x-1} = 12$, (d) $2^{x+1} = 0.5$, (e) $0.9^{2x+3} = 4.5$,

(f) $2^{2x-1} = 13$, (g) $e^{2x} = 4$, (h) $e^{x-1} = 4$

8 The amount N of a radioactive substance decays with time t according to the equation $N = N_0\,e^{-\lambda t}$. What will be the time taken for the amount of the radioactive material to drop to half its initial value?

9 The resistance R of an electrical conductor increases with the temperature θ, being given by the equation $R = R_0\,e^{\alpha\theta}$, where R_0 and α are constants. Determine the value of α if $R_0 = 500\ \Omega$ and R is $520\ \Omega$ at a temperature of 300°C.

Problems

1 Write the following equations in logarithmic form:

(a) $2.4 = 10^x$, (b) $2.5 = 10^x$, (c) $0.68 = 10^x$, (d) $1.3 = e^x$, (e) $55.6 = e^x$,

(f) $5.8 = e^x$, (g) $0.35 = e^x$

2 Write the following equations in exponential form:

(a) $x = \lg_{10} 34$, (b) $x = \lg_{10} 0.54$, (c) $x = \lg_{10} 345$, (d) $x = \ln 37$,

(e) $x = \ln 0.86$, (f) $x = \ln 397$

3 Determine the values of x, to three decimal places, for the following:

(a) $2.4 = 10^x$, (b) $2.5 = 10^x$, (c) $0.68 = 10^x$, (d) $1.3 = e^x$, (e) $55.6 = e^x$,

(f) $5.8 = e^x$, (g) $0.35 = e^x$, (h) $x = \lg 34$, (i) $x = \lg_{10} 0.54$, (j) $x = \lg_{10} 345$,

(k) $x = \ln 37$, (l) $x = \ln 0.86$, (m) $x = \ln 397$

4 Evaluate the following:

(a) $\ln(12 \times 0.019)$, (b) $\lg_{10}(13 \times 57)$, (c) $\ln(193/27)$,

(d) $\lg_{10}(0.019/0.178)$, (e) $\ln 24.6^5$, (f) $\lg_{10} 18.2^5$

5 Determine x in the following equations:

(a) $1.67 = \ln\left(\frac{4.56}{x}\right)$, (b) $6.78 = \ln\left(\frac{x}{234}\right)$, (c) $0.189 = \ln(5x)$,

(d) $0.013 = \ln(2x)$

6 Determine t in the following equations:

(a) $3 = 5\ \mathrm{e}^{-t/3}$, (b) $0.1 = 10\ \mathrm{e}^{-4t}$, (c) $12 = 100(1 - \mathrm{e}^{-t/4})$,

(d) $0.6 = 2(1 - \mathrm{e}^{-t/5})$

7 The current i, in amperes, in an electrical circuit varies with time t, in seconds, and is given by the equation $i = I(1 - \mathrm{e}^{-t/CR})$. If $I = 10$ A, what will be the value of t when $i = 1$ A if $CR = 5$ s.

8 Determine x in the following equations:

(a) $3^x = 100$, (b) $5^x = 17$, (c) $2^{3x+1} = 5$, (d) $4^{4x-5} = 0.56$,

(e) $10^{4x+5} = 0.056$, (f) $\mathrm{e}^{3x} = 2\ \mathrm{e}^x$, (g) $\mathrm{e}^{-(2x+1)} = 4$

9 The atmospheric pressure p varies with the height h above sea level, being given by the equation $p = p_0\ \mathrm{e}^{-h/C}$, where p_0 and C are constants. If $p_0 = 1.01 \times 10^5$ Pa and $p = 9.96 \times 10^4$ Pa at $h = 1000$ m, determine C.

10 The voltage drop v across an inductor in an electrical circuit varies with time t and is given by the equation $v = 10\ \mathrm{e}^{-Rt/L}$. What will be the time taken for the voltage to drop to 4 V if $R = 100\ \Omega$ and $L = 0.001$ H?

11 When a hot object cools from a temperature θ_0 above its surroundings, then its temperature θ varies with time t according to the equation $\theta = \theta_0\ \mathrm{e}^{-kt}$. Determine k if for $\theta_0 = 500$°C, the object cools to 100°C in a time of 3000 s.

12 Determine the value of T_1 when $T_2 = 20$, $\mu = 0.5$ and $\theta = 1.2$ rad, given that $\ln\left(\frac{T_1}{T_2}\right) = \mu\theta$.

6 Straight-line motion

6.1 Introduction

In the earlier chapters in this section, the algebraic techniques developed have been illustrated by showing their use in engineering problems. In this chapter the emphasis is on the engineering with the algebraic techniques used to support the development of an engineering topic. The aim is to put the algebraic techniques in an engineering context. The engineering topic in this chapter is the development and use of the equations for straight-line motion. The next chapter considers another engineering topic, basic electrical circuit analysis. Both these topics are followed up in later sections when illustrating the use of other mathematical techniques.

6.1.1 Terms

The discussion in this chapter is restricted to motion in a straight line. The following are basic terms used in the discussion:

1. *Velocity* is the rate at which the distance of an object along a straight line varies with time. The term *displacement* is generally used for the distance when measured in a straight line between the initial and final positions, the term *distance* being used when the path followed could be curved and the distance is measured along the curve rather than the stright line distance between the initial and final positions. In the situations considered in this chapter where the motion is restricted to a straight line, the terms distance and displacement mean the same thing.
2. The *average velocity* over some time interval is given by

$$\text{average velocity} = \frac{\text{displacement}}{\text{time interval}}$$

3. A *uniform velocity* occurs when equal distances are covered in the same straight line direction in equal intervals of time, however small the time interval.
4. *Acceleration* is the rate of change of velocity with time. The term *retardation* is often used if the velocity is decreasing.
5. The *average acceleration* over some time interval is given by

$$\text{average acceleration} = \frac{\text{change of velocity}}{\text{time interval}}$$

   Because with a retardation we have a final velocity less than the initial velocity, the average retardation is a negative quantity. A positive value for the acceleration indicates that the velocity is increasing.
6. A *uniform acceleration* occurs when the velocity changes by equal amounts in equal intervals of time, however small the time interval.

6.2 Equations of motion

Consider an object moving along a straight line with a uniformly accelerated motion. If u is the initial velocity, i.e. at time $t = 0$, and v the velocity after some time t, then the change in velocity in the time interval t is $(v - u)$. Hence the acceleration a is

$$a = \frac{v - u}{t}$$

Multiplying both sides of the equation by t gives $at = v - u$. Hence the equation can be manipulated to give

$$v = u + at \qquad [1]$$

If the object, in its straight line motion, covers a distance s in a time t, then the average velocity in that time interval is s/t. With an initial velocity of u and a final velocity of v at the end of the time interval, the average velocity is

$$\text{average velocity} = \frac{u + v}{2}$$

Hence

$$\frac{s}{t} = \frac{u + v}{2}$$

Multiplying both sides of the equation by t gives

$$s = \left(\frac{u + v}{2}\right)t \qquad [2]$$

Substituting for v by using the equation $v = u + at$ gives

$$s = \left(\frac{u + (u + at)}{2}\right)t = \left(\frac{2u + at}{2}\right)t$$

Hence

$$s = ut + \tfrac{1}{2}at^2 \qquad [3]$$

Consider the equation $v = u + at$. Squaring both sides of this equation gives

$$v^2 = (u + at)^2 = u^2 + 2uat + a^2t^2$$

We can rewrite this equation as

$$v^2 = u^2 + 2a(ut + \tfrac{1}{2}at^2)$$

Hence

$$v^2 = u^2 + 2as \qquad [4]$$

The equations [1], [2], [3] and [4] are referred to as the equations for straight-line motion. The following examples illustrate their use in solving engineering problems.

Example

An object moves in a straight line with a uniform acceleration. If it starts from rest and takes 12 s to cover 100 m, what is the acceleration?

We have $u = 0$, $s = 100$ m, $t = 12$ s and are required to obtain a. The equation containing these terms is equation [3], i.e. $s = ut + \frac{1}{2}at^2$. Thus

$$100 = 0 + \tfrac{1}{2}a \times 12^2 = 0 + 72a$$

Rearranging this equation to give $72a = 100$, then dividing both sides by 72, gives

$$a = \frac{100}{72} = 1.4 \text{ m/s}^2$$

Example

A package slides down a slope with a constant acceleration. Its velocity increases from 100 mm/s to 320 mm/s in a distance of 600 mm. What is the acceleration?

The question gives $u = 100$ mm/s, $v = 320$ mm/s, $s = 600$ mm and we are required to obtain a. The equation containing these terms is [4], i.e. $v^2 = u^2 + 2as$. We thus have

$$320^2 = 100^2 + 2a \times 600$$

Transposing terms, we can rewrite this equation as

$$1200a = 320^2 - 100^2 = 102\,400 - 10\,000 = 92\,400$$

Hence, dividing both sides of the equation by 1200 gives

$$a = \frac{92\,400}{1\,200} = 77 \text{ mm/s}^2$$

Example

A car travelling at 25 m/s brakes and slows down with a uniform retardation of 1.2 m/s^2. How long will it take to come to rest?

A retardation is a negative acceleration in that the final velocity is less than the starting velocity. Thus we have $u = 25$ m/s, a final velocity of

$v = 0$, retardation $a = -1.2$ m/s^2 and are required to obtain t. Thus, since the equation containing these terms is [1], i.e. $v = u + at$, then

$$0 = 25 + (-1.2)t = 25 - 1.2t$$

Transposing the $1.2t$ to the other side of the equation gives $1.2t = 25$. Dividing both sides of the equation by 1.2 then gives

$$t = \frac{25}{1.2} = 20.8 \text{ s}$$

Example

A car is moving with a velocity of 10 m/s. It then accelerates at 0.2 m/s^2 for 100 m. What will be the time taken for the 100 m to be covered?

We have $u = 10$ m/s, $a = 0.2$ m/s^2 and $s = 100$ m and are required to obtain t. The equation containing these terms is equation [3], i.e. $s = ut + \frac{1}{2}at^2$. Thus

$$100 = 10t + \tfrac{1}{2} \times 0.2t^2$$

We can write this equation as

$$0.1t^2 + 10t - 100 = 0$$

This is a quadratic equation. Using the formula

$$t = \frac{-b \pm \sqrt{b^2 - 4ac}}{2a}$$

then

$$t = \frac{-10 \pm \sqrt{100 - 4 \times 0.1(-100)}}{2 \times 0.1} = \frac{-10 \pm \sqrt{140}}{0.2} = \frac{-10 \pm 11.83}{0.2}$$

Hence $t = -50 - 59.2 = -109.2$ s or $t = -50 + 59.2 = 9.2$ s. Since the negative time has no significance, the answer is 9.2 s. The answer can be checked by substituting it back in the original equation.

Revision

1 The following all refer to an object moving in a straight line with uniform acceleration.

(a) Initially at rest, acceleration 2 m/s^2 for 8 s. Determine the distance travelled in that time.
(b) Initial velocity 3 m/s, acceleration 2 m/s^2 for 6 s. Determine the velocity at the end of that time.

(c) Initial velocity 3 m/s, final velocity 5 m/s after 2 s. Determine the distance travelled in that time.
(d) Initial velocity 3 m/s, distance travelled 20 m, final velocity 7 m/s. Determine the time taken.
(e) Initial velocity 10 m/s, acceleration –4 m/s^2, final velocity 2 m/s. Determine the distance travelled.
(f) Initial velocity 10 m/s, acceleration –4 m/s^2, final velocity 2 m/s. Determine the time taken.
(g) Acceleration 0.5 m/s^2 for 10 s, final velocity 10 m/s. Determine the initial velocity.

2 A train is moving with a velocity of 10 m/s. It then accelerates at a uniform rate of 2.5 m/s^2 for 8 s. What is the velocity after the 8 s?

3 An object starts from rest and moves with a uniform acceleration of 2 m/s^2 for 20 s. What is the distance moved by the object?

4 A cyclist is moving with a velocity of 1 m/s. He/she then accelerates at 0.4 m/s^2 for 100 m. What will be the time taken for the 100 m?

5 Rearrange the equation $v^2 = u^2 + 2as$ to give (a) s in terms of the other quantities, (b) a in terms of the other quantities, (c) u in terms of the other quantities.

6 A car accelerates from 7.5 m/s to 22.5 m/s at 2 m/s^2. What is (a) the time taken, (b) the distance travelled during this acceleration?

6.2.1 Simultaneous equations

The following example and revision problems involve straight-line motion when simultaneous equations can occur.

Example

A stationary car A is passed by car B moving with a uniform velocity of 15 m/s. Two seconds later, car A starts moving with a constant acceleration of 1 m/s^2 in the same direction. How long will it take for car A to draw level with car B?

Both the cars will have travelled the same distance s when they have drawn level. We will measure time t from when car B initially passes car A. For car B, since u = 15 m/s and a = 0, we can write for equation [3], i.e. $s = ut + \frac{1}{2}at^2$,

$$s = 15t + 0$$

For car A, since u = 0 and a = 1 m/s^2, we can write for equation [3],

$$s = 0 + \tfrac{1}{2} \times 1 \times (t - 2)^2$$

We write $(t - 2)$ for the time since car A is in motion for a time 2 s less than t. These two simultaneous equations can be solved by substituting for s in the second equation. Thus

$$15t = 0.5(t-2)^2 = 0.5(t^2 - 4t + 4) = 0.5t^2 - 2t + 2$$

Transposing the $15t$ to the other side of the equation gives the quadratic equation

$$0.5t^2 - 17t + 2 = 0$$

Using the formula for the quadratic equation

$$t = \frac{-b \pm \sqrt{b^2 - 4ac}}{2a} = \frac{17 \pm \sqrt{17^2 - 4 \times 0.5 \times 2}}{2 \times 0.5} = 17 \pm \sqrt{285}$$

Hence $t = 17 + 16.9 = 33.9$ s or $t = 17 - 16.9 = 0.1$ s. With the data given, this second answer is not feasible and so the time is 33.9 s after car B has passed car A.

Revision problems

7 Car A, travelling with a uniform velocity of 25 m/s, overtakes a stationary car B. Two seconds later, car B starts and accelerates at 6 m/s^2. How far will B have to travel before it catches up A?

8 Car A and car B are parked along the same straight road with car B 27 m in front of A. Car A moves off with a uniform acceleration of 1 m/s^2 and then 6 s later car B starts off with a uniform acceleration of 1.5 m/s^2. How long will it take, from car A starting off, for the cars to draw level?

6.3 Motion under the action of gravity

All freely falling objects in a vacuum fall with the same uniform acceleration directed towards the surface of the earth as a result of a gravitational force acting between the object and the earth. This acceleration is termed the *acceleration due to gravity* g. For most practical purposes, the acceleration due to gravity at the surface of the earth is taken as being 9.81 m/s^2. The equations for motion of a falling object are those for motion in a straight line with the acceleration as g.

When an object is thrown vertically upwards it suffers an acceleration directed towards the surface of the earth. An acceleration directed in the opposite direction to which an object is moving is a retardation, i.e. a negative acceleration since it results in a final velocity less than the initial velocity. The result is that the object slows down. The object slows down until its velocity upwards eventually becomes zero, it then having attained its greatest height above the ground. Then the object reverses the direction of its motion and falls back towards the earth, accelerating as it does. Figure 6.1 illustrates these points.

	Direction of motion		
At greatest height H velocity = 0			Fall starts at height H with initial velocity = 0
	↑	↓	
Velocity decreasing	↑	↓	Velocity increasing
Retardation of $-g$	↑	↓	Acceleration of $+g$
	↑	↓	
Ground, h = 0	↑	↓	Back at ground, h = 0

Figure 6.1 *Vertical motion under gravity*

The equations for motion due to gravity are those for straight-line motion with uniform acceleration, the acceleration being g. Care has to be taken with the signs used with velocities, displacements and the acceleration to ensure that all the directions involved for the elements in an equation are the same. Thus when an object is falling then the velocities, displacements and accelerations are all measured in a downard direction. This means that g is a positive quantity since the acceleration is in that direction. When an object is moving upwards then the velocities, displacements and accelerations must all be in an upward direction. Thus for the acceleration we must use $-g$.

Example

A stone is dropped down a vertical shaft and reaches the bottom in 5 s. How deep is the shaft? Take g as 9.8 m/s^2.

We have $u = 0$, $t = 5$ s and $a = g = 9.8$ m/s^2. We have to determine the distance fallen, h. Thus, using equation [3], i.e. $s = ut + \frac{1}{2}at^2$, we obtain

$$h = 0 + \frac{1}{2} \times 9.8 \times 5^2 = 122.5 \text{ m}$$

This equation is consistent with regard to directions since the acceleration and h are both in the same direction, i.e. downwards.

Example

An object is thrown vertically upwards with a velocity of 8.0 m/s. What will be the greatest height reached and the time taken to reach that height? Take g as 9.8 m/s^2.

We have $u = 8.0$ m/s, $a = -g = -9.8$ m/s^2 and, at the greatest height H, $v = 0$. We have initially to determine H. The velocities and displacement are measured in an upward direction and so the acceleration due to gravity is negative. Thus, using equation [4], i.e. $v^2 = u^2 + 2as$, we have

$$0 = 8.0^2 + 2(-9.8)H$$

Thus $0 = 64 - 19.6H$. Transposing the $19.6H$ term to the other side of the equation gives $19.6H = 64$, and so

$$H = \frac{64}{19.6} = 3.27 \text{ m}$$

To determine t we can use equation [1], i.e. $v = u + at$. Thus

$$0 = 8.0 + (-9.8)t$$

Thus $8.0 - 9.8t = 0$ and, transposing the $-9.8t$ to the other side of the equation, gives $9.8t = 8.0$, and so

$$t = \frac{8.0}{9.8} = 0.82 \text{ s}$$

Revision

9 A stone is thrown vertically upwards with an initial velocity of 15 m/s. What will be the distance from the point of projection and the velocity after 1 s? Take g as 9.8 m/s^2.

10 A stone is thrown vertically upwards with an initial velocity of 5 m/s. What will be the time taken for it to reach the greatest height? Take g as 9.8 m/s^2.

11 A stone falls off the edge of a cliff. With what velocity will it hit the beach below if the height of the cliff above the beach is 50 m? Take g as 9.8 m/s^2.

Problems

1 A conveyor belt is used to move an object along a production line with an acceleration of 1.5 m/s^2. What is the velocity after it has moved 3 m?

2 A train starts from rest and moves with a uniform acceleration so that it takes 300 s to cover 9000 m. What is the acceleration?

3 Sound travels in air at 0°C at 330 m/s. How far will the sound of a shot travel in 5 s?

4 A car is initially at rest. It then accelerates at 2 m/s^2 for 6 s. What will be the velocity after that time?

5 A car accelerates at a uniform rate from 15 m/s to 35 m/s in 20 s. How far does the car travel in this time?

6 An object has an initial velocity of 10 m/s and is accelerated at 3 m/s^2 for a distance of 800 m. What is the time taken to cover this distance?

7 Rearrange the equation $v = u + at$ to give (a) a in terms of the other quantities, (b) t in terms of the other quantities, (c) u in terms of the other quantities.

8 Rearrange the equation $s = ut + \frac{1}{2}at^2$ to give (a) a in terms of the other quantities, (b) u in terms of the other quantities.

9 What is the velocity attained by an object falling from rest through a distance of 4.9 m? Take g as 9.8 m/s^2.

10 An object is thrown vertically upwards with an initial velocity of 8 m/s. How long will it take for the object to return to the same point from which it was thrown? Take g as 9.8 m/s^2.

11 A fountain projects water vertically to a height of 5 m. What is the velocity with which the water must be leaving the fountain nozzle? Take g as 9.8 m/s^2.

12 A stone is thrown vertically upwards with an initial velocity of 9 m/s. What is the greatest height reached by the stone and the time taken? Take g as 9.8 m/s^2.

13 An object is thrown vertically upwards with a velocity of 34.2 m/s. How long will it take to reach a height, on its way up and its way back down, of 49 m above the point of projection? Take g as 9.8 m/s^2.

14 An object is thrown vertically upwards with a velocity of 20 m/s from an initial height h above the ground. It takes 5 s from the time of being projected upwards before the object hits the gound. Determine h. Take g as 9.8 m/s^2.

15 An object is thrown vertically upwards with a velocity of 30 m/s and, from the same point at the same time, another object is thrown vertically downwards with a velocity of 30 m/s. How far apart will the objects be after 3s? Take g as 9.8 m/s^2.

16 A ball is held 2.0 m above the ground and then released. The ball hits the ground and then rebounds with half the speed it had prior to the impact. Find the greatest height the ball reaches after bouncing. Take g as 9.8 m/s^2.

7 Electrical circuits

7.1 Introduction

In the earlier chapters in this section, the algebraic techniques developed have been illustrated by showing their use in engineering problems. In this chapter the emphasis is on the engineering with the algebraic techniques used to support the development of engineering topics. The aim is to put the algebraic techniques in an engineering context. The engineering topic discussed in this chapter is basic electrical circuit analysis, the discussion being restricted to d.c. circuits containing resistors. This discussion is followed up in later sections.

7.1.1 Terms

The following are the basic terms used in this discussion of electrical circuits, and circuits discussed in later chapters in this book:

1. *Current* This has the unit of ampere or amp (A) and is the rate of movement of charge. When there is a charge Q moved in a time t then the average current I over that time is given by $I = Q/t$.
2. *Charge* The unit of charge is the coulomb (C) and is the quantity of charge passing a point in a circuit when a current of 1 A flows for 1 s.
3. *Power* When there is a current through a conductor then charge is continuously being moved through it. Energy is continually required to keep the charge moving. The rate at which this energy is required is called *power P*. Thus if energy W is required over a time t then $P = W/t$. The unit of power is the watt (W), one watt being a rate of energy use of 1 joule per second.
4. *Voltage* or *potential difference* This is the energy required to move a unit charge between two points and has the unit of volt (V). Thus if V is the potential difference between two points in a circuit then the energy W required to move a charge Q between the points is given by the equation $V = W/Q$. Thus we can write

   $$P = \frac{W}{t} = \frac{Q}{t} \times \frac{W}{Q} = IV$$

   A potential difference of 1 V is said to exist between two points of a conducting wire carrying a constant current of 1 A when the power dissipated between these points is equal to 1 W.
5. *Resistance* The electrical *resistance*, R, of a circuit element is the property it has of impeding the flow of electrical current and is defined by the equation

   $$R = \frac{V}{I}$$

where V is the potential difference across an element when I is the current through it. The unit of resistance is the ohm (Ω) when the potential difference is in volts and the current in amps.

Often in dealing with electrical circuits the values of currents, voltages, powers, etc. are expressed in multiples or submultiples of their units by the addition of a prefix letter to the unit. For example, a current may be written as 10 mA (milliamperes) or a voltage as 2 kV (kilovolts). The following are some common multiples.

Multiplication factor	Prefix	
1 000 000 000 = 10^9	giga	G
1 000 000 = 10^6	mega	M
1000 = 10^3	kilo	k
100 = 10^2	hecto	h
10 = 10	deca	da
0.1 = 10^{-1}	deci	d
0.01 = 10^{-2}	centi	c
0.001 = 10^{-3}	milli	m
0.000 001 = 10^{-6}	micro	μ
0.000 000 001 = 10^{-9}	nano	n
0.000 000 000 001 = 10^{-12}	pico	p

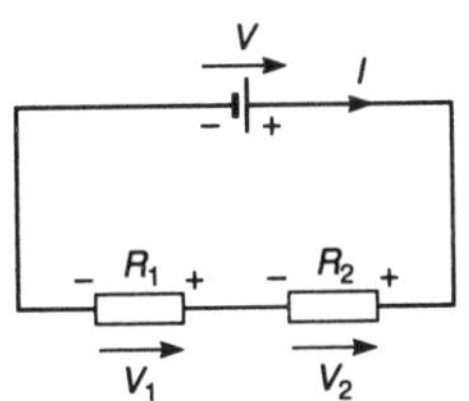

Figure 7.1 *Circuit notation*

The convention adopted with electrical circuits is to indicate the direction of the current in a circuit conductor by an arrow on the conductor (figure 7.1). Current is taken as flowing out of the positive plate of a cell, round the circuit and back into the negative plate, the arrow for current indicating this direction. The direction of the potential difference between two points in a circuit is indicated by an arrow between the points and generally drawn parallel to the conductor. The arrow for the potential difference points towards the point which is taken to be more positive. Think of charge flowing down the potential hill with a cell pumping the charge back up to the top of the hill again. The current arrow points down the hill and the potential difference arrows up the hill.

7.2 Resistors in circuits

For many resistors, if the temperature does not change, the potential difference across a resistor is proportional to the current through it. Thus V/I = a constant and so, since V/I is the resistance R we can write

$$\frac{V}{I} = R$$

Multiplying both sides of the equation by I gives

$$V = RI$$

This is called *Ohm's law*. Thus circuit elements which obey Ohm's law have a constant value of resistance which does not change when the current changes. Also, since the power $P = IV$, we can write

$$P = IV = I^2R = \frac{V^2}{R}$$

Example

If the current through a resistor of resistance 10 Ω is 0.2 A, what will be (a) the potential difference across it and (b) the power dissipated in it?

(a) Using Ohm's law,

$$V = RI = 10 \times 0.2 = 20 \text{ V}$$

(b) The power P is given by

$$P = I^2R = 0.2^2 \times 10 = 0.4 \ \Omega$$

Example

Rearrange the equation $P = I^2R$ to give the current I in terms of the other quantities.

We can write the equation as $I^2R = P$. Dividing both sides of the equation by R gives

$$I^2 = \frac{P}{R}$$

Then taking the square root of both sides gives

$$I = \sqrt{\frac{P}{R}}$$

Revision

1 Rearrange the following equation to give V in terms of the other quantities.

$$P = \frac{V^2}{R}$$

2 What is the resistance of a resistor if the potential difference across it is 3 V and the current through it is 0.2 A?

3 If the current through a 10 Ω resistor is 0.5 A, what is the potential difference across it?

4 If the current through a resistor of resistance 100 Ω is 20 mA, what will be (a) the potential difference across it and (b) the power dissipated in it?

7.2.1 Resistors in series and parallel

When circuit elements are in *series*, as the resistors are in figure 7.2, then we have the same current I through each element. Thus the potential difference V_1 across resistor R_1 is IR_1. The potential difference V_2 across resistor R_2 is IR_2. The potential difference V across the series arrangement is the sum of the potential differences, i.e.

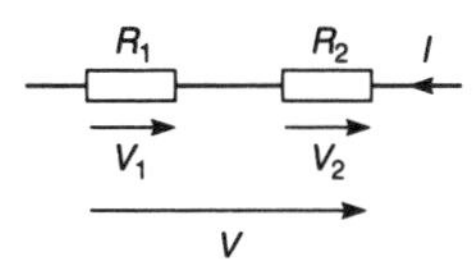

Figure 7.2 *Resistors in series*

$$V = V_1 + V_2 = IR_1 + IR_2 = I(R_1 + R_2)$$

Dividing both sides of the equation by I gives

$$\frac{V}{I} = R_1 + R_2$$

We could replace the two resistors by a single, equivalent, resistor R giving the same current I and overall potential difference V if we have $R = V/I$. Hence

$$R = R_1 + R_2$$

For resistors in series the total resistance R is the sum of their resistances.

When circuit elements are in *parallel*, as the resistors are in figure 7.3, then the same potential difference occurs across each element. Since charge does not accumulate at a junction, then the current entering a junction must equal the sum of the current leaving it, i.e. $I = I_1 + I_2$. But for resistor R_1 we have $V = I_1R_1$ and for resistor R_2 we have $V = I_2R_2$. Thus, substituting these values in the above equation gives

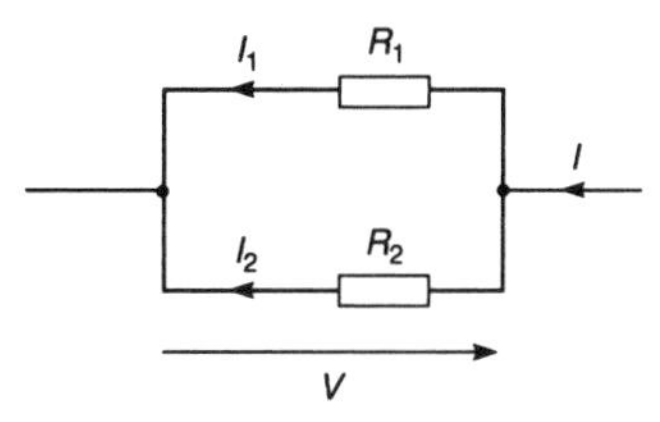

Figure 7.3 *Resistors in parallel*

$$I = \frac{V}{R_1} + \frac{V}{R_2} = V\left(\frac{1}{R_1} + \frac{1}{R_2}\right)$$

Dividing both sides of the equation by V gives

$$\frac{I}{V} = \frac{1}{R_1} + \frac{1}{R_2}$$

We could replace the two resistors by a single, equivalent, resistor R if we have $R = I/V$. Hence

$$\frac{1}{R} = \frac{1}{R_1} + \frac{1}{R_2}$$

The reciprocal of the equivalent resistance is equal to the sum of the reciprocals of the parallel resistances.

Example

A circuit consists of four resistors in series, they having resistances of 2 Ω, 10 Ω, 5 Ω and 8 Ω. A voltage of 12 V is applied across them. Determine (a) the total resistance, (b) the circuit current, (c) the power dissipated in each resistor, (d) the total power dissipation.

(a) The total resistance is the sum of the four resistances, i.e.

$$R = R_1 + R_2 + R_3 + R_4$$

and so we have $R = 2 + 10 + 5 + 8 = 25\ \Omega$

(b) The current I through the circuit is given by $V = IR$. Dividing both sides of this equation by R gives

$$\frac{V}{R} = I$$

Hence

$$I = \frac{V}{R} = \frac{12}{25} = 0.48 \text{ A or } 480 \text{ mA}$$

(c) The power developed in each resistance is given by $P = I^2R$, the current I being the same through each resistor. Thus, for each resistor:

$$P = 0.48^2 \times 2 = 0.46 \text{ W}$$

$$P = 0.48^2 \times 10 = 2.30 \text{ W}$$

$$P = 0.48^2 \times 5 = 1.15 \text{ W}$$

$$P = 0.48^2 \times 8 = 1.84 \text{ W}$$

(d) The total power developed can be obtained, either by adding together the powers dissipated by each resistor or determining the power dissipated by the equivalent resistor. Thus

$$P = 0.46 + 2.30 + 1.15 + 1.84 = 5.75 \text{ W}$$

or $P = I^2R = 0.48^2 \times 25 = 5.76$ W. The slight difference in the answers is due to rounding errors.

Example

A circuit consists of two resistors of 5 Ω and 10 Ω in parallel with a voltage of 2 V applied across them. Determine (a) the total circuit resistance, (b) the current through each resistor, (c) the total current through the circuit.

(a) The total resistance R is given by

$$\frac{1}{R} = \frac{1}{R_1} + \frac{1}{R_2} = \frac{1}{5} + \frac{1}{10} = \frac{3}{10}$$

We thus have

$$R = \frac{10}{3} = 3.3\ \Omega$$

(b) The same potential difference V will be across each resistor. Since $V = IR$, then if we divide both sides of this equation by R we obtain

$$I = \frac{V}{R}$$

Thus the currents are

$$I = \frac{V}{R_1} = \frac{2}{5} = 0.4 \text{ A or } 400 \text{ mA}$$

$$I = \frac{V}{R_2} = \frac{2}{10} = 0.2 \text{ A or } 200 \text{ mA}$$

(c) The total current entering the arrangement can be obtained by either adding the currents through each of the branches or determining the current taken by the equivalent resistance. Thus, adding the currents through the branches gives

$$I = 0.4 + 0.2 = 0.6 \text{ A or } 600 \text{ mA}$$

or, considering the current taken by the equivalent resistance,

$$I = \frac{V}{R} = \frac{2}{3.3} = 0.61 \text{ A or } 610 \text{ mA}$$

Rounding errors account for the slight differences in these answers.

Revision

5 Calculate the equivalent resistance of three resistors of 5 Ω, 10 Ω and 15 Ω which are connected in (a) series, (b) parallel.

6 Three resistors of 15 kΩ, 20 kΩ and 24 kΩ are connected in parallel. What is (a) the total resistance and (b) the power dissipated if a voltage of 24 V is applied across the circuit?

7 A circuit consists of two resistors in parallel. The total resistance of the parallel arrangement is 4 Ω. If one of the resistors has a resistance of 12 Ω, what will be the resistance of the other one?

8 A circuit consists of two resistors in series. If they have resistances of 4 Ω and 12 Ω, what will be (a) the circuit current and (b) the potential difference across each resistor when a potential difference of 4 V is applied across the circuit?

9 A circuit consists of two resistors in parallel. If they have resistances of 4 Ω and 12 Ω, what will be (a) the current through each resistor and (b) the total power dissipated when a potential difference of 4 V is applied across the circuit?

7.2.2 Series–parallel circuits

Many circuits are more complicated than just the simple series or parallel connections of resistors and have varied combinations of both series and parallel connected resistors. In general, such circuits are analysed by beginning with the innermost parts of the circuit and block-by-block reducing the elements to single equivalent resistances, before finally simplifying the circuit to just one equivalent resistance. The following example illustrates the procedure.

Example

Determine the total resistance of the circuits shown in figure 7.4.

(a) The circuit consists of a parallel arrangement of 4 Ω and 2 Ω which is then in series with resistance of 6 Ω. For the two resistors in parallel, the equivalent resistance is given by

$$\frac{1}{R} = \frac{1}{4} + \frac{1}{2} = \frac{3}{4}$$

Hence

$$R = \frac{4}{3} = 1.33\ \Omega$$

This is in series with the 6 Ω resistor and so the total resistance is given by

$$R = 1.33 + 6 = 7.33\ \Omega.$$

(b) The circuit consists of 2 Ω and 4 Ω in series. Thus the equivalent resistance R is

$$R = R_1 + R_2 = 2 + 4 = 6\ \Omega$$

This is then in parallel with 3 Ω. Thus the total resistance is given by

$$\frac{1}{R} = \frac{1}{3} + \frac{1}{6} = \frac{3}{6}$$

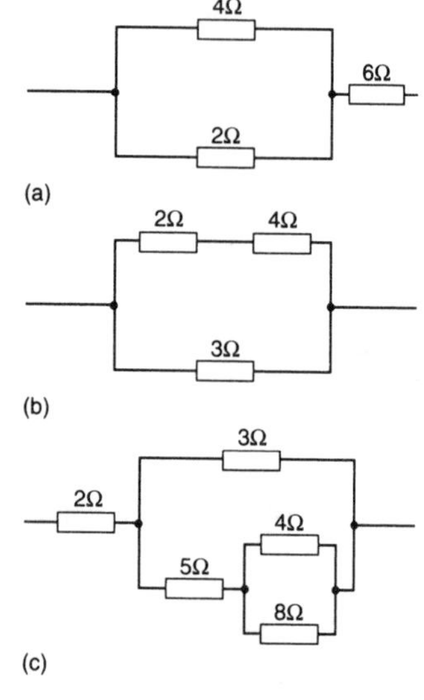

Figure 7.4

and so $R = \frac{6}{3} = 2\,\Omega$

(c) The 4 Ω and the 8 Ω resistors are in parallel and can be replaced by an equivalent resistance R, where

$$\frac{1}{R} = \frac{1}{4} + \frac{1}{8} = \frac{3}{8}$$

and so $R = 8/3 = 2.67\ \Omega$. This is in series with the 5 Ω resistor and so their combined resistance is $5 + 2.67 = 7.67\ \Omega$. This is in parallel with 3 Ω and so the combined resistance is given by

$$\frac{1}{R} = \frac{1}{7.67} + \frac{1}{3} = \frac{3 + 7.67}{7.67 \times 3} = \frac{10.67}{23.01}$$

Thus the combined resistance is

$$R = \frac{23.01}{10.67} = 2.16\,\Omega$$

This is in series with the 2 Ω resistance and so the total resistance of the circuit $2.16 + 2 = 4.16\ \Omega$.

Revision

10 Determine the total resistances of the circuits and the circuit currents I shown in figure 7.5 .

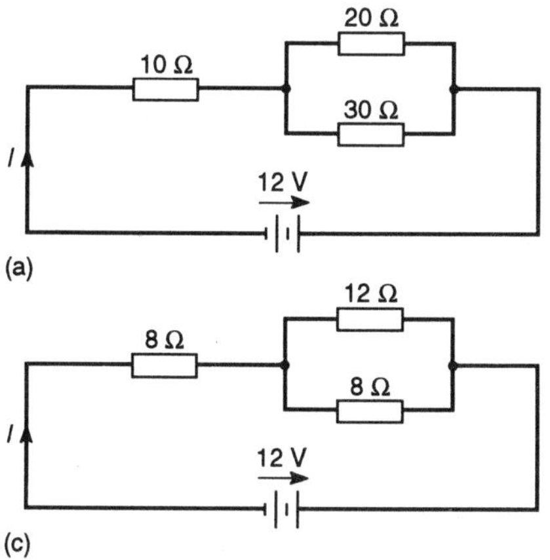

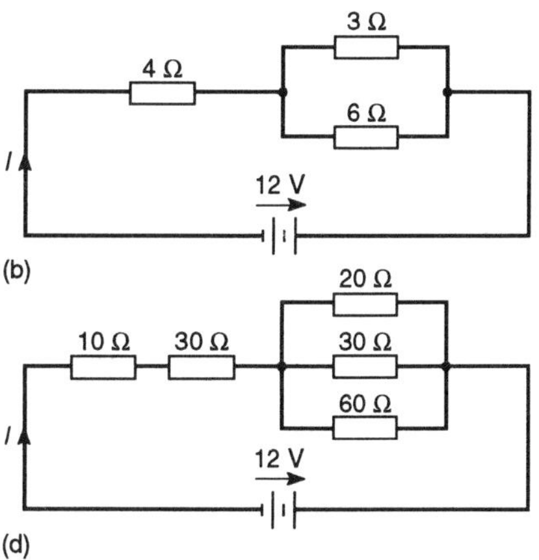

Figure 7.5

11 A resistor of 16 Ω is connected in series with an arrangement of a 12 Ω resistor in parallel with a 6 Ω resistor. If the supply voltage to the circuit is 12 V, what is (a) the total circuit resistance, (b) the current through each of the resistors, (c) the total power dissipated?

12 A resistor of 10 Ω is connected in series with an arrangement of a 20 Ω resistor in parallel with a 30 Ω resistor. If the supply voltage is 12 V, what is (a) the total circuit resistance, (b) the potential difference across each resistor, (c) the total power dissipated?

7.3 Kirchhoff's laws

Kirchhoff's current law states that at any node in an electrical circuit, the current entering it equals the current leaving it. The term *node* is used for a point in an electrical circuit at which conductors join to form a junction. Thus for the node shown in figure 7.6(a), we must have

$$I_1 = I_2 + I_3 + I_4$$

This law is often stated as: the algebraic sum of the currents flowing towards a node is zero. The algebraic sum is the sum obtained when account is taken of the direction of the currents, with perhaps currents into a node being taken as positive and currents leaving it as negative. Then for the node in figure 7.6(a) we have

$$I_1 - I_2 - I_3 - I_4 = 0$$

Kirchhoff's voltage law states that around any closed path in a circuit, the algebraic sum of the voltages across all the components and the algebraic sum of the applied voltages is zero. A closed path, often called a *loop*, is exactly what is implied. If we start at one node and follow the circuit through perhaps a number of nodes we end up at the starting node without encountering a node more than once. Figure 7.6(b) shows such a closed path. The term 'algebraic sum' means that we have to take account of the directions of the potential differences and applied voltages when we add them. The potential difference across a resistor R when a current I flows through it is IR and is in the opposite direction to the current. Thus moving round the loop in figure 7.6(b) in a clockwise direction gives

$$E - IR_1 - IR_2 - IR_3 = 0$$

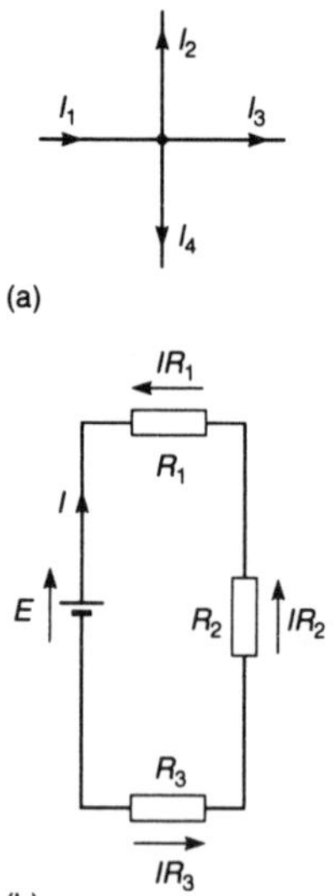

Figure 7.6 *Using Kirchhoff's laws*

We can analyse a network by using both Kirchhoff's current law and Kirchhoff's voltage law and obtaining a set of simultaneous equations. The first of the examples that follows illustrates this. However, we can simplify the analysis and effectively only use the voltage law if we use what are termed *mesh currents*. The term *mesh* is used for a loop which does not contain any other loop. Thus for the circuit shown in figure 7.7, there are three loops ABCF, CDEF and ABCDEF but only the first two are meshes. For each mesh a mesh current is defined, such a current being considered to circulate round the mesh. The same direction of circulation must be taken for all the mesh currents in a circuit, the usual convention being for all mesh

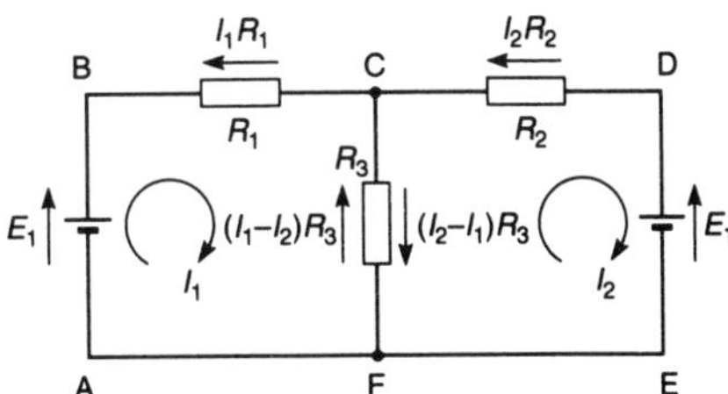

Figure 7.7 *Mesh currents*

currents to circulate in a clockwise direction. Consider mesh 1. The current through R_1 is the mesh current I_1. The current through resistor R_2, which is common to the two meshes, is the algebraic sum of the two mesh currents, i.e. $I_1 - I_2$ in the direction of I_1. Thus, applying Kirchhoff's voltage law to the mesh gives

$$E_1 - I_1R_1 - (I_1 - I_2)R_2 = 0$$

For mesh 2 we have a current of I_2 through R_3 and a current of $(I_2 - I_1)$, in the direction of I_2, through R_2. Thus applying Kirchhoff's voltage law to this mesh gives

$$-E_2 - I_2R_2 - (I_2 - I_1)R_2 = 0$$

Thus, with a two-mesh circuit, we end up with two simultaneous equations. With a three-mesh circuit we would have obtained three simultaneous equations, with a four-mesh circuit four simultaneous equations.

Example

Determine the current through the 20 Ω resistor in the circuit shown in figure 7.8

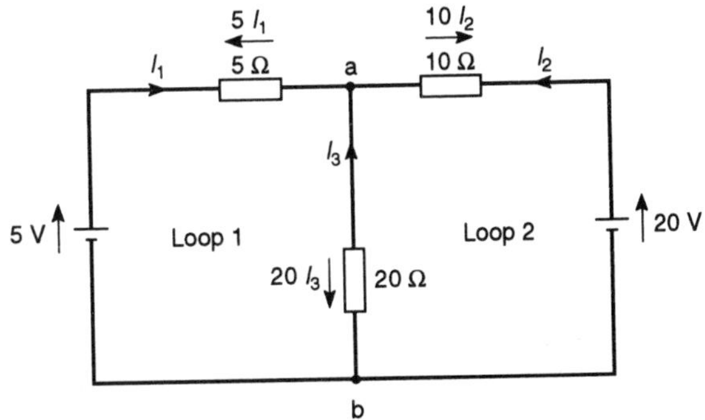

Figure 7.8

Consider node a in the circuit. With the currents as labelled, Kirchhoff's current law gives

$$I_1 + I_2 + I_3 = 0 \qquad [1]$$

Now apply Kirchhoff's voltage law to loop 1. The current through the 5 Ω resistor is I_1 and so the potential difference across the resistor, in the opposite direction to the current, is $5I_1$. The current through the 20 Ω resistor is I_3 and so the potential difference across the resistor, in the opposite direction to the current, is $20I_3$. Thus for loop 1 we can write

$$5 - 5I_1 + 20I_3 = 0 \qquad [2]$$

Now consider loop 2. The current through the 10 Ω resistor is I_2 and so the potential difference across the resistor, in the opposite direction to the current, is $10I_2$. The current through the 20 Ω resistor is I_3 and so the potential difference across the resistor, in the opposite direction to the current, is $20I_3$. Thus for loop 2 we can write

$$10I_2 - 20 - 20I_3 = 0 \qquad [3]$$

We thus have three simultaneous equations with three unknowns.

Adding 5 times equation [1] to equation [2] gives

$$\begin{array}{rl} & 5I_1 + 5I_2 + 5I_3 = 0 \\ \text{plus} & \underline{5 - 5I_1 \qquad + 20I_3 = 0} \\ & 5 \qquad + 5I_2 + 25I_3 = 0 \end{array}$$

If we multiply this equation by 2 and subtract equation [3] from it then we obtain

$$\begin{array}{rl} & 10 + 10I_2 + 50I_3 = 0 \\ \text{minus} & \underline{-20 + 10I_2 - 20I_3 = 0} \\ & +30 \qquad + 70I_3 = 0 \end{array}$$

Hence $70I_3 = -30$ and so $I_3 = -30/70 = -0.43$ A. The minus sign is because the direction of the current is in the opposite direction to that indicated by the arrow for the current direction in the figure.

Example

Determine, using mesh analysis, the current through the 20 Ω resistor in the circuit shown in figure 7.9.

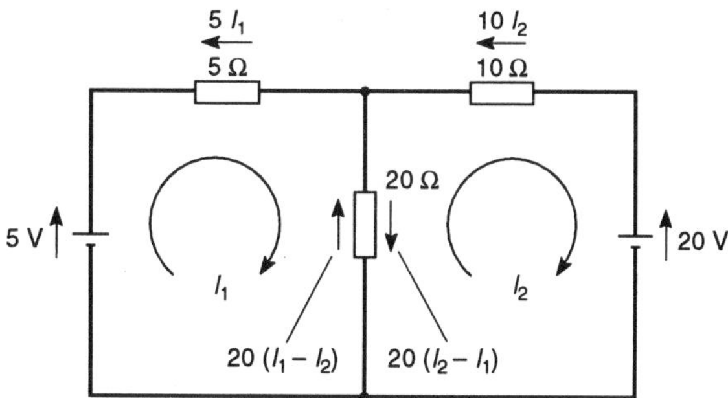

Figure 7.9

This question is a repeat of the previous example but tackled by mesh analysis. With a current of I_1 through the 5 Ω resistor then the potential difference across that resistor is $5I_1$. For the 20 Ω resistor, the current through it is $(I_1 - I_2)$ and so the potential difference across that resistor is $20(I_1 - I_2)$. For mesh 1, applying Kirchhoff's voltage law, we thus obtain

$$5 - 5I_1 - 20(I_1 - I_2) = 0$$

This can be rewritten as

$$5 = 25I_1 - 20I_2 \qquad [4]$$

For mesh 2, the current through the 10 Ω resistor is I_2 and so the potential difference across it is $10I_2$. The current through the 20 Ω resistor is $(I_2 - I_1)$ and so the potential difference across that resistor is $20(I_2 - I_1)$. Thus, applying Kirchhoff's voltage law to mesh 2,

$$-10I_2 - 20 - 20(I_2 - I_1) = 0$$

This can be rewritten as

$$20 = 20I_1 - 30I_2 \qquad [5]$$

We now have in equations [4] and [5] a pair of simultaneous equations. Multiplying equation [4] by 4 and subtracting from it five times equation [5] gives

$$\begin{array}{rrcl} & 20 = & 100I_1 & - \ 80I_2 \\ \text{minus} & \underline{100 =} & \underline{100I_1} & \underline{- \ 150I_2} \\ & -80 = & 0 & + \ 70I_2 \end{array}$$

Thus $I_2 = 80/70 = -1.14$ A. Substituting this value in equation [4] gives $5 = 25I_1 - 20(-1.14)$ and so $-17.8 = 25I_1$ and so $I_1 = -0.71$ A. The minus signs indicate that the currents are in the opposite directions to those indicated in the figure. The current through the 20 Ω resistor, which is $(I_1 - I_2)$ in the direction of I_1, is thus $-0.71 - (-1.14) = 0.43$ A.

Revision

13 Determine the currents labelled as I in the circuits given in figure 7.10.

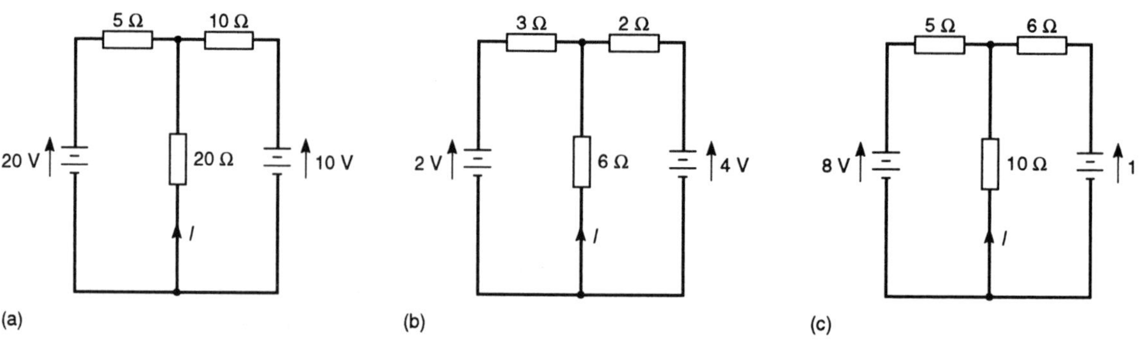

Figure 7.10

Problems

1 An electrical lamp filament dissipates 20 W when the current through it is 0.10 A. What is (a) the potential difference across the lamp and (b) its resistance at that current?

2 A 100 Ω resistor is specified as being able to be used up to a maximum power dissipation of 250 mW. What should be the maximum current used with the resistor?

3 If the current through a resistor of resistance 100 Ω is 50 mA, what will be (a) the potential difference across it and (b) the power dissipated in it?

4 If the current through a resistor of resistance 1 kΩ is 0.02 A, what will be (a) the potential difference across it and (b) the power dissipated in it?

5 Resistors of resistances 20 Ω, 30 Ω and 50 Ω are connected in series across a d.c. supply of 20 V. What is the current in the circuit and the potential difference across each resistor?

6 Determine the power dissipated in each resistor and the total power dissipated when a voltage of 10 V is applied across (a) a series-connected (b) a parallel-connected pair of resistors with resistances of 20 Ω and 50 Ω.

7 A circuit consists of two parallel-connected resistors of 2.2 kΩ and 3.9 kΩ in series with a 1.5 kΩ resistor. If the supply voltage connected to the circuit is 20 V, what is the current drawn from the supply?

8 A circuit consists of three parallel-connected resistors of 330 Ω, 560 Ω and 750 Ω in series with a 800 Ω resistor. If the supply voltage connected to the circuit is 12 V, what are the voltages across the parallel resistors and the series resistor?

9 Three resistors of 6 Ω, 12 Ω and 24 Ω are connected in parallel across a 12 V voltage supply. What is the total resistance and the current through each resistor?

10 A circuit consists of a resistor of 8 Ω in series with a an arrangement of another 8 Ω resistor in parallel with a 12 Ω resistor. If the voltage input to the circuit is 24 V, what will be the current taken from the voltage source?

11 A circuit consists of a resistor of 2 Ω in series with an arrangement of three parallel resistors, these having resistances of 4 Ω, 10 Ω and 20 Ω. If the voltage input to the circuit is 10 V, what will be the current through the 2 Ω resistor?

12 Determine the equivalent resistances of the circuits shown in figure 7.11.

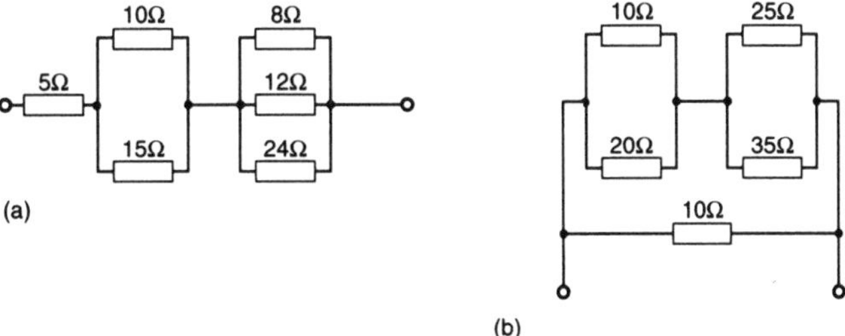

Figure 7.11

13 Resistors with resistances of 10 Ω, 20 Ω and 30 Ω are connected (a) in series, (b) in parallel across a 12 V supply. Determine the current taken from the supply in each case.

14 Determine, using Kirchhoff's laws, the current I in each of the circuits shown in figure 7.12.

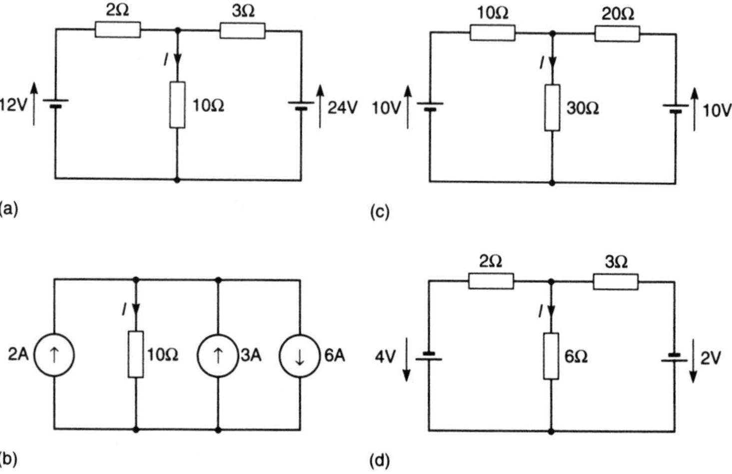
2Ω
3Ω
I
12V
10Ω
24V
(a)
10Ω
20Ω
I
10V
30Ω
10V
(c)
I
2A
10Ω
3A
6A
(b)
2Ω
3Ω
I
4V
6Ω
2V
(d)

Figure 7.12

Section B
Trigonometry

The aims of this section are to enable the reader to:

- Calculate trigonometric ratios.
- Use trigonometric identities.
- Use and manipulate trigonometric formulae.
- Determine the angles, sides and areas of triangles.
- Solve engineering problems involving trigonometric ratios.
- Represent and manipulate vector quantities.
- Solve engineering problems involving vectors.
- Represent and manipulate phasors.
- Solve engineering problems involving phasors.

This section is concerned with the development of the skills to use trigonometry in the solution of engineering problems, in particular those involved in the use of vectors and phasors. It is assumed that the reader is able to manipulate algebraic expressions, e.g. has completed chapter 1, and is able to plot graphs. Thus the reader might like to refer to chapter 12.

Chapter 8 assumes some familiarity with angles, triangles and the Pythagoras theorem, though the topics are revised in that chapter. Chapter 9 assumes that chapter 8 has been completed. Chapter 10 applies trigonometric concepts to the solution of problems involving vectors. It assumes that chapters 8 and 9 have been completed. Chapter 11 applies trigonometric concepts to an introduction to the solution of alternating current problems via the use of phasors. It assumes that chapter 8 has been completed and links in with some of the basic concepts of vectors given in chapter 10.

8 Trigonometric ratios

8.1 Introduction

Trigonometry is concerned with the relationships between the sides and angles of triangles. Such relationships are involved in a wide range of engineering problems. For example, in considering the equilibrium of a structure the angles at which forces are applied have to be taken into account and this can be done by the use of trigonometry. Quantities, such as forces, where both the direction and the size are necessary to determine their effect are termed vector quantities. Another example involving trigonometry is the description of alternating currents and voltages by equations involving sines or cosines and their representation by phasors.

In this opening section, the units used for angles are reviewed and then later sections develop the trigonometric ratios and their inter-relationships. The chapter concludes with a discussion of the trigonometric ratios for any magnitude angle and graphs of the trigonometric ratios plotted against angle.

8.1.1 Degrees and radians

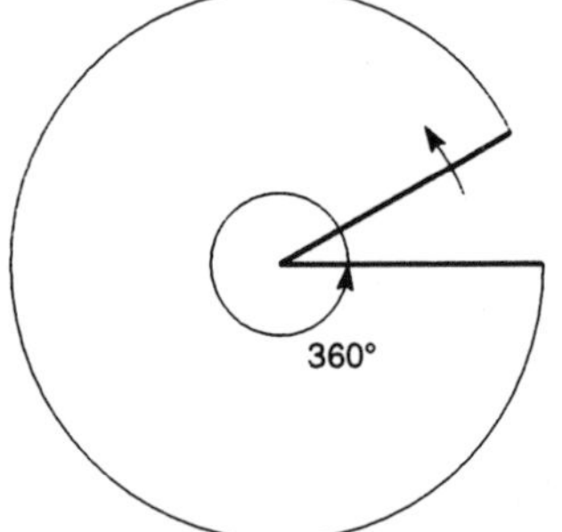

Figure 8.1 *Degrees*

The basic units in which angles are measured are *degrees* and *radians*. With degrees, one complete rotation of a radial line (figure 8.1) is defined as being a rotation through 360°. A rotation through a quarter of a revolution is a rotation through 90°, such an angle being referred to as a ***right angle***. With radians, when the tip of the radial line has moved through a distance equal to the length of the line, i.e. the radius, then the angle swept out is said to be 1 radian (figure 8.2), the radian being defined as the angle subtended at the centre of a circle by an arc whose length is equal to the radius. Since the circumference of a circle is $2\pi r$, then the number of radians swept out by the rotating radial line in moving through one complete circle is 2π radians. Thus

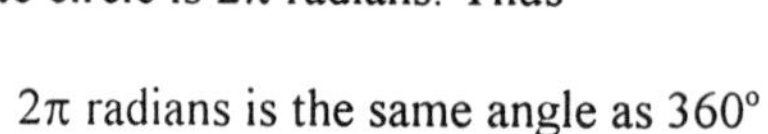

2π radians is the same angle as 360°

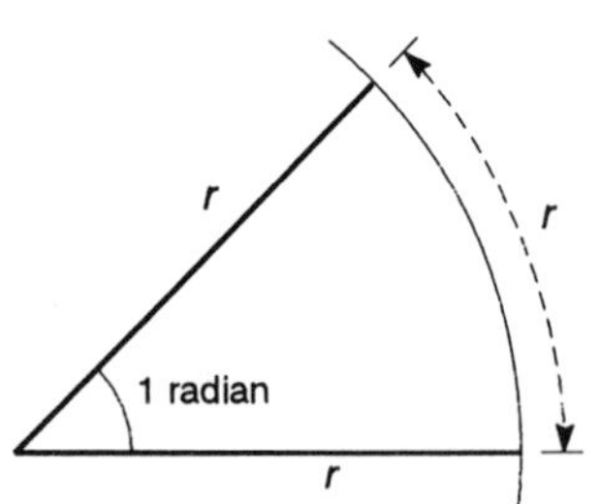

Figure 8.2 *The radian*

Hence

$$1 \text{ radian} = \frac{360°}{2\pi} = 57.3°$$

A rotation through a quarter of a revolution, i.e. a right angle, is a rotation through a quarter of 2π rad and hence is $\pi/2$ rad. Thus 90° is the same as $\pi/2$ rad.

Example

Convert the following angles in degrees into angles in radian measure: (a) 30°, (b) 60°, (c) 90°, (d) 180°, (e) 270°.

(a) Since 360° is 2π radians, then 30° is $\frac{30}{360} \times 2\pi = \frac{\pi}{6} = 0.52$ rad.

(b) Since 360° is 2π radians, then 60° is $\frac{60}{360} \times 2\pi = \frac{\pi}{3} = 1.05$ rad.

(c) Since 360° is 2π radians, then 90° is $\frac{90}{360} \times 2\pi = \frac{\pi}{2} = 1.57$ rad.

(d) Since 360° is 2π radians, then 180° is $\frac{180}{360} \times 2\pi = \pi = 3.14$ rad.

(e) Since 360° is 2π radians, then 270° is $\frac{270}{360} \times 2\pi = \frac{3\pi}{2} = 4.71$ rad.

Example

Convert the following angles in radians into angles in degrees:
(a) 0.80 rad, (b) 1.2 rad.

(a) Since 2π radians is 360°, then 0.80 radians is $\frac{0.8}{2\pi} \times 360 = 45.8°$.

(b) Since 2π radians is 360°, then 1.2 radians is $\frac{1.2}{2\pi} \times 360 = 68.8°$.

Revision

1 Convert the following angles in degrees into angles in radian measure:
(a) 65°, (b) 125°, (c) 245°, (d) 20°, (e) 119°, (f) 312°

2 Convert the following angles in radians into angles in degrees:
(a) 0.3 rad, (b) 1.9 rad, (c) 3.5 rad, (d) $\pi/6$ rad, (e) $3\pi/7$ rad, (f) $\pi/5$ rad

8.2 Trigonometric ratios

Consider the two right-angled triangles shown in figure 8.3 with the same angle θ. For such triangles the ratios of corresponding sides are the same. For example, for each triangle the ratio opposite side/hypotenuse has the same value, the ratio opposite side/adjacent side has the same value, etc. The ratios between two sides of a right-angled triangle occur frequently and so are given special names. The ratios are defined as:

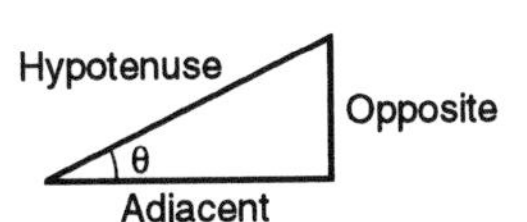

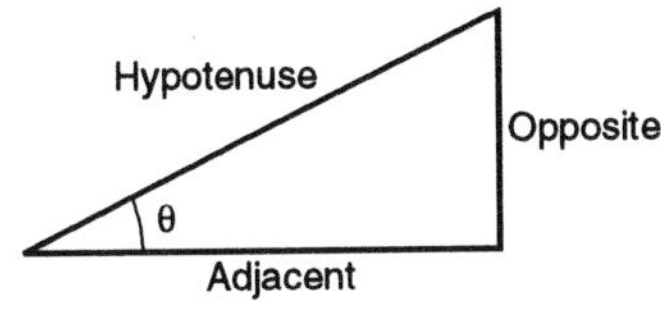

Figure 8.3 *Similar triangles*

1 sine of angle θ, written as sin θ

$$\sin\theta = \frac{\text{opposite}}{\text{hypotenuse}}$$

2 cosine of angle θ, written as cos θ

$$\cos\theta = \frac{\text{adjacent}}{\text{hypotenuse}}$$

3 tangent of angle θ, written as tan θ

$$\tan\theta = \frac{\text{opposite}}{\text{adjacent}}$$

4 cosecant of angle θ, written as cosec θ

$$\operatorname{cosec}\theta = \frac{\text{hypotenuse}}{\text{opposite}}$$

5 secant of angle θ, written as sec θ

$$\sec\theta = \frac{\text{hypotenuse}}{\text{adjacent}}$$

6 cotangent of angle θ, written as cot θ

$$\cot\theta = \frac{\text{adjacent}}{\text{opposite}}$$

Example

For the triangle shown in figure 8.4, what are the values of (a) sin θ, (b) cos θ, (c) tan θ?

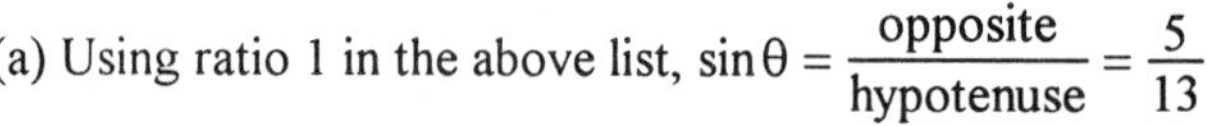

(a) Using ratio 1 in the above list, $\sin\theta = \frac{\text{opposite}}{\text{hypotenuse}} = \frac{5}{13}$

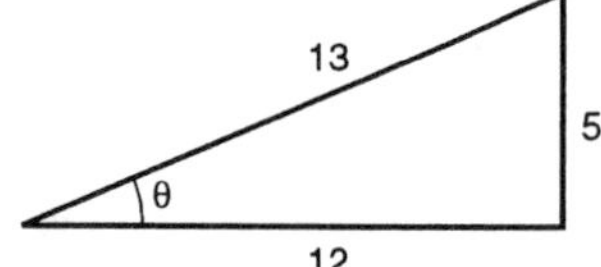

Figure 8.4

(b) Using ratio 2 in the above list, $\cos\theta = \frac{\text{adjacent}}{\text{hypotenuse}} = \frac{12}{13}$

(c) Using ratio 3 in the above list, $\tan\theta = \frac{\text{opposite}}{\text{adjacent}} = \frac{5}{12}$

Revision

3 For the triangle shown in figure 8.5, determine:

(a) sin θ, (b) cos θ, (c) tan θ, (d) cosec θ, (e) sec θ, (f) cot θ.

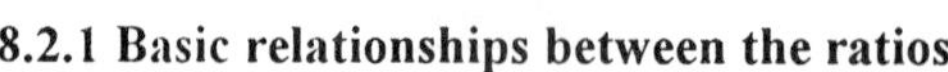

8.2.1 Basic relationships between the ratios

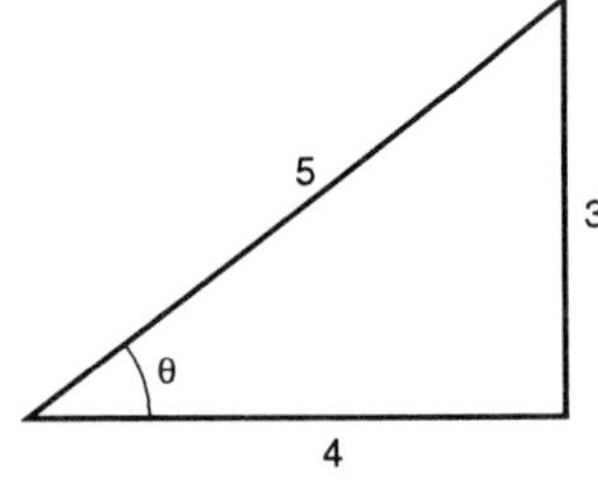

Figure 8.5

From the definitions of the trigonometric ratios given above it can be seen that:

$$\operatorname{cosec}\theta = \frac{1}{\sin\theta},\quad \sec\theta = \frac{1}{\cos\theta},\quad \cot\theta = \frac{1}{\tan\theta}$$

Also we can write

$$\tan\theta = \frac{\text{opposite}}{\text{adjacent}} = \frac{\text{opposite}}{\text{hypotenuse}} \times \frac{\text{hypotenuse}}{\text{adjacent}}$$

Thus

$$\tan\theta = \frac{\sin\theta}{\cos\theta}$$

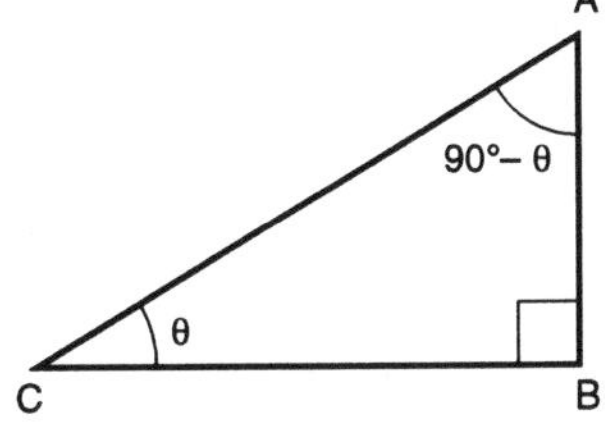

Figure 8.6 *Angles in triangle*

Since the angles in a triangle add up to 180° and one of them in a right angled triangle is 90°, then the angles in such a triangle are always 90°, θ and (90° – θ). Thus if we consider, for a right-angled triangle, the sine of the angle (90° – θ) then we have, with the sides labelled as in figure 8.6,

$$\sin(90^\circ - \theta) = \frac{BC}{AC}$$

But BC/AC is the cosine of angle θ. Thus

$$\sin(90^\circ - \theta) = \cos\theta$$

If we consider the cosine of the angle (90° – θ) then

$$\cos(90^\circ - \theta) = \frac{AB}{AC}$$

But AB/BC is the sine of angle θ. Thus

$$\cos(90^\circ - \theta) = \sin\theta$$

Example

If cos 40° = 0.766, what is sec 40°?

Since sec 40° = 1/cos 40°, then sec 40° = 1/0.766 = 1.305.

Example

If cos 40° = 0.766, what is sin 50°?

Since 40° + 50° = 90°, then sin 50° = cos (90° – 50°) = cos 40°. Thus sin 50° = 0.766.

Revision

4 If sin 20° is 0.342, what are (a) cos 70° and (b) cosec 20°?

5 If cos 45° is 0.707, what are (a) sin 45° and (b) sec 45°?

6 If sin 30° = 0.500 and cos 30° = 0.866, what are (a) tan 30° and (b) cot 30°?

8.2.2 Using a calculator for the trigonometric ratios

The values of the trigonometric ratios for particular angles can be obtained by the use of a calculator. Thus for example to find sin 50°, with my calculator I press the keys for 50 to appear in the display, then the key for sin. The result then appears as 0.7660... Note that the display when carrying out such a calculation displays the symbol deg, indicating that the number 50 is assumed to be in degrees. If the angle is in radians, e.g. sin 0.40, with my calculator I press the mode key until rad appears in the display, then 0.40, then sin. The result then appears as 0.3894...

Sometimes we require to determine the value of an angle that gives a particular trigonometric ratio. For example, we might need to find the angle that gives a sine of 0.6, i.e. we have $\sin\theta = 0.6$ and need to find θ. To do this we use the calculator key for $\sin^{-1}$. This symbol indicates that we are carrying out the inverse operation to finding the sine of an angle. The key sequence is thus: press the keys for 0.6, then press the key for $\sin^{-1}$. The result is 36.9°. The keys marked as $\sin^{-1}$, $\cos^{-1}$ and $\tan^{-1}$ are used to carry out the operations of finding the angles that give particular sines, cosine or tangents.

Example

Using a calculator, determine the values to three decimal places of the following:

(a) sin 25°, (b) cos 78°, (c) tan 35°, (d) cosec 50°, (e) sec 40°, (f) cot 70°, (g) sin 0.6, (h) cos 0.2, (i) tan 1.3, (j) cosec 1.1, (k) sec 0.5, (l) cot 0.6.

The following are the key sequences with my calculator.

(a) Press keys for 25, then key for sin. The result is 0.423.

(b) Press keys for 78, then key for cos. The result is 0.208.

(c) Press keys for 35, then key for tan. The result is 0.700.

(d) We can utilise here the fact that cosec 50° = 1/sin 50°. Thus the sequence is: press keys for 50, key for sin, then key for $1/x$. The result is 1.305.

(e) We utilise here the fact that sec 40° = 1/cos 40°. Thus the sequence is: press keys for 40, key for cos, then key for $1/x$. The result is 1.305.

(f) We utilise here the fact that cot 70° = 1/tan 70°. Thus the sequence is: press keys for 70, key for tan, then key for $1/x$. The result is 0.364.

(g) Press key for mode to give rad in the display, keys for 0.6, then key for sin. The result is 0.565.

(h) Press key for mode to give rad in the display, keys for 0.2, then key for cos. The result is 0.980.

(i) Press key for mode to give rad in the display, keys for 1.3, then key for tan. The result is 3.602.

(j) We utilise the fact that cosec 1.1 = 1/sin 1.1. Thus the sequence is: press key for mode to give rad in the display, keys for 1.1, key for sin, then key for $1/x$. The result is 1.122.
(k) We utilise the fact that sec 0.5 = 1/cos 0.5. Thus the sequence is: press key for mode to give rad in the display, keys for 0.5, key for cos, then key for $1/x$. The result is 1.139.
(l) We utilise the fact that cot 0.6 = 1/tan 0.6. Thus the sequence is: press key for mode to give rad in the display, keys for 0.6, key for tan, then key for $1/x$. The result is 1.462.

Example

Use a calculator to determine the following angles:

(a) the angle that gives the sine of 0.8,
(b) the angle that gives the cosine of 0.3,
(c) the angle that gives the tangent of 0.7.

The key sequences are:
(a) Press the keys for 0.8, then the key for $\sin^{-1}$. The result is 53.1°.
(b) Press the keys for 0.3, then the key for $\cos^{-1}$. The result is 72.5°.
(c) Press the keys for 0.7, then the key for $\tan^{-1}$. The result is 35.0°.

Revision

7 Using a calculator, determine the values to three decimal places of the following trigonometric ratios:

(a) sin 55°, (b) cos 15°, (c) tan 60°, (d) cosec 35°, (e) sec 80°, (f) cot 25°, (g) sin 0.9, (h) cos 1.1, (i) tan 0.3, (j) cosec 0.9, (k) sec 0.35, (l) cot 0.45.

8 Use a calculator to determine, to one decimal place, the following angles:

(a) the angle with a sine of 0.85, (b) the angle with a cosine of 0.35, (c) the angle with a tangent of 1.5, (d) the angle with a sine of 0.9, (e) the angle with a sine of 0.21, (f) the angle with a cosecant of 1.9, (g) the angle with a cotangent of 1.2, (h) the angle with a secant of 1.3.

8.2.3 Problems involving the trigonometric ratios

The following are examples of problems involving the trigonometric ratios.

Example

Determine the lengths x, to two decimal places, in the triangles shown in figure 8.7.

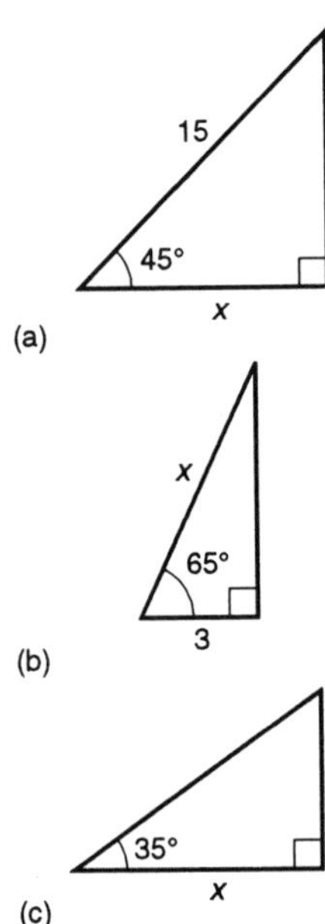

Figure 8.7

(a) We are given the hypotenuse and asked to obtain the adjacent side. Hence, using the cosine we can write

$$\frac{x}{15} = \cos 45°$$

and so $x = 15 \cos 45° = 15 \times 0.707 = 10.61$.

(b) We are given the adjacent and asked to obtain the hypotenuse. Hence, using the secant we can write

$$\frac{x}{3} = \sec 65°$$

and so $x = 3 \sec 65° = \dfrac{3}{\cos 65°} = 7.10$.

(c) We are given the opposite and asked to obtain the adjacent. Hence, using the cotangent we can write

$$\frac{x}{4} = \cot 35°$$

and so $x = 4 \cot 35° = \dfrac{4}{\tan 35°} = 5.71$.

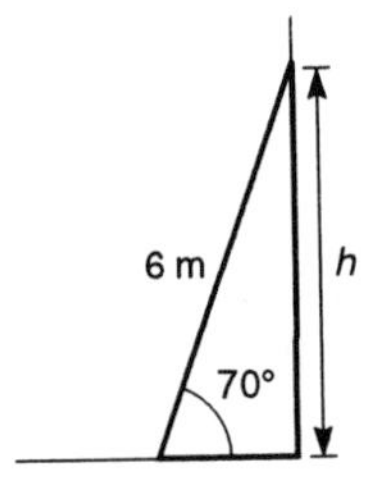

Figure 8.8

Example

A ladder of length 6 m leans against a wall, making an angle of 70° with the level ground. How high up the wall is the upper end of the ladder?

Figure 8.8 shows the situation. Thus

$$\frac{h}{6} = \sin 70°$$

and so $h = 6 \times \sin 70° = 5.6$ m.

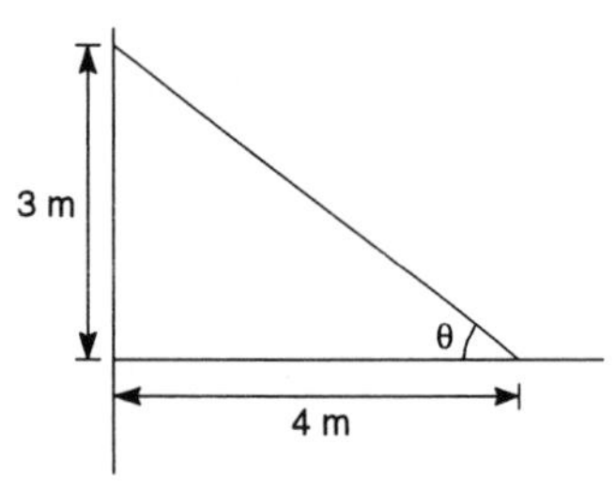

Figure 8.9

Example

For the supported beam shown in figure 8.9, what angle does the supporting stay make with the beam?

For the angle θ we have

$$\tan\theta = \frac{3}{4}$$

We thus need to find the angle which has a tangent of 0.75. We can do this by means of a calculator. The key sequence is: press the keys for 0.75, then press the key for $\tan^{-1}$. The result is 36.9°.

Example

A tower stands on level ground. The angle of elevation of the top of the tower from a point on the ground some distance away is 40°. On moving 30 m towards the tower the elevation becomes 60°. What is the height of the tower?

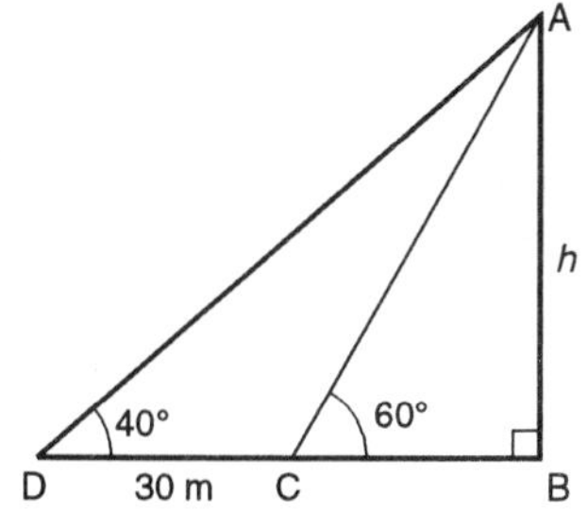

Figure 8.10

Figure 8.10 shows the situation, AB representing the tower. For the triangle ABD we can write

$$\frac{h}{\text{BD}} = \tan 40°$$

Thus BD = h cot 40°. For the triangle ABC we can write

$$\frac{h}{\text{BC}} = \tan 60°$$

Thus BC = h cot 60°. But CD = BD – BC. Thus

$$30 = h \cot 40° - h \cot 60° = h(\cot 40° - \cot 60°)$$

Hence

$$h = \frac{30}{\cot 40° - \cot 60°} = \frac{30}{1.192 - 0.577} = 48.8 \text{ m}$$

Revision

9 Solve for x, to two decimal places, in the following right-angled triangles:

(a) Hypotenuse 4, opposite x, angle 46°,
(b) Hypotenuse 10, adjacent x, angle 55°,
(c) Adjacent 10, opposite x, angle 70°,
(d) Adjacent 12, hypotenuse x, angle 58°,
(e) Opposite 4, hypotenuse x, angle 25°,
(f) Opposite 12, adjacent x, angle 67°.

10 The angle of elevation of the top of a flagpole from a point on the ground level with its base and 20 m from it is 40°. What is the height of the flagpole? Note that the angle of elevation is the upward angle from the horizontal.

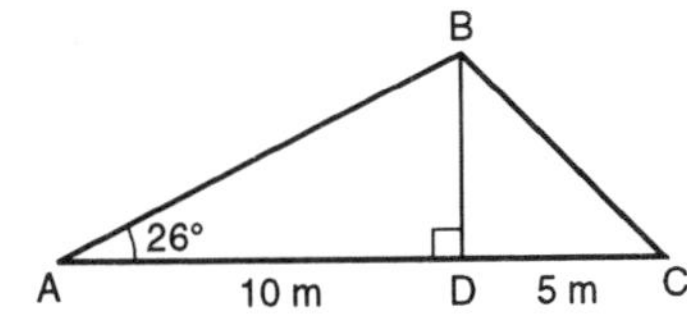

Figure 8.11

11 A pendulum of length 200 mm is swinging. Calculate the angle to the vertical it will be when the pendulum bob is displaced horizontally from its vertical position by 50 mm.

12 Figure 8.11 shows a roof truss for a building. Determine the lengths AB, BD and BC.

13 A road up an incline rises 12 m for every 100 m of roadway. What angle does the road make with the horizontal?

14 B is a point 420 m due east of A and C is another point which is 310 m from A in a direction 50° east of north. Determine the distance and bearing of C from B. Note that the bearing of an object from a point of observation is the angle that the line through the two points makes with the north and south line, or east and west line, through the point of observation.

15 When seen from the window of a building 100 m from a tower, the angle of elevation of the top of the tower is 26° and the angle of depression of the bottom of the tower is 10°. What is the height of the tower? Note that the angle of depression is the angle through which a horizontal line must be lowered.

8.3 Trigonometric relationships

The *Pythagoras theorem* states that, for a right-angled triangle, the square of the hypotenuse is equal to the sum of the squares of the other two sides. Thus, for the triangle shown in figure 8.12, we can write

$$AB^2 + BC^2 = AC^2$$

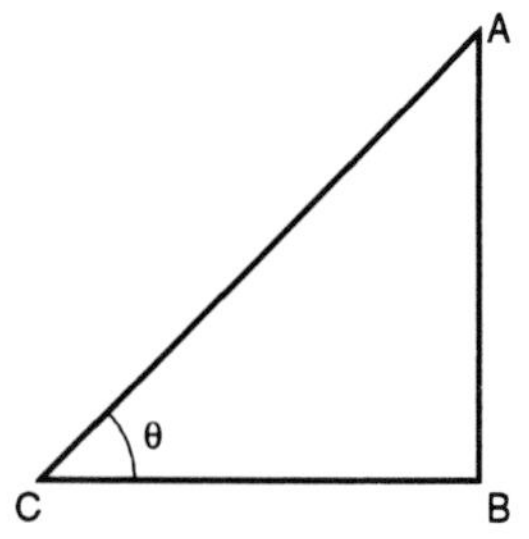

Figure 8.12 *The Pythagoras theorem*

If we divide both sides of the equation by AC^2 we obtain

$$\left(\frac{AB}{AC}\right)^2 + \left(\frac{BC}{AC}\right)^2 = 1$$

But AB/AC = sin θ and BC/AC = cos θ. Hence we have the very commonly used relationship of

$$\sin^2\theta + \cos^2\theta = 1$$

$(\sin\theta)^2$ is written as $\sin^2\theta$ and $(\cos\theta)^2$ as $\cos^2\theta$.

We can obtain two further relationships from this equation. If we divide the $\sin^2\theta + \cos^2\theta = 1$ equation by $\cos^2\theta$ we obtain

$$\left(\frac{\sin\theta}{\cos\theta}\right)^2 + 1 = \left(\frac{1}{\cos\theta}\right)^2$$

and so, since sin θ/cos θ = tan θ and 1/cos θ = sec θ, we obtain the relationship

$$\tan^2\theta + 1 = \sec^2\theta$$

If we divide the $\sin^2\theta + \cos^2\theta = 1$ equation by $\sin^2\theta$ we obtain

$$1 + \left(\frac{\cos\theta}{\sin\theta}\right)^2 = \left(\frac{1}{\sin\theta}\right)^2$$

and so, since $\cos\theta/\sin\theta = \cot\theta$ and $1/\sin\theta = \text{cosec}\,\theta$, we obtain:

$$1 + \cot^2\theta = \text{cosec}^2\theta$$

Example

Show that $\dfrac{\sec\theta + \tan\theta}{1 + \sin\theta} = \sec\theta$.

We can write the expression as

$$\frac{\sec\theta + \tan\theta}{1 + \sin\theta} = \frac{\dfrac{1}{\cos\theta} + \dfrac{\sin\theta}{\cos\theta}}{1 + \sin\theta} = \frac{\dfrac{1 + \sin\theta}{\cos\theta}}{1 + \sin\theta} = \frac{1}{\cos\theta} = \sec\theta$$

Hence the relationship is confirmed.

Example

Show that $\dfrac{\cos\theta}{1 - \sin\theta} + \dfrac{\cos\theta}{1 + \sin\theta} = 2\sec\theta$.

We can write the equation with a common denominator as

$$\frac{\cos\theta}{1 - \sin\theta} + \frac{\cos\theta}{1 + \sin\theta} = \frac{\cos\theta(1 + \sin\theta) + \cos\theta(1 - \sin\theta)}{(1 - \sin\theta)(1 + \sin\theta)}$$

Multiplying out the brackets gives

$$\frac{2\cos\theta}{1 - \sin^2\theta}$$

As $\sin^2\theta + \cos^2\theta = 1$, then this can be written as

$$\frac{2\cos\theta}{\cos^2\theta} = \frac{2}{\cos\theta} = 2\sec\theta$$

Hence the relationship is confirmed.

Revision

16 Simplify the following:

(a) $\dfrac{\text{cosec}\,\theta}{\cot\theta + \tan\theta}$, (b) $\dfrac{\tan\theta + 1}{\cot\theta + 1}$, (c) $\dfrac{\sin^2\theta\cot\theta}{\cos\theta}$, (d) $\sin\theta\cos\theta\tan\theta$

17 Show that $\dfrac{\cos\theta\cot\theta\sec^2\theta}{\text{cosec}\,\theta} = 1$

18 Show that $\tan\theta + \cot\theta = \sec\theta\,\text{cosec}\,\theta$.

19 Show that $1 - 2\sin^2\theta = 2\cos^2\theta - 1$.

20 Show that $\dfrac{1}{1 - \sin^2\theta} = \sec^2\theta$.

8.4 Ratios for angles

We can use the Pythagoras theorem to find the values of the sine, cosine and tangent for the commonly encountered angles of 0°, 30°, 45°, 60° and 90°.

Consider a right-angled triangle with the angle θ zero. The opposite size must be zero, with the adjacent side and the hypotenuse both having the same value. Thus

$$\sin 0° = 0, \quad \cos 0° = 1, \quad \tan 0° = 0$$

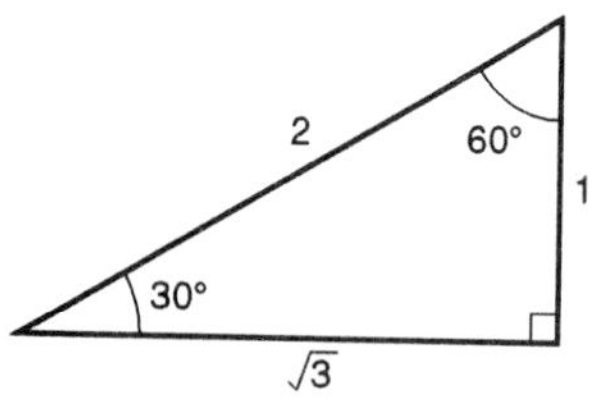

Figure 8.13 *Right-angled triangle with 30° and 60° angles*

Consider a right-angled triangle containing 30° and 60° angles. Since sin 30° = 0.5, then we can draw such a triangle with an opposite side of 1 for the 30° angle and a hypotenuse of 2 (figure 8.13). The adjacent side must then have a length of

$$\text{adjacent side} = \sqrt{2^2 - 1^2} = \sqrt{3}$$

Thus we have

$$\sin 30° = \frac{1}{2}, \quad \cos 30° = \frac{\sqrt{3}}{2}, \quad \tan 30° = \frac{1}{\sqrt{3}}$$

For the 60° angle in the triangle we have

$$\sin 60° = \frac{\sqrt{3}}{2}, \quad \cos 60° = \frac{1}{2}, \quad \tan 60° = \frac{\sqrt{3}}{1} = \sqrt{3}$$

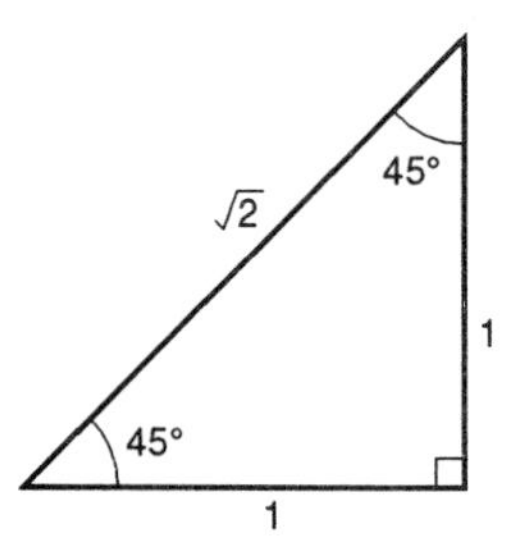

Figure 8.14 *Right-angled triangle with 45° angles*

Consider a right-angled triangle containing two 45° angles (figure 8.14). Since we have tan 45° = 1 then the opposite and adjacent sides can be 1. The Pythagoras theorem thus gives a hypotenuse of length

$$\text{hypotenuse} = \sqrt{1^2 + 1^2} = \sqrt{2}$$

Thus we have

$$\sin 45° = \frac{1}{\sqrt{2}}, \quad \cos 45° = \frac{1}{\sqrt{2}}, \quad \tan 45° = 1$$

Consider a right-angled triangle with the angle θ equal to 90°. For this we must have the adjacent side equal to zero and the opposite side and hypotenuse the same value. Thus

$$\sin 90° = 1, \quad \cos 90° = 0, \quad \tan 90° \text{ infinite}$$

Table 8.1 summarises the above results for these angles of 0°, 30°, 45°, 60° and 90°.

Table 8.1 Common angles between 0° and 90°

θ	sin θ	cos θ	tan θ
0°	0	1	0
30°	$\frac{1}{2}$	$\frac{\sqrt{3}}{2}$	$\frac{1}{\sqrt{3}}$
45°	$\frac{1}{\sqrt{2}}$	$\frac{1}{\sqrt{2}}$	1
60°	$\frac{\sqrt{3}}{2}$	$\frac{1}{2}$	$\sqrt{3}$
90°	1	0	infinity

Example

For the triangle shown in figure 8.15, using the values given in the above table, determine AB if AC = 12 m.

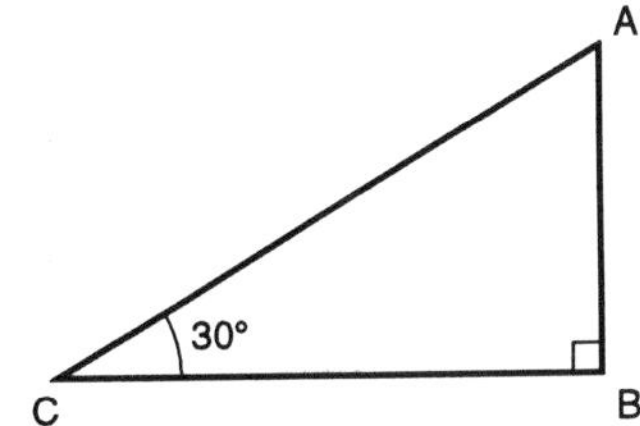

Figure 8.15

We can write

$$\frac{AB}{AC} = \sin 30°$$

Since sin 30° = 1/2 then AB = $\frac{1}{2}$AC. Thus AB = 6 m.

Example

Show that cos 60° cos 30° + sin 60° sin 30° = $\frac{\sqrt{3}}{2}$.

Using the values given in the above table, then

$$\cos 60° \cos 30° + \sin 60° \sin 30° = \frac{1}{2} \times \frac{\sqrt{3}}{2} + \frac{\sqrt{3}}{2} \times \frac{1}{2} = \frac{\sqrt{3}}{2}$$

The relationship is thus confirmed.

Revision

21 For the triangle shown in figure 8.15, use the values of the ratios given in the above table without working out the square roots.

(a) If AB = 12 m, determine BC.

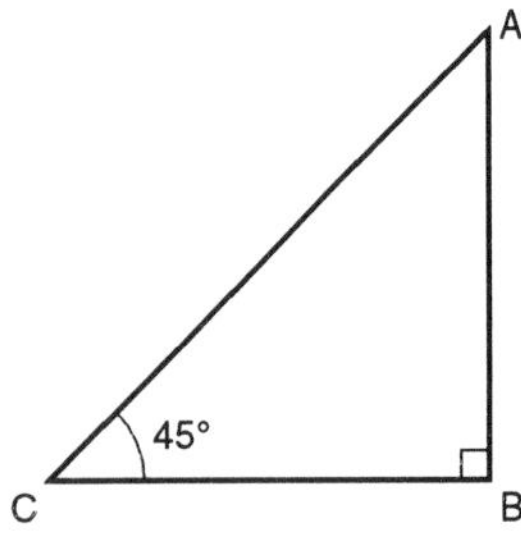

Figure 8.16

(b) If BC = 4 m, determine AC.
(c) If AC = 10 m, determine BC.

22 For the triangle shown in figure 8.16, use the values of the ratios given in the above table without working out the square roots.

(a) If AB = 10 m, determine BC.
(b) If AC = 6 m, determine AB.
(c) If BC = 5 m, determine AB.

23 Show that $\sin^2 30° + \sin^2 45° + \sin^2 60° = \frac{3}{2}$.

8.4.1 Ratios for angles of any magnitude

In section 8.2, the sine, cosine and tangent trigonometric ratios were defined for acute angles, i.e. angles less than 90°. In this section we consider a more general definition which will enable us to consider the trigonometric ratios for angles greater than 90°. The general definitions are consistent with the definitions for angles less than 90°.

Consider an angle θ formed by a line OP rotating anticlockwise about O from an initial line OX (figure 8.17). Whatever the size of the angle θ, we will define the trigonometric ratios as:

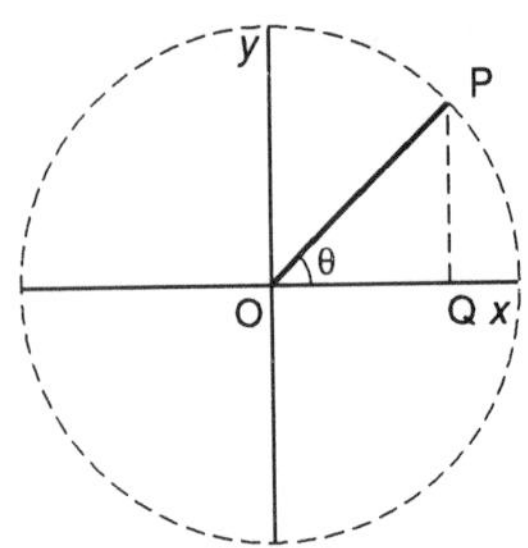

Figure 8.17 *Defining the ratios*

$$\sin\theta = \frac{\text{vertical displacement of P}}{\text{OP}}$$

$$\cos\theta = \frac{\text{horizontal displacement of P}}{\text{OP}}$$

$$\tan\theta = \frac{\text{vertical displacement of P}}{\text{horizontal displacement of P}}$$

We consider O to be at the centre of a system of coordinates, with distances measured to the right of O being positive, distances to the left negative. Distances measured upwards from O are positive and distances downwards are negative. This is the conventional system of coordinates used with graphs (see chapter 12) with O as the origin.

When the line OP has rotated through an angle greater than 0° and less than 90° and is in the top right-hand quadrant, as in figure 8.17, then we have for the three basic trigonometric ratios

$$\sin\theta = \frac{\text{QP}}{\text{OP}},\quad \cos\theta = \frac{\text{OQ}}{\text{OP}},\quad \tan\theta = \frac{\text{QP}}{\text{OQ}}$$

The displacements QP and OQ are both positive quantities (the radius OP is defined as always being positive) so that, with angles between 0° and 90°, we have with each of the above ratios positive quantities divided by positive

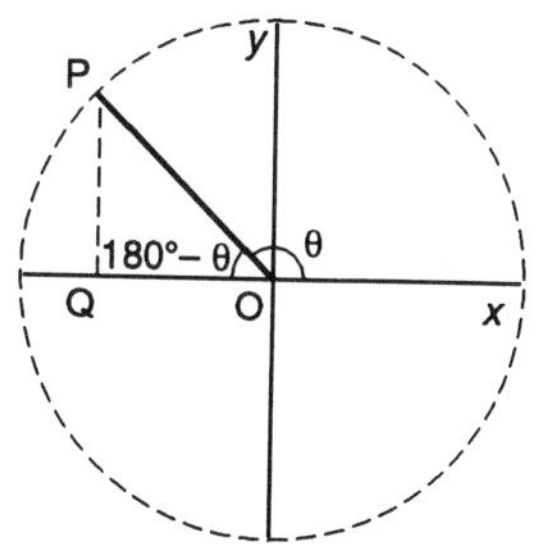

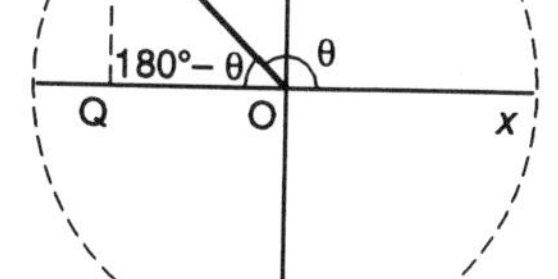

Figure 8.18 *Angles between 90° and 180°*

quantities and so the sine, cosine and tangent of such angles are all designated as being positive.

Now consider the situation when the line OP is in the top left-hand quadrant (figure 8.18). To reach this position the line has rotated through more than 90° but less than 180°. Thus, since we have

$$\sin\theta = \frac{QP}{OP},\ \cos\theta = \frac{OQ}{OP},\ \tan\theta = \frac{QP}{OQ}$$

then for angles between 90° and 180°, as QP is positive and OQ negative, the sine is positive, the cosine negative and the tangent negative.

For angles in this quadrant, QP/OP could be the sine of θ or, when considering just the triangle OPQ, the sine of the angle (180° – θ). Thus

$$\sin\theta = \sin(180° - \theta)$$

For angles in this quadrant, OQ/OP could be the cosine of θ or, when considering just the triangle OPQ, the cosine of the angle (180° – θ). Thus

$$\cos\theta = -\cos(180° - \theta)$$

For angles in this quadrant, QP/OQ could be the tangent of θ or, when considering just the triangle OPQ, the tangent of the angle (180° – θ). Thus

$$\tan\theta = -\tan(180° - \theta)$$

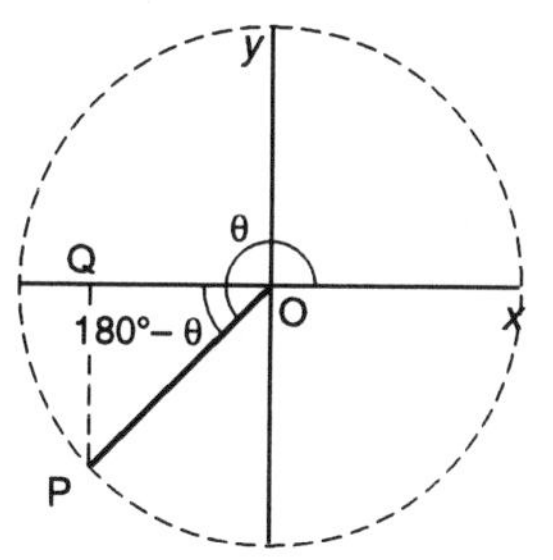

Figure 8.19 *Angles between 180° and 270°*

Now consider the situation when the line OP is in the bottom left-hand quadrant (figure 8.19). To reach this position the line has rotated through more than 180° but less than 270°. Thus, since we have

$$\sin\theta = \frac{QP}{OP},\ \cos\theta = \frac{OQ}{OP},\ \tan\theta = \frac{QP}{OQ}$$

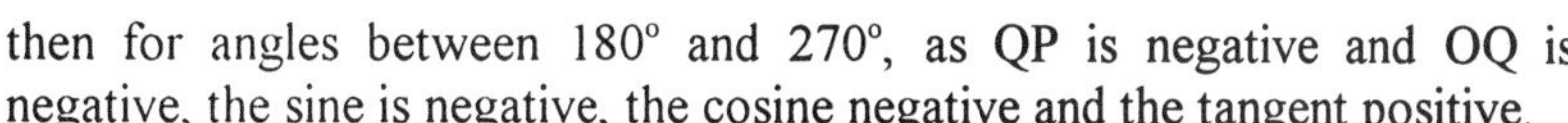

then for angles between 180° and 270°, as QP is negative and OQ is negative, the sine is negative, the cosine negative and the tangent positive.

For angles in this quadrant, QP/OP could be the sine of θ or, when considering just the triangle OPQ, the sine of the angle (θ – 180°). Thus

$$\sin\theta = -\sin(\theta - 180°)$$

For angles in this quadrant, OQ/OP could be the cosine of θ or, when considering just the triangle OPQ, the cosine of the angle (θ – 180°). Thus

$$\cos\theta = -\cos(\theta - 180°)$$

For angles in this quadrant, QP/OQ could be the tangent of θ or, when considering just the triangle OPQ, the tangent of the angle (θ – 180°). Thus

$$\tan\theta = \tan(\theta - 180°)$$

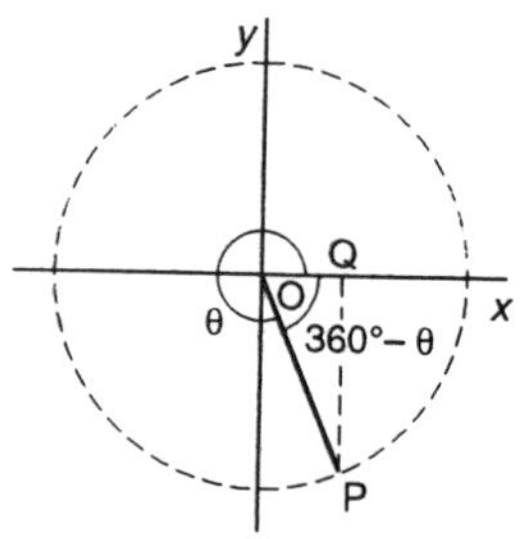

Figure 8.20 *Angles between 270° and 360°*

Now consider the situation when the line **OP** is in the bottom right-hand quadrant (figure 8.20). To reach this position the line has rotated through more than 270° but less than 360°. Thus, since we have

$$\sin\theta = \frac{QP}{OP},\ \cos\theta = \frac{OQ}{OP},\ \tan\theta = \frac{QP}{OQ}$$

then for angles between 270° and 360°, as **QP** is negative and **OQ** is positive, the sine is negative, the cosine positive and the tangent negative.

For angles in this quadrant, QP/OP could be the sine of θ or, when considering just the triangle OPQ, the sine of the angle (360° – θ). Thus

$$\sin\theta = -\sin(360° - \theta)$$

For angles in this quadrant, OQ/OP could be the cosine of θ or, when considering just the triangle OPQ, the cosine of the angle (360° – θ). Thus

$$\cos\theta = \cos(360° - \theta)$$

For angles in this quadrant, QP/OQ could be the tangent of θ or, when considering just the triangle OPQ, the tangent of the angle (360° – θ). Thus

$$\tan\theta = -\tan(360° - \theta)$$

In summary, we have the following pattern of signs for the trigonometric ratios:

Angles between:	
0° and 90° (0 and π/2 rad)	All the ratios are positive
90° and 180° (π/2 and π rad)	Only the sine is positive
180° and 270° (π and 3π/2 rad)	Only the tangent is positive
270° and 360° (3π/2 and 2π rad)	Only the cosine is positive

sin+, cos–, tan– S	sin+, cos+, tan+ A
sin–, cos–, tan+ T	sin–, cos+, tan– C

Figure 8.21 *Signs of ratios*

Figure 8.21 summarises the situation with regard to the signs of sines, cosines and tangents of angles between 0° and 360°. A mnemonic that might be used to remember this is, when working anticlockwise round the quadrants,

positively **A**ll **S**ilver **T**ea **C**ups

A = all in the first quadrant, **S** = sine in the second quadrant, **T** = tangent in the third quadrant, **C** = cosine in the fourth quadrant

Example

Determine, to three decimal places, the values of: (a) sin 200°, (b) cos 115°, (c) tan 295°, (d) cos 320°.

If we use a calculator to determine these angles then the signs should automatically come right.
(a) With a calculator the key sequence is: press the keys for 200, then press the key for sin. The result is –0.342
(b) With a calculator the key sequence is: press the keys for 115, then press the key for cos. The result is –0.423.
(c) With a calculator the key sequence is: press the keys for 295, then press the key for tan. The result is –2.145.
(d) With a calculator the key sequence is: press the keys for 320, then press the key for cos. The result is 0.766.

Alternatively we could have just used the values of the trigonometric ratios for angles between 0° and 90°.
(a) Since $\sin 200° = -\sin (200° - 180°)$, then $\sin 200° = -\sin 20°$ and so is –0.342.
(b) Since $\cos 115° = -\cos (180° - 115°)$, then $\cos 115° = -\cos 65°$ and so is –0.423.
(c) Since $\tan 295° = -\tan (360° - 295°)$, then $\tan 295° = -\tan 65°$ and so is –2.145.
(d) Since $\cos 320° = \cos (360° - 320°)$, then $\cos 320° = \cos 40°$ and so is 0.766.

Example

Determine, to three decimal places the values of (a) sin 1.8, (b) cos 2.1, (c) tan 4.6.

(a) This angle is in radian measure. Thus the key sequence is: press the key for mode to give the display in rad, press the keys for 1.8, then press the key for sin. The result is 0.974.
(b) This angle is in radian measure. Thus the key sequence is: press the key for mode to give the display in rad, press the keys for 2.1, then press the key for cos. The result is –0.505.
(c) This angle is in radian measure. Thus the key sequence is: press the key for mode to give the display in rad, press the keys for 4.6, then press the key for tan. The result is 8.860.

Example

If $\sin 60° = \frac{\sqrt{3}}{2}$, determine $\sin (180° - 60°)$.

We can use $\sin \theta = \sin (180° - \theta)$. Thus $\sin (180° - 60°) = \frac{\sqrt{3}}{2}$.

Example

Determine the angles between 0° and 360° having the trigonometric ratios of (a) $\sin \theta = 0.7$, (b) $\cos \theta = -0.4$, (c) $\sin \theta = -0.3$.

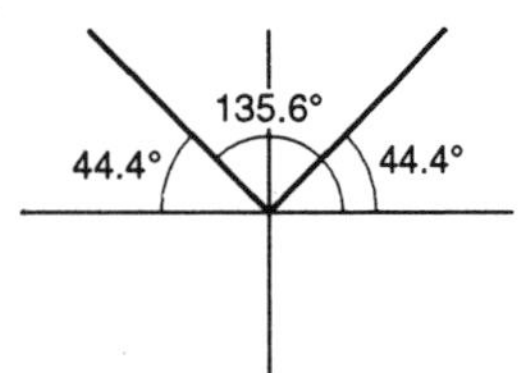

Figure 8.22

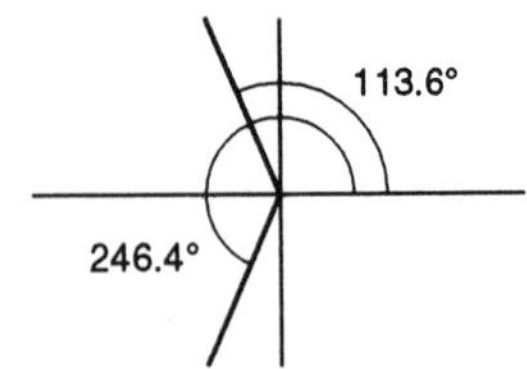

Figure 8.23

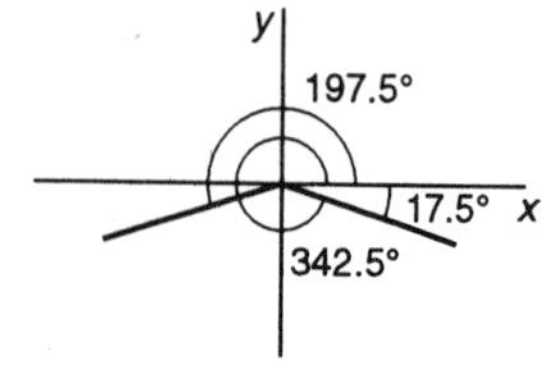

Figure 8.24

(a) Using a calculator the key sequence can be: press the keys for 0.7, then the key for $\sin^{-1}$. The result given by the calculator is 44.4°. But sines are positive in the first and second quadrants (figure 8.22). Thus (180° – 44.4°) also gives $\sin\theta = 0.7$. Hence the angles are 44.4° and 135.6°. You can check that these angles give a sine of 0.7 by using the calculator.

(b) Using a calculator the key sequence can be: press the keys for 0.4, press the key for +/–, then the key for $\cos^{-1}$. The result given by the calculator is 113.6°. But cosines are negative in the second and third quadrants (figure 8.23). Thus (180° + {180° – 113.6°}) also gives $\cos\theta = -0.4$. Hence the angles are 113.6° and 246.4°. You can check that these angles give a cosine of –0.4 by using the calculator.

(c) Using a calculator the key sequence can be: press the keys for 0.3, press the key for +/–, then the key for $\sin^{-1}$. The result given by the calculator is –17.5°. Positive angles are given between the *x*-axis and the rotating line when it rotates in an anticlockwise direction. A negative angle occurs when the rotation is in clockwise direction. Thus this angle represents (360° – 17.5°) = 342.5°. Sines are negative in the third and fourth quadrants (figure 8.24). Thus the angle in the third quadrant is (180° + 17.5°). Hence the angles are 197.5° and 342.5°. You can check that these angles give a sine of –0.3 by using the calculator.

Revision

24 Determine, to three decimal places, the values of the following:

(a) sin 125°, (b) cos 145°, (c) tan 170°, (d) sin 200°, (e) cos 220°, (f) tan 260°, (g) sin 290°, (h) cos 300°, (i) tan 340°, (j) sin 345°, (k) sin 2.5, (l) cos 3.1, (m) tan 4.2, (n) sin 3.8.

25 If $\sin 30° = \frac{1}{2}$, determine sin (180° – 30°).

26 If $\cos 45° = \dfrac{\sqrt{2}}{2}$, determine cos (180° – 45°).

27 Determine the angles having the following trigonometric ratios:

(a) $\tan\theta = 1.7$, (b) $\sin\theta = 0.8$, (c) $\cos\theta = -0.7$, (d) $\sin\theta = -0.5$, (e) $\cos\theta = 0.8$

8.4.2 Graphs of trigonometric ratios

Table 8.2 gives values of the trigonometric ratios as the angles increase from 0° to 360° and 0 to 2π radians and figure 8.25 shows the graphs of the ratios. A particular value of a trigonometric ratio can be given by more than one angle. For example, a sine of 0.71 is given by angles of 45° and 135°. Indeed if we consider angles greater than 360° there are even more angles which will give this value. Note that $\sin\theta$ and $\cos\theta$ are the same basic graph, one being just the other displaced by 90° ($\pi/2$ rad). The tan θ graph goes towards + infinity and returns from – infinity at 90° and 270°. Since the graphs repeat themselves at regular intervals they are called *periodic functions*. The sine and cosine curves are continuous and repeat themselves every 360° or 2π rad.

Table 8.2 Values of trigonometric ratios

Angle θ		sin θ	cos θ	tan θ
degrees	radians			
0	0	0	1	0
30	$\pi/6$	0.5	0.87	0.58
45	$\pi/4$	0.71	0.71	1
60	$\pi/3$	0.87	0.50	1.73
90	$\pi/2$	1	0	± infinity
120	$2\pi/3$	0.87	–0.50	–1.73
135	$3\pi/4$	0.71	–0.71	–1
150	$5\pi/6$	0.50	–0.87	–0.58
180	π	0	–1	0
210	$7\pi/6$	–0.5	–0.87	0.58
225	$5\pi/4$	–0.71	–0.71	1
240	$4\pi/3$	–0.87	–0.50	1.73
270	$3\pi/2$	–1	0	± infinity
300	$5\pi/3$	–0.87	0.5	–1.73
315	$7\pi/4$	–0.71	0.71	–1
330	$11\pi/6$	–0.50	0.87	–0.58
360	2π	0	1	0

Revision

28 What angles between 0° and 360° have (a) sin θ = 0.5, (b) cos θ = –0.5, (c) tan θ = 1?

Problems

1 Determine, to three decimal places, the values of the following:

(a) sin 23°, (b) cos 15.1°, (c) tan 19.5°, (d) cosec 34°, (e) sec 59°, (f) cot 57°, (g) sin 0.50, (h) cos 1.21, (i) tan 0.56, (j) cosec 0.85, (k) sec 1.05, (l) cot 1.23, (m) sin 140°, (n) cos 150°, (o) tan 130°, (p) sin 190°, (q) cos 195°, (r) tan 230°, (s) sin 280°, (t) cos 290°, (u) tan 315°, (v) sin 320°, (w) cos 350°, (x) sin 261.2°, (y) sec 301.8°, (z) cot 342.3°.

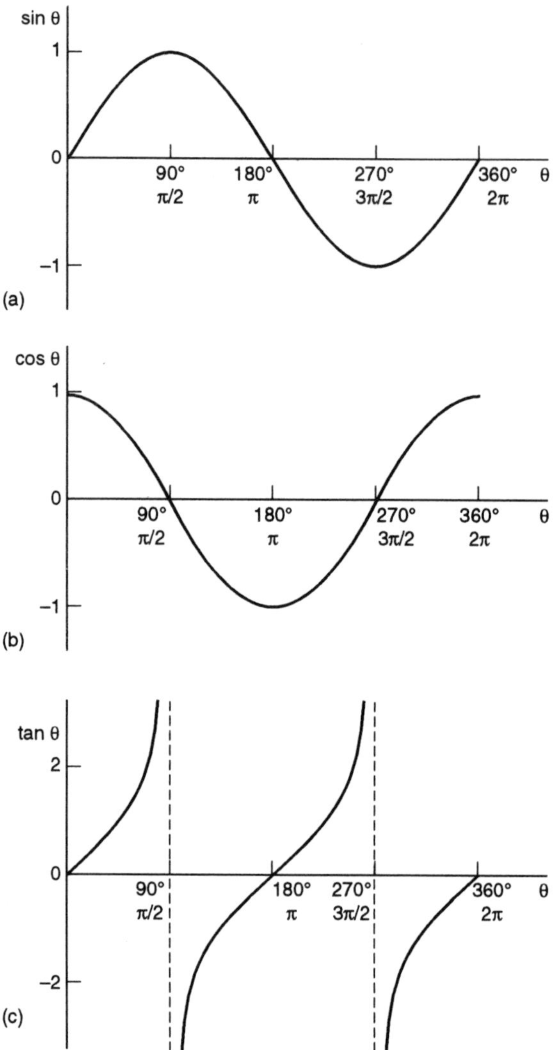

Figure 8.25 *(a) sin θ, (b) cos θ, (c) tan θ*

2 Determine using a calculator the values between 0° and 90°, to one decimal place, of the following angles:

(a) The angle which has a sine of 0.78,
(b) The angle which has a cosine of 0.15,
(c) The angle which has a tangent of 1.9.

3 Solve for x, to two decimal places, in the following right-angled triangles:

(a) Adjacent 10, opposite x, angle 68°,
(b) Adjacent 4, hypotenuse x, angle 55°,
(c) Opposite 6, adjacent x, angle 23°,
(d) Adjacent 12, opposite x, angle 58°,
(e) Hypotenuse 4, adjacent x, angle 70°,
(f) Hypotenuse 2, opposite x, angle 38°.

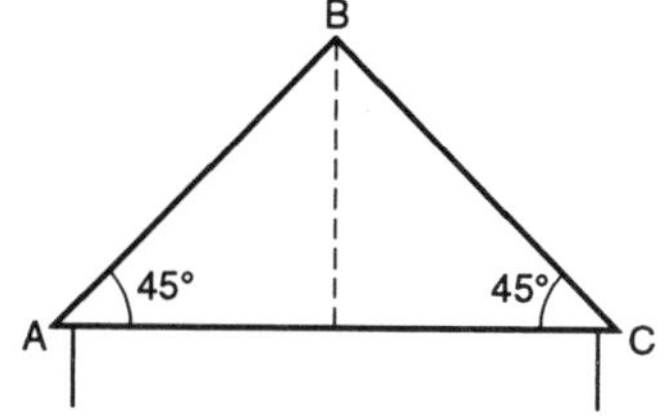

Figure 8.26

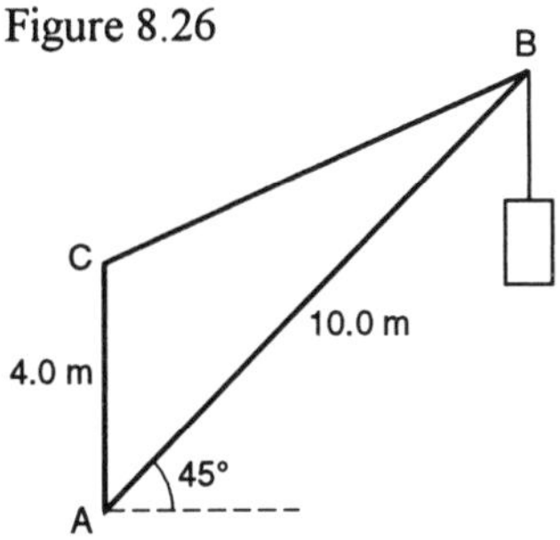

Figure 8.27

4 What is the height of a flagpole if the angle of elevation of the top of the pole from a point on the level ground 10 m from its base is 70°?

5 A ladder of length 6 m leans against a vertical wall with its base on horizontal ground. If the ladder makes an angle of 70° with the ground, how far is the bottom end from the wall?

6 The roof of a house is as shown in figure 8.26. What is the length AB if AC = 9.0 m?

7 A and B are two towns 30 km apart. If B is on a bearing 50° east of north from A, how far east is B from A?

8 A guy wire of length 5 m to a pole rising from level ground makes an angle of 62° with the ground. How high above the ground is the wire attached to the pole?

9 In figure 8.27, AB represents the jib of a crane and BC the tie. If AB is 10.0 m and BC is 4.0 m, what is the length of the tie?

10 Observer A sees an aeroplane as directly overhead. Observer B, who is 300 m due east of A, sees the plane as being at an angle of elevation of 70°. What is the height of the plane?

11 A ramp has to be built with a slope of 15° to enable a heavy machine to be slid up it to a height of 1.3 m. What must be the length of the ramp needed to achieve this?

12 A roof has to span 12 m with one slope pitched at 40° to the horizontal and the other at 50°. What will be the height of the ridge above the eaves?

13 Show that:

(a) $\sec^2\theta - \dfrac{\sin^2\theta}{\cos^2\theta} = 1$, (b) $(\sin\theta + \cos\theta)^2 + (\sin\theta - \cos\theta)^2 = 2$,

(c) $1 - 2\cos^2\theta = 2\sin^2\theta - 1$, (d) $\dfrac{1 - \tan^2\theta}{1 + \tan^2\theta} = \cos^2\theta - \sin^2\theta$,

(e) $(1 + \tan^2\theta)\cos^2\theta = 1$, (f) $\dfrac{2\tan\theta}{1 + \tan^2\theta} = 2\sin\theta\cos\theta$,

(g) $\dfrac{1}{\sqrt{1-\sin^2\theta}} = \sec\theta$, (h) $\sin\theta - \sin^3\theta = \dfrac{\sin\theta}{\sec^2\theta}$,

(i) $(1 + \cot\theta)^2 + (1 - \cot\theta)^2 = 2\operatorname{cosec}^2\theta$, (j) $\cot\theta = \sqrt{\dfrac{\cos^2\theta}{1-\cos^2\theta}}$

14 The horizontal distance x and the vertical distance y travelled in a time t by a projectile when fired with a velocity v at an angle θ to the horizontal are given by

$$x = vt\cos\theta$$

$$y = vt\sin\theta - \tfrac{1}{2}gt^2$$

Substitute for t and so show that

$$y = x\tan\theta - \frac{gx^2}{2v^2}(\tan^2\theta + 1)$$

15 Show that $\dfrac{\tan 60° - \tan 30°}{1 + \tan 60° \tan 30°} = \dfrac{1}{\sqrt{3}}$.

16 Determine the angles between 0° and 360°, to one decimal place, having the trigonometric ratios:

(a) $\sin\theta = 0.35$, (b) $\cos\theta = -0.55$, (c) $\cos\theta = 0.60$, (d) $\tan\theta = 0.10$, (e) $\tan\theta = -1.5$, (f) $\sin\theta = -0.48$, (g) $\cot\theta = -1.24$, (h) $\sec\theta = -2.5$, (i) $\operatorname{cosec}\theta = -2.5$, (j) $\operatorname{cosec}\theta = 1.5$

17 Determine, to two decimal places, the values of the following:

(a) $\sin \pi/6$, (b) $\tan \pi/2$, (c) $\cos 3\pi/2$, (d) $\sin 5\pi/6$, (e) $\tan 7\pi/4$, (f) $\cos 7\pi/6$, (g) $\tan 7\pi/6$

9 Solving triangles

9.1 Introduction

In chapter 8 problems were solved which involved right-angled triangles. But how can we solve problems involving triangles which do not have a 90° angle? This chapter is about such triangles and the sine rule and the cosine rule which can be used with problems involving such triangles. Applications of such rules in engineering are discussed in chapter 10. This chapter also includes a consideration of the areas of triangles.

9.2 The sine rule

Consider the triangle ABC shown in figure 9.1. This triangle does not contain a right angle. The notation system that is used for the angles and sides of triangles is that the angles are labelled as being A at apex A, B at apex B and C at apex C while the side opposite angle A is labelled as being of length a, that opposite angle B of length b, and that opposite angle C of length c.

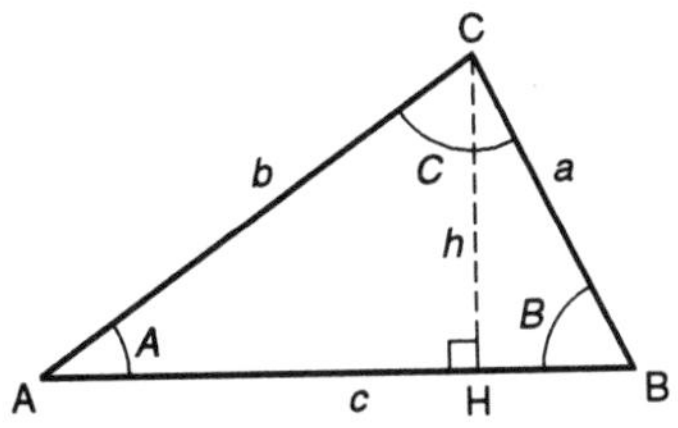

Figure 9.1 *The sine rule*

If we draw a line from C which is at right angles to AB then we divide the triangle into two right-angled triangles. Suppose this line has a length of h. Then for triangle AHC we can write

$$\sin A = \frac{h}{b}$$

and so $h = b \sin A$. For triangle BHC we can write

$$\sin B = \frac{h}{a}$$

and so $h = a \sin B$. Thus we have $h = b \sin A = a \sin B$, and so

$$\frac{a}{\sin A} = \frac{b}{\sin B}$$

Similarly, by drawing a line from A at right angles to BC we can show that

$$\frac{b}{\sin B} = \frac{c}{\sin C}$$

Thus the *sine rule* is stated as

$$\frac{a}{\sin A} = \frac{b}{\sin B} = \frac{c}{\sin C}$$

This rule may be used with triangles to find the values of unknown sides and angles when one side and any two angles are given, or two sides and an angle not included between those sides are given.

Example

For the triangle shown in figure 9.2, determine the length of the side AC.

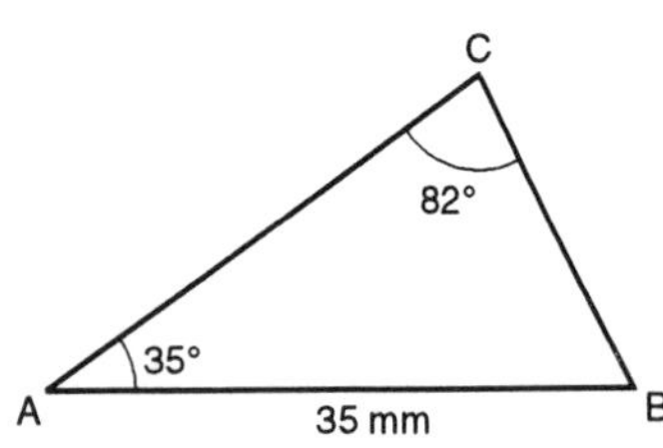

Figure 9.2

In the triangle we have angle A = 35°, angle C = 82°, side c = 35 mm and are required to determine b. Since all the angles in a triangle add up to 180° then angle B = 180° – 35° – 82° = 63°. Thus, using the sine rule, we have

$$\frac{b}{\sin B} = \frac{c}{\sin C}$$

then

$$\frac{b}{\sin 63°} = \frac{35}{\sin 82°}$$

Hence

$$b = \frac{35}{\sin 82°} \times \sin 63° = 31.5 \text{ mm}$$

Thus AC = 31.5 mm.

Example

For the triangle shown in figure 9.3, determine the angle B.

In the triangle we have angle A = 70°, side b = 35 mm, side a = 45 mm and are required to determine angle B. Thus, using the sine rule, we have

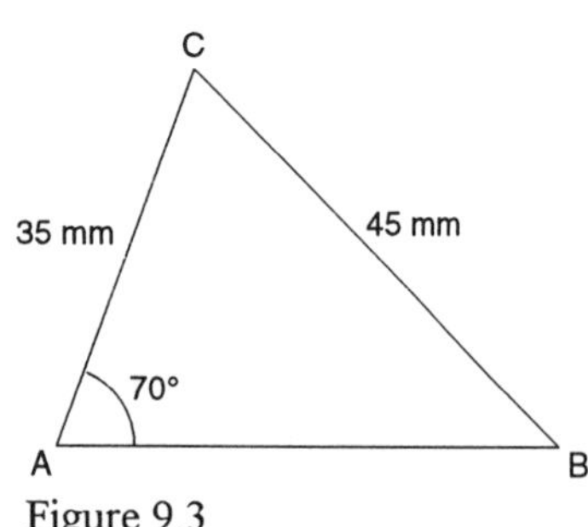

Figure 9.3

$$\frac{a}{\sin A} = \frac{b}{\sin B}$$

and so

$$\frac{45}{\sin 70°} = \frac{35}{\sin B}$$

Thus

$$\sin B = \frac{35 \times \sin 70°}{45} = 0.731$$

Hence B = 47.0°.

Example

A radio aerial is 20 m high and stands on the roof of a building. The angle of elevation of the top of the aerial is 48° and from the same point

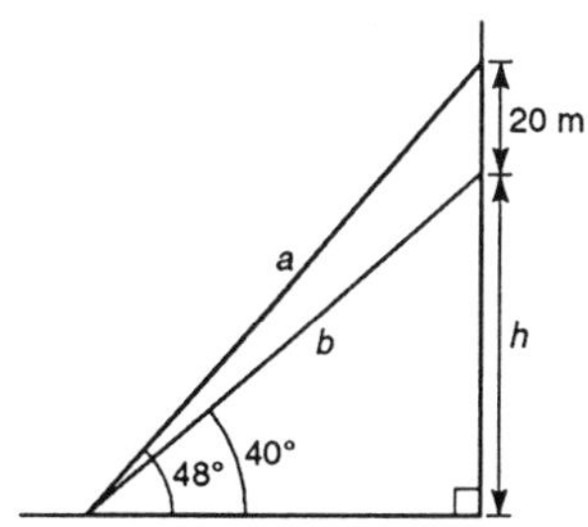

Figure 9.4

the angle of elevation of the base of the aerial is 40°. What is the height of the building?

Figure 9.4 shows the situation, with h representing the height of the building. For the triangle with sides of a, b and 20 m we can use the sine rule to write

$$\frac{b}{\sin(90° - 48°)} = \frac{20}{\sin(48° - 40°)}$$

Thus

$$b = \frac{20 \times \sin 42°}{\sin 8°} = 96.2 \text{ m}$$

For the right-angled triangle we now have

$$\frac{h}{b} = \sin 40°$$

and so the height of the tower $h = 96.2 \sin 40° = 61.8$ m.

Revision

1 Determine the required side or angle in each of the following triangles:

(a) Angle $B = 47°$, angle $C = 78°$, side $a = 340$ mm, determine side c,
(b) Angle $A = 67°$, angle $C = 51°$, side $c = 150$ mm, determine side a,
(c) Angle $A = 80°$, side $a = 200$ mm, side $b = 120$ mm, determine angle B,
(d) Angle $A = 40°$, side $a = 30$ mm, side $c = 37$ mm, determine angle C,
(e) Angle $C = 35°$, side $a = 36$ mm, side $c = 30$ mm, determine side b,
(f) Angle $A = 68°$, angle $C = 30°$, side $c = 26$ mm, determine side b,
(g) Angle $A = 80°$, angle $B = 70°$, side $b = 80$ mm, determine side a,
(h) Angle $B = 55°$, side $a = 70$ mm, side $b = 80$ mm, determine angle A.

2 A roof has a span of 8 m with eaves that slope on one side at 42° to the horizontal and on the other side at 35° to the horizontal. Determine the lengths of the sloping parts of the roof.

3 A tower stands on the top of a slope which rises by 1 m for every 12 m of slope. From a position 120 m down the slope, the angle of elevation of the top of the tower is 18°. Determine the height of the tower.

9.2.1 The ambiguous solution

When given two sides of a triangle and the angle opposite one of the sides, there is the possibility that using the sine rule will lead to two solutions.

This is because a positive value for a sine can mean an angle between 0° and 90° or 90° and 180°.

For example, suppose we have angle A = 25°, side a = 28 mm and side b = 37 mm. Using the sine rule we can find angle B. Thus

$$\frac{28}{\sin 25°} = \frac{37}{\sin B}$$

Thus

$$\sin B = \frac{37 \times \sin 25°}{28} = 0.558$$

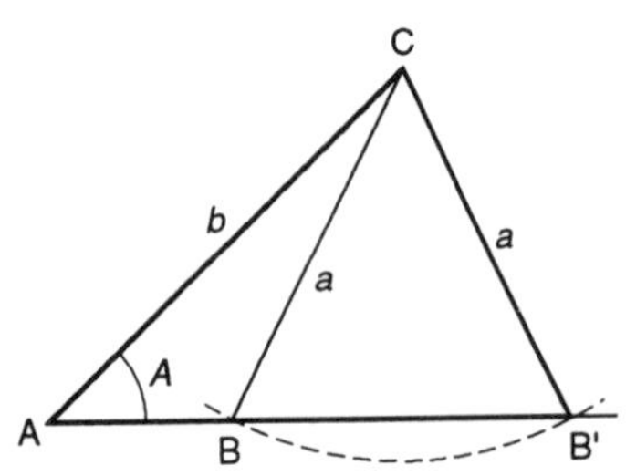

Figure 9.5 *Two solutions*

The angle B which would give a positive sine of 0.558 is either 33.9° or 146.1°. Are both these values feasible? Figure 9.5 shows the possible triangles we could have. With one case we have the triangle ABC, with angle C = 180° – 25° – 146.1° = 8.9°, and with the other the triangle AB'C, with angle C = 180° – 25° – 33.9° = 121.1°. Thus both the solutions are feasible and there are two possible triangles which fit the supplied data and which can be drawn.

Sometimes, though there might be two possible values for angle B, one of the answers leads to an impossible situation, such as the sum of the angles $A + B + C$ being greater than 180°, and so one of the answers is then not feasible. Two triangles cannot be drawn on the basis of the data, only one. To illustrate this, consider the determination of the triangle, or triangles, which fit the data given of angle A = 37°, side a = 46 mm and side b = 21 mm. Angle B is given by the sine rule as

$$\frac{46}{\sin 37°} = \frac{21}{\sin B}$$

Thus

$$\sin B = \frac{21 \times \sin 37°}{46} = 0.275$$

The angles which give this sine are 15.9° and 164.1°. Since we must have $A + B + C$ = 180°, then with the first answer for B we have angle C as 180° – 37° – 15.9° = 127.1° and with the other answer for B we have angle C as 180° – 37° – 164.1° = –21.1°. This last answer is not feasible and so we have only one solution, namely angle B is 15.9°.

Revision

4 Determine the required angles in the following triangles:

(a) Angle A = 25°, side a = 18 mm, side b = 38 mm, determine angles B and C,
(b) Angle A = 42°, side a = 160 mm, side b = 200 mm, determine angles B and C.

9.3 The cosine rule

Consider the triangle ABC in figure 9.6. From apex C a line is drawn at right angles to side AB, intersecting it at point H. Let BH = x, and so we then have AH = $c - x$. Applying the Pythagoras theorem to the triangle ACH gives

$$b^2 = h^2 + (c - x)^2$$

Figure 9.6 *The cosine rule*

Applying the Pythagoras theorem to the triangle BCH gives

$$a^2 = h^2 + x^2$$

Subtracting the second equation from the first one gives

$$b^2 - a^2 = (c - x)^2 - x^2 = c^2 - 2cx + x^2 - x^2$$

Thus

$$b^2 = a^2 + c^2 - 2cx$$

But, with triangle BCH we have $\cos B = x/a$. Hence $x = a \cos B$ and so

$$b^2 = a^2 + c^2 - 2ac \cos B$$

This is one form of what is termed the *cosine rule*. By repeating the above with lines drawn from A to intersect BC at right angles and from B to intersect AC at right angles, we can also obtain

$$a^2 = b^2 + c^2 - 2bc \cos A$$

and

$$c^2 = a^2 + b^2 - 2ab \cos C$$

The rule can be stated as: the square of a side is equal to the sum of the squares of the other two sides minus twice the product of those sides times the cosine of the angle between them.

This rule may be used to find the values of unknown sides and angles when two sides and the angle included between those sides, or the three sides are given.

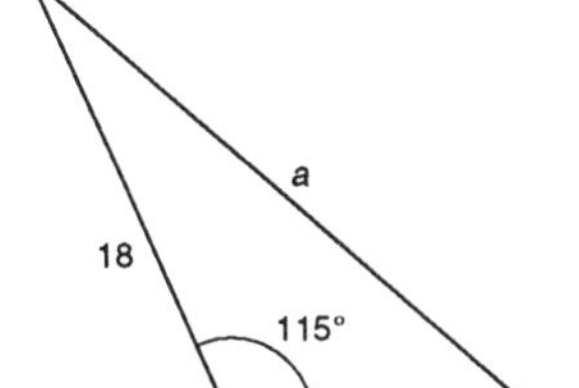

Figure 9.7

Example

Determine the side a for the triangle given in figure 9.7.

Using the cosine law we have $a^2 = b^2 + c^2 - 2bc \cos A$, and so

$$a^2 = 12^2 + 18^2 - 2 \times 12 \times 18 \cos 115°$$

Because the cosine of 115° is negative, we have

$$a^2 = 144 + 324 - (-182.6) = 650.6$$

and so $a = 25.5$.

Example

Determine the angles in a triangle which has sides of lengths $a = 13$, $b = 15$ and $c = 17$.

Using the cosine rule we can write $a^2 = b^2 + c^2 - 2bc \cos A$, hence

$$\cos A = \frac{b^2 + c^2 - a^2}{2bc} = \frac{15^2 + 17^2 - 13^2}{2 \times 15 \times 17} = 0.676$$

Hence $A = 47.5°$. Similarly for angle B, $b^2 = a^2 + c^2 - 2ac \cos B$ and so

$$\cos B = \frac{a^2 + c^2 - b^2}{2ac} = \frac{13^2 + 17^2 - 15^2}{2 \times 13 \times 17} = 0.527$$

Hence $B = 58.2°$. Since $A + B + C = 180°$, then $C = 180° - 47.5° - 58.2°$ and so is 74.3°.

Revision

5 Determine the required angle or length in each of the following triangles:

(a) Angle $A = 60°$, side $b = 14$, side $c = 10$, determine side a,
(b) Side $a = 5$, side $b = 6$, side $c = 7$, determine angle A,
(c) Angle $C = 75°$, side $a = 40$, side $b = 60$, determine side c,
(d) Angle $B = 110°$, side $a = 9.0$, side $c = 11.5$, determine side b,
(e) Side $a = 90$, side $b = 75$, side $c = 65$, determine angle A,
(f) Angle $C = 120°$, side $a = 8$, side $b = 10$, determine side c,
(g) Angle $C = 80°$, side $a = 30$ mm, side $b = 20$ mm, determine side c.
(h) Angle $A = 102°$, side $b = 50$ mm, side $c = 30$ mm, determine side a.

6 An object is supported by two chains connected from the same point on the object to points 3 m apart on the horizontal ceiling. The chain lengths are 1.8 m and 2.1 m. What is the angle between the chains at their connecting point on the object?

7 A parallelogram has sides of length 120 mm and 70 mm and the longer diagonal has a length of 150 mm. What are the angles of the parallelogram?

8 A 600 mm length of wire is bent into the shape of a triangle. If two of the sides have lengths of 220 mm and 180 mm, what are the angles in the triangle?

9.4 The areas of triangles

The area of any triangle is given by

$$\text{area} = \tfrac{1}{2}\ \text{base} \times \text{height}$$

Thus, for the triangle shown in figure 9.8, the area is $\frac{1}{2}ch$. Since, for triangle ACH, $\sin A = h/b$ we can substitute using $h = b \sin A$ and so write the area as

$$\text{area} = \tfrac{1}{2}bc \sin A$$

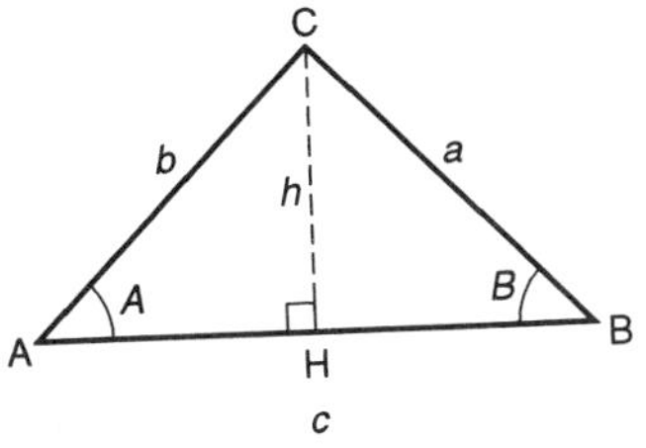

Figure 9.8 *Area of a triangle*

Alternatively we can consider triangle BCH and use $\sin B = h/a$ and so obtain

$$\text{area} = \tfrac{1}{2}ac \sin B$$

We could alternatively have considered the triangle area in terms of the height from BC as the base and so obtained

$$\text{area} = \tfrac{1}{2}ab \sin C$$

In general

$$\text{area} = \tfrac{1}{2} \times (\text{product of two sides}) \times (\text{sine of angle between those sides})$$

Example

Determine the area of the triangle shown in figure 9.9.

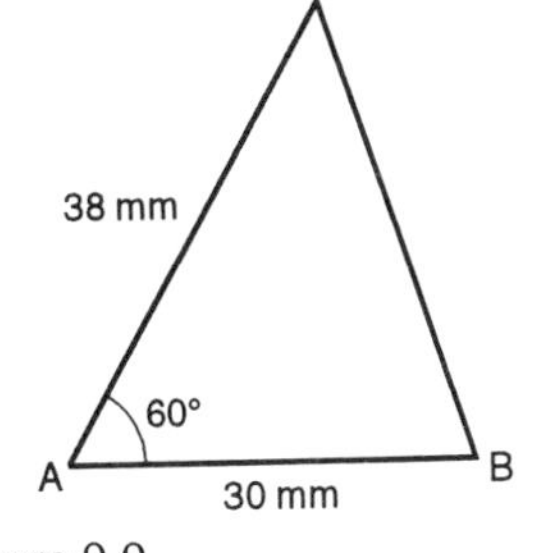

Figure 9.9

The data given on the figure is for two sides and the angle between those sides. Hence we can use the equation

$$\text{area} = \tfrac{1}{2}bc \sin A = \tfrac{1}{2} \times 38 \times 30 \times \sin 60° = 493.6\ \text{mm}^2$$

Example

Determine the area of the quadrilateral shown in figure 9.10.

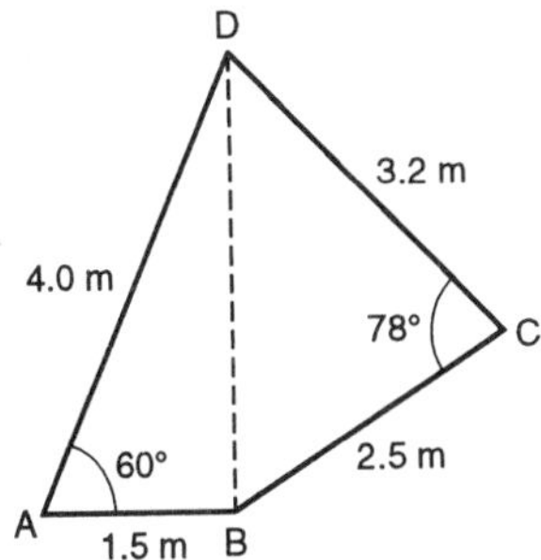

Figure 9.10

The quadrilateral can be considered to be made up of two triangles. For the triangle ABD, the area is

$$\text{area} = \tfrac{1}{2} \times 4 \times 1.5 \times \sin 60° = 2.60\ \text{m}^2$$

For the triangle BCD, the area is

$$\text{area} = \tfrac{1}{2} \times 2.5 \times 3.2 \times \sin 78° = 3.91\ \text{m}^2$$

Thus the total area is $2.60 + 3.91 = 6.51\ \text{m}^2$.

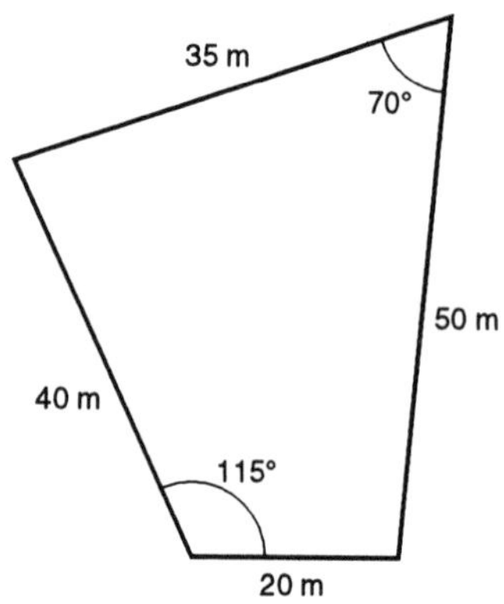

Figure 9.11

Revision

9 Determine the areas of each the following triangles:

(a) Angle $A = 45°$, side $b = 20$ mm, side $c = 50$ mm,
(b) Angle $C = 50°$, side $a = 2$ m, side $b = 3$ m,
(c) Angle $B = 110°$, side $a = 50$ mm, side $c = 140$ mm.
(d) Angle $A = 80°$, side $b = 4$ m, side $c = 3$ m,
(e) Angle $C = 38°$, side $a = 40$ mm, side $b = 60$ mm,
(f) Angle $C = 40°$, side $a = 80$ mm, side $b = 50$ mm,
(g) Angle $A = 145°$, side $b = 30$ mm, side $c = 40$ mm.

10 Determine the area of the quadrilateral shown in figure 9.11.

9.4.1 Area in terms of the semi-perimeter

If the three sides of a triangle are known but no angle, then an alternative equation can be used to determine the area of the triangle. Let us take the area of a triangle (e.g. that in figure 9.8) as being given by $\frac{1}{2}bc \sin A$. If we square both sides of the equation we obtain

$$(\text{area})^2 = \left(\tfrac{1}{2}bc\sin A\right) \times \left(\tfrac{1}{2}bc\sin A\right) = \tfrac{1}{4}b^2c^2\sin^2 A$$

But $\sin^2 A + \cos^2 A = 1$ (see section 8.3). Thus we can write

$$(\text{area})^2 = \tfrac{1}{4}b^2c^2(1 - \cos^2 A)$$

We can write this as

$$(\text{area})^2 = \tfrac{1}{4}b^2c^2(1 - \cos A)(1 + \cos A)$$

Using the cosine law for triangles we can write for $\cos A$

$$\cos A = \frac{b^2 + c^2 - a^2}{2bc}$$

Hence

$$(\text{area})^2 = \tfrac{1}{4}b^2c^2\left(1 - \frac{b^2 + c^2 - a^2}{2bc}\right)\left(1 + \frac{b^2 + c^2 - a^2}{2bc}\right)$$

$$= \tfrac{1}{4}b^2c^2\left(\frac{2bc - b^2 - c^2 + a^2}{2bc}\right)\left(\frac{2bc + b^2 + c^2 - a^2}{2bc}\right)$$

$$= \tfrac{1}{16}\left[a^2 - (b-c)^2\right]\left[(b+c)^2 - a^2\right]$$

$$= \tfrac{1}{16}(a - b + c)(a + b - c)(b + c + a)(b + c - a)$$

Let $a + b + c = 2s$, i.e. s is half the length of the perimeter of the triangle. Then subtracting $2b$ from both sides of this equation gives

$$a - b + c = 2(s - b)$$

If we had subtracted $2a$ we would have obtained

$$b + c - 2a = 2(s - a)$$

and if we had subtracted $2c$

$$a + b - c = 2(s - c)$$

Thus we obtain

$$(\text{area})^2 = \tfrac{1}{16} \times 2(s - b) \times 2(s - c) \times 2s \times 2(s - a)$$

Hence

$$\text{area} = \sqrt{s(s - a)(s - b)(s - c)}$$

Example

Determine the area of a triangle which has sides of $a = 6$ m, $b = 5$ m and $c = 9$ m.

We have $2s = a + b + c = 6 + 5 + 9 = 20$. Hence $s = 10$. Thus

$$\text{area} = \sqrt{10(10 - 6)(10 - 5)(10 - 9)} = \sqrt{10 \times 4 \times 5 \times 1} = \sqrt{200}$$

Hence the area is 14.1 m^2.

Revision

11 Determine the areas of the following triangles:

(a) Sides $a = 3$ m, $b = 4$ m, $c = 5$ m,
(b) Sides $a = 6$ m, $b = 8$ m, $c = 10$ m,
(c) Sides $a = 7$ m, $b = 10$ m, $c = 12$ m,
(d) Sides $a = 5$ cm, $b = 7$ cm, $c = 8$ cm,
(e) Sides $a = 7$ cm, $b = 8$ cm, $c = 9$ cm,
(f) Sides $a = 5$ cm, $b = 6$ cm, $c = 7$ cm,
(g) Sides $a = 4$ cm, $b = 5$ cm, $c = 7$ cm.

12 Determine the area of a parallelogram which has adjacent sides of lengths 2.4 cm and 3.2 cm and the length of the shortest diagonal is 2.0 cm.

Problems

1 Determine, using the sine rule, the required sides and angles for each of the following triangles:

(a) Angle A = 68°, angle C = 70°, side c = 42 mm, determine angle B, sides a and b,
(b) Angle A = 58°, side a = 180 mm, side b = 150 mm, determine angles B and C, side c,
(c) Angle C = 25°, angle B = 80°, side b = 10 cm, determine angle A and sides a and c,
(d) Angle = 10°, angle C = 60°, side b = 50 mm, determine angle B and sides a and c,
(e) Angle B = 50°, side b = 30 mm, side c = 20 mm, determine angles A and C and side a,
(f) Angle A = 70°, angle C = 40°, side c = 100 mm, determine angle B and sides a and b,
(g) Angle A = 62°, angle C = 62°, side b = 50 mm, determine angle B and sides a and c,
(h) Angle A = 53°, angle B = 62°, side a = 125 mm, determine angle C and sides b and c,
(i) Angle A = 80°, side a = 165 mm, side b = 130 mm, determine angles B and C and side c,
(j) Angle A = 30°, side a = 50 mm, side b = 80 mm, determine angles B and C and side c.

2 A roof has a span of 9 m with eaves that slope on one side at 44° to the horizontal and on the other side at 36° to the horizontal. Determine the lengths of the sloping parts of the roof.

3 A wall is leaning outwards at an angle to the vertical of 15°. A ladder of length 2 m is placed against the wall with its base on level ground 1 m from the base of the wall. How high up the wall does the ladder reach?

4 A vertical pole of length 6 m stands alongside a road which slopes away from it with a constant slope. The pole casts a shadow which is 14 m long directly downhill when the angle of elevation of the sun is 56°. What is the angle of the road with the horizontal?

5 Determine, using the cosine rule, the required sides and angles for each of the following triangles:

(a) Side a = 12, side b = 63, side c = 52, determine angles A, B and C,
(b) Angle A = 57.5°, side b = 78, side c = 97, determine side a and angles B and C,
(c) Side a = 9, side b = 13, side c = 17, determine angles A, B and C,
(d) Side a = 7, side b = 5, side c = 3, determine angles A, B and C,
(e) Side a = 4, side b = 7, side c = 5, determine angles A, B and C,
(f) Angle A = 80°, side b = 7, side c = 8, determine side a and angles B and C,

(g) Angle A = 66°, side b = 40 mm, side c = 50 mm, determine angles B and C and side a,
(h) Side a = 40 mm, side b = 50 mm, side c = 70 mm, determine angles A, B and C,
(i) Angle C = 74°, side a = 100 mm, side b = 50 mm, determine angles A and B and side c.

6 A plane flies for 50 km in a straight line from its start point before turning through 20° and flying in a straight line on the new bearing for 70 km. What distance is the plane from the start?

7 A parallelogram has adjacent sides of lengths 55 mm and 82 mm with an angle between them of 68°. What is the length of the diagonal through this corner of the parallelogram?

8 A parallelogram has adjacent sides of lengths 30 mm and 50 mm with an angle between them of 144°. What is the length of the diagonal through this corner of the parallelogram?

9 Determine the areas of each of the following triangles:

(a) Angle A = 85°, side b = 40 mm, side c = 10 mm,
(b) Angle B = 130°, side a = 40 mm, side c = 60 mm,
(c) Angle C = 52°, side a = 2 m, side b = 5 m,
(d) Angle A = 55°, side b = 12 m, side c = 4 m,
(e) Side a = 4 cm, side b = 7 cm, side c = 9 cm,
(f) Side a = 2 m, side b = 2 m, side c = 2 m,
(g) Side a = 5 cm, side b = 12 cm, side c = 13 cm.

10 A parallelogram of area 30 cm^2 has adjacent sides of lengths 6 cm and 9 cm. Determine the angles of the parallelogram.

11 A parallelogram has adjacent sides of lengths 9 cm and 5 cm. If the shortest diagonal has a length of 7 cm, what is the area of the parallelogram?

10 Vectors

10.1 Introduction

With many engineering problems there are certain quantities for which not only their size but their direction is important. For example, if we want to specify the location of some town in relation to, say, London then we can do this by stating how far away from London it is and in what direction. In the case of forces, both the direction of a force and its size need to be considered if we want to predict the effect of the force on an object. Those quantities for which both the size and direction have to be specified are termed *vectors*. Quantities for which only size is relevant are termed *scalar quantities*. This chapter is an introduction to the problems involving vector quantities, considering how vector quantities can be added or subtracted and how a vector quantity can be replaced by two equivalent vector quantities. The study of vectors offers an example of the usefulness of trigonometry in engineering.

Example

Which of the following italicised terms have a direction associated with their specification and so are vector quantities?

(a) The *distance* that I can walk in an hour is 6 km.
(b) The *force* due to friction acting on a block sliding across the floor is 4 N and is in a direction opposite to its direction of motion.
(c) The surface *area* of a block is 4 m^2.
(d) The *displacement* of town A from town B is 40 km in a direction 50° east of north.

(a) This is a scalar quantity since there is no direction needed to specify the distance.
(b) This is a vector quantity since the force is specified by both a size and a direction.
(c) This is a scalar quantity since there is no direction needed to specify the area.
(d) This is a vector quantity since the displacement is specified by both size and direction.

Revision

1 Which of the following italicised terms have a direction associated with their specification and so are vector quantities?

(a) The *mass* of an object is 5 kg.
(b) The *force* acting on an object due to gravity is 19.6 N vertically towards the ground.

(c) The *temperature* of an object is 40°C.
(d) The atmospheric *pressure* is about 10^5 Pa.
(e) The *velocity* of a car is 100 km/hour due north along the motorway.
(f) A car has a top *speed* of 150 km/hour.

10.1.1 Representation of vectors

To specify a scalar quantity all we need to do is give a single number to represent its size. To specify a vector quantity we need to indicate both size and direction. The term *magnitude* tends to be used for the size of vector quantities. To represent vector quantities on a diagram we use arrows. The length of the arrow is chosen according to some scale to represent the magnitude of the vector and the direction of the arrow, with reference to some reference direction, the direction of the vector. For example, to represent a displacement of 30 km in a north-east direction from A to B we might use the arrow shown in figure 10.1. The length of the arrow is scaled to represent the magnitude of 30 km and the direction of the arrow is 45° to the horizontal to represent the north-east direction, the horizontal being the reference direction for the east-west direction.

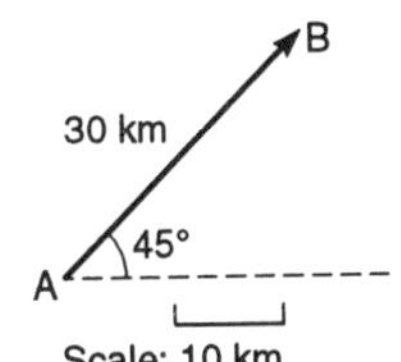

Figure 10.1 *A vector*

We can indicate that a line from A to B is a vector by writing it as $\overrightarrow{AB}$, the direction of the arrow indicating that the direction is from A to B. With a direction from B to A we would have used $\overrightarrow{BA}$. An alternative notation is however more usually used. Vector quantities, such as that represented by the line $\overrightarrow{AB}$, are denoted by using a bold letter such as **a**, or when hand-written by underlining the symbol $\underline{a}$. When we are referring to $\overrightarrow{BA}$ we would use −**a**, or $-\underline{a}$, with the minus sign being used to indicate that it is in the opposite direction to **a**, or $\underline{a}$.

The term *modulus of a vector* is often used for the magnitude of a vector **a** and is represented by |**a**| or just writing *a*. The direction of a vector with reference to some reference direction is specified by an angle. This angle is often called the *argument of the vector*. For the displacement of 30 km in a north-east direction from A to B represented graphically in figure 10.1, we have a modulus of 30 km and an argument of 45°.

For two vectors to be equal they must represent the same quantity and have the same magnitude and direction. Thus the two vectors represented in figure 10.2 by $\overrightarrow{AB}$ and $\overrightarrow{CD}$ are equal. For two forces, vector quantities, to be equal then they must have the same magnitude and direction. We cannot say that a force of 2 N is equal to another force of 2 N without knowing that both forces are acting in the same direction. Thus a force of 2 N acting in a northerly direction is the same as another force of 2 N acting in a northerly direction. We can represent these forces as the identical vectors **F** and **F**. A force of 2 N in a northerly direction is however different from a force of 2 N acting in a southerly direction. In this case the magnitudes are the same but the directions are directly opposite. We can represent these forces as the vectors **F** and −**F**.

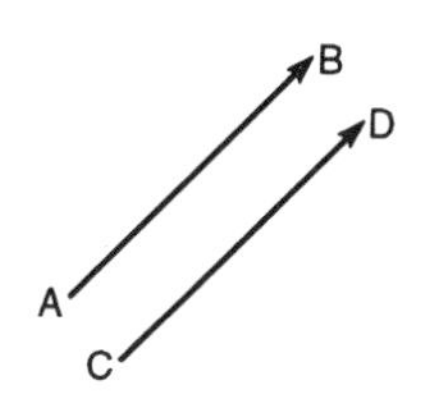

Figure 10.2 *Two equal vectors*

Example

An object slides down a slope which is at 20° to the horizontal. Draw a diagram to represent the vectors for the vertical force of 30 N acting on the object as a result of gravity and the frictional force acting up the plane and opposing the motion of 10 N.

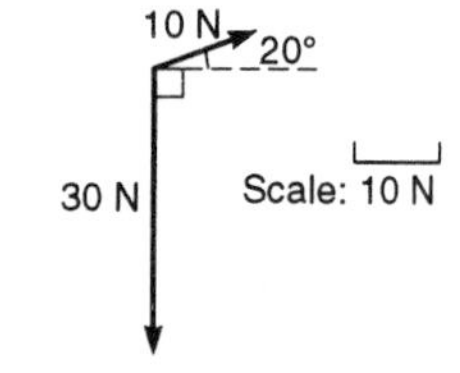

Figure 10.3

There is a force of 30 N in a vertical direction and a force of 10 N at an angle of 20° to the horizontal. We can thus represent the two vector quantities by an arrow with a length proportional to 30 N in the vertical direction and an arrow of length proportional to 10 N at 20° to the horizontal. Figure 10.3 shows the result.

Revision

2 Draw diagrams to represent the following vectors:

(a) An object acted on by a force of 10 N in a horizontal direction and a force of 5 N at an angle of 30° to the horizontal direction and upwards.
(b) A projectile which has a velocity of 2 m/s in the horizontal direction and a velocity of 3 m/s in the vertically upward direction.
(c) An object which has been given a displacement of 200 mm to the right and 400 mm in a direction at right angles to this.

3 Which of the vectors shown in figure 10.4 are equal?

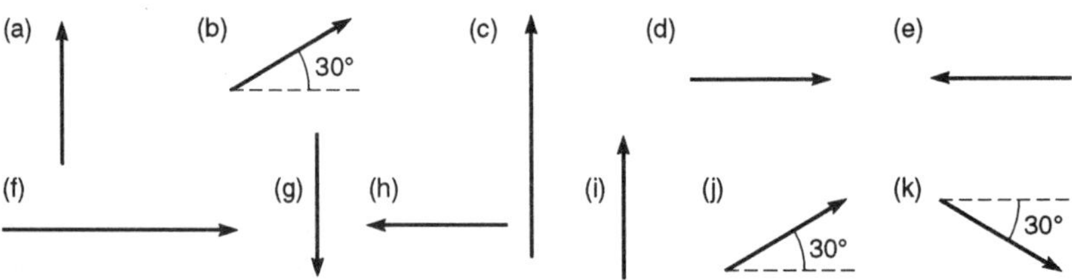

Figure 10.4

10.2 Addition of vectors

When we add two scalar quantities, say a mass of 2 kg added to a mass of 3 kg, we can write 2 + 3 = 5 and so the addition gives a mass of 5 kg. With vectors we have to consider both the magnitude and direction of each vector in carrying out the addition.

The rule for adding two vectors is called the *triangle rule*. This can be stated as: to add two vectors **a** and **b** we place the tail of the arrow representing one vector at the head of the arrow representing the other and then the line that forms the third side of the triangle represents the vector which is the sum of **a** and **b**. Figure 10.5 shows such a triangle. Note that the directions of **a** and **b** go in one sense round the triangle and the sum goes in the opposite direction. The vector representing the sum of **a** and **b** is termed the *resultant*. We could replace the two vectors **a** and **b** by the vector for the sum and have the same effect.

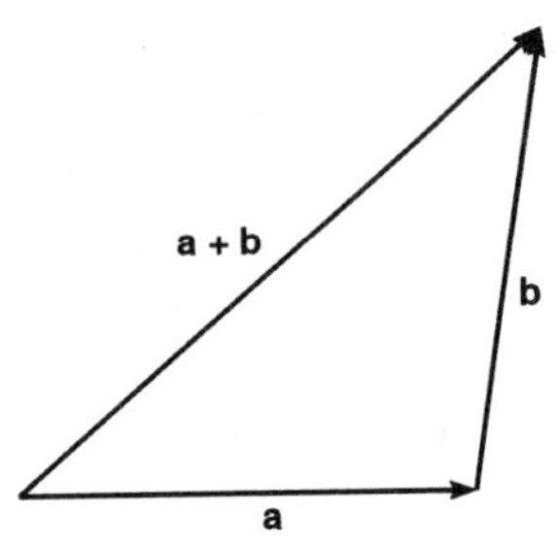

Figure 10.5 *The triangle rule*

Note that in engineering you may come across the *triangle of forces* defined as the condition for equilibrium of three forces acting at the same point and in the same plane on some object. This may be stated as: if the three forces are represented in magnitude and direction by arrows $\mathbf{F}_1$, $\mathbf{F}_2$ and $\mathbf{F}_3$, then these arrows when taken in the order of the forces must form a triangle if the three forces are to be in equilibrium (figure 10.6). This is just the triangle rule. If we consider two forces $\mathbf{F}_1$ and $\mathbf{F}_2$ then the sum of their vectors $\mathbf{F}_1 + \mathbf{F}_2$ is the third side of the triangle. This vector must be of the same magnitude but in the opposite direction if there is to be no net force acting on the object. Thus $\mathbf{F}_1 + \mathbf{F}_2$ is $-\mathbf{F}_3$.

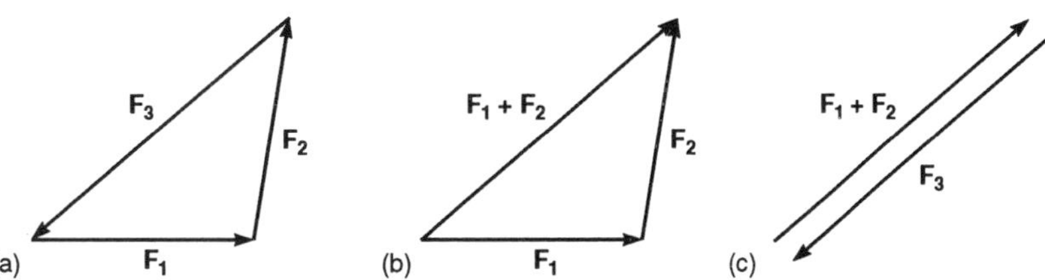

Figure 10.6 *Triangle of forces*

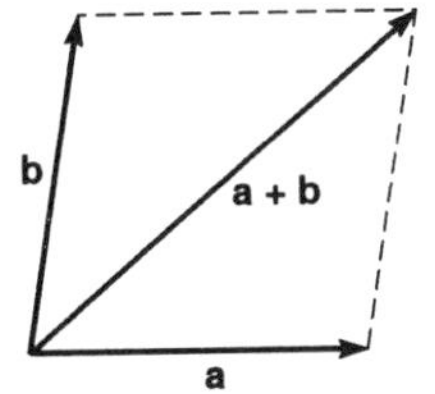

Figure 10.7 *The parallelogram rule*

An alternative and equivalent rule to the triangle rule for determining the sum of two vectors is the *parallelogram rule*. This can be stated as: if we place the tails of the arrows representing the two vectors **a** and **b** together and complete a parallelogram, then the diagonal of that parallelogram drawn from the junction of the two tails represents the sum of the vectors **a** and **b**. Figure 10.7 illustrates this. The triangle rule is just the triangle formed between the diagonal and two adjacent sides of the parallelogram.

Example

The vector **a** has a magnitude of 10 and an argument of 30°. The vector **b** has a magnitude of 5 and an argument of 100°. Determine **a** + **b**.

We can represent **a** by an arrow with a length which represents the magnitude 10 at an angle of 30° to some reference direction. Using the triangle rule then we draw the vector **b** with a length representing 5 at an angle of 100° to the same reference direction and with its tail at the arrow end of **a**. The sum of the two vectors, i.e. the resultant, is then the vector indicated by completing the triangle. Figure 10.8 shows the triangle. We could determine this resultant by using a scale drawing. Alternatively we could use the cosine rule (see section 9.3). This rule is: the square of a side is equal to the sum of the squares of the other two sides minus twice the product of those sides times the cosine of the angle between them. Since the angle opposite the resultant of **a** + **b** is (180° − 100°) + 30° = 110°, then

$$(\text{resultant})^2 = 5^2 + 10^2 - 2 \times 5 \times 10 \cos 110° = 25 + 100 + 34.2$$

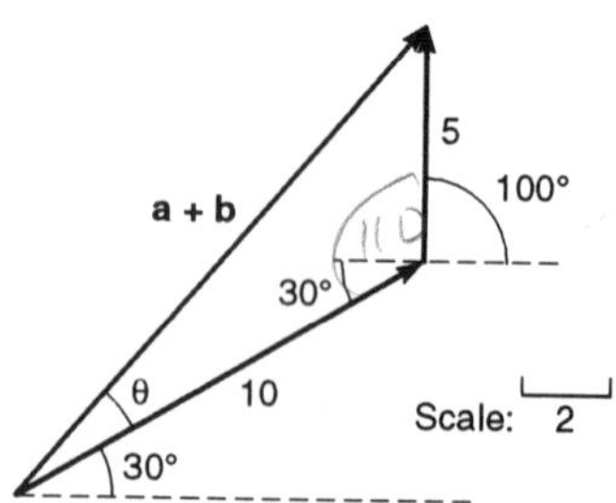

Figure 10.8 *Example*

Hence the $(\text{resultant})^2 = 159.2$ and the resultant has a magnitude of 12.6. The angle this resultant makes with **a** can be determined by means of the sine rule. Thus

$$\frac{12.6}{\sin 110°} = \frac{5}{\sin\theta}$$

Hence

$$\sin\theta = \frac{5 \times \sin 110°}{12.6} = 0.373$$

Hence $\theta = 21.9°$. Thus the resultant is at an angle of $21.9° + 30° = 51.9°$ to the reference axis.

Example

An object is acted on by two forces, of magnitudes 5 N and 4 N, at an angle of 60° to each other. What is the resultant force on the object?

Figure 10.9 shows how these two vectors can be represented by the sides of a parallelogram.

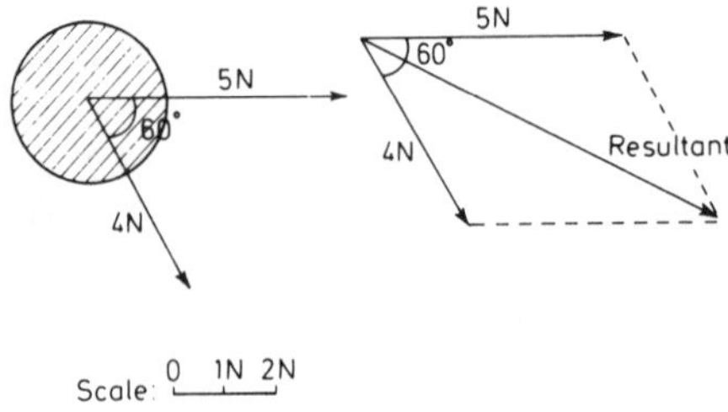

Figure 10.9 *Example*

The resultant can be determined from a scale drawing from measurements of its length and angle or by calculation. We have a triangle with adjacent sides of 5 and 4 and an angle of $180° - 60° = 120°$ between them. We need to determine the length of the other side. The cosine rule (see section 9.3) can be used. Thus

$$(\text{resultant})^2 = 5^2 + 4^2 - 2 \times 5 \times 4 \cos 120° = 25 + 16 + 20$$

Hence the resultant is $\sqrt{61} = 7.8$ N. We can determine the angle θ between the resultant and the 5 N force by the use of the sine rule (see section 9.2). Thus

$$\frac{4}{\sin\theta} = \frac{7.8}{\sin 120°}$$

Hence

$$\sin\theta = \frac{4 \times \sin 120°}{7.8} = 0.444$$

Hence θ = 26.4°.

Example

Figure 10.10(a) shows an object supported by two wires. If the weight of the object is 25 N, what are the forces in the two wires when the object is in equilibrium?

When the object is in equilibrium then the resultant of the two forces in the wires must be opposite and equal to the weight of 25 N. For the weight we know the magnitude and direction. For the forces in the two wires we only know the directions. Figure 10.10(b) shows the triangle of forces produced by utilising this information. We could obtain the forces in the wires from a scale diagram of the triangle of forces. Alternatively we can use the sine rule (see section 9.2). The angle between the two wires is, from figure 10.10(a), 180° – 30° – 60°. Thus

$$\frac{T_1}{\sin(90° - 60°)} = \frac{25}{\sin(180° - 30° - 60°)}$$

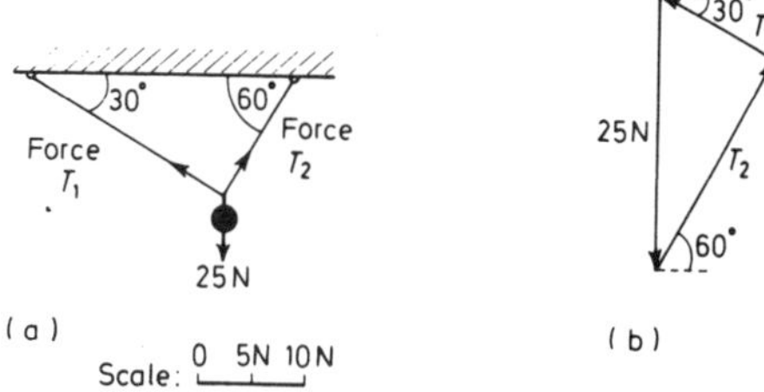

Figure 10.10

Thus

$$T_1 = \frac{25 \times \sin 30^\circ}{\sin 90^\circ} = 12.5\ \text{N}$$

For the other string,

$$\frac{T_2}{\sin(90^\circ - 30^\circ)} = \frac{25}{\sin(180^\circ - 30^\circ - 60^\circ)}$$

Thus

$$T_2 = \frac{25 \sin 60^\circ}{\sin 90^\circ} = 21.7\ \text{N}$$

Example

A tool is given a displacement of 100 mm in the x direction and 60 mm in a direction which is at right angles to the x direction. What is the resultant displacement?

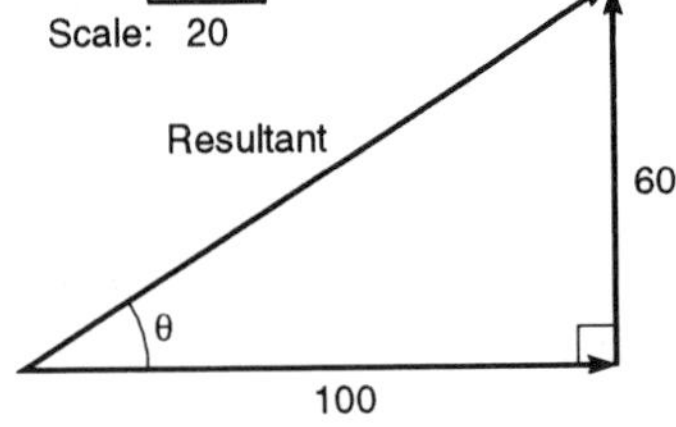

Figure 10.11

Displacement is a vector quantity and thus the resultant displacement is found by vector means. We can thus use the triangle rule to determine the sum of the two vectors. We draw an arrow with a length scaled to represent 100 mm. The arrow to represent the second displacement is then drawn with its tail starting from the arrow head of the first in a direction at right angles to it and with a length scaled to represent 60 mm. The resultant is the third side of the triangle. Figure 10.11 shows the triangle. We can calculate the resultant by the use of the Pythagoras theorem. Thus

$$(\text{resultant})^2 = 60^2 + 100^2 = 3600 + 10\,000$$

Hence the resultant has a magnitude of 116.6 mm. The resultant is at an angle θ where $60/100 = \tan\theta$. Hence $\theta = 31.0^\circ$.

Example

A ship heads on a course due south at 10 km/h. It is driven off course by a current flowing in a north-easterly direction at 6 km/h. What is the resultant velocity?

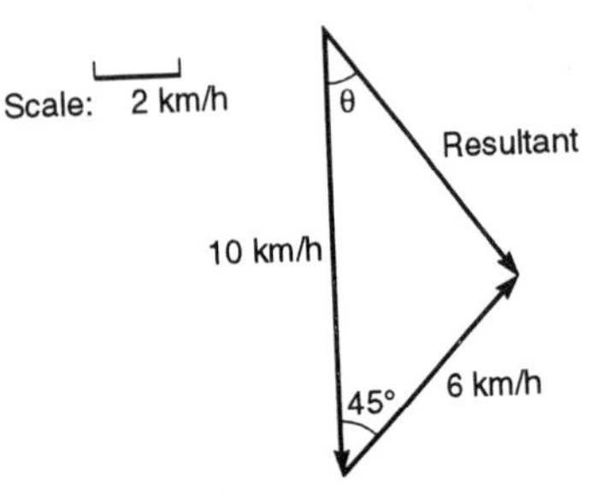

Figure 10.12

The ship has two velocities, one in the southerly direction and one in the north-easterly direction. The resultant velocity is the vector sum of the two. If it helps, you can consider the problem in terms of what the resultant displacement of the ship would be if in one hour it has a displacement of 10 km in the southerly direction and 6 km in the north-easterly direction. We can use the triangle rule to determine the resultant. Figure 10.12 shows the result. The 10 km/h velocity is represented by an arrow of length scaled to represent 10 km in a direction which represents south. From the end of that arrow an arrow is drawn to represent the 6 km/h velocity. This will be at 45° to the southerly direction in order to give the north-easterly direction. The resultant is then the third side of the triangle. We can determine the resultant from a scale diagram or by calculation using the cosine rule (section 9.3). Thus

$$(\text{resultant})^2 = 10^2 + 6^2 - 2 \times 10 \times 6 \cos 45° = 100 + 36 - 84.85$$

Hence the resultant has a magnitude of 7.2 km/h. We can use the sine rule (section 9.2) to determine the angle θ between the resultant and the 10 km/h velocity:

$$\frac{7.2}{\sin 45°} = \frac{6}{\sin \theta}$$

Thus

$$\sin \theta = \frac{6 \times \sin 45°}{7.2} = 0.589$$

Hence θ = 36.1°. Thus the resultant velocity is 7.2 km/h in a direction which is 36.1° east of south.

Revision

4 If **a** is a vector of magnitude 4 in an easterly direction and **b** a vector of magnitude 3 in a southerly direction, what is the resultant, i.e. the vector sum **a** + **b**?

5 If **a** is a vector of magnitude 5 in an easterly direction and **b** a vector of magnitude 4 in a north-easterly direction, what is the resultant, i.e. the vector sum **a** + **b**?

6 Determine the resultant forces acting on objects subject to the following forces acting at a point on the object:

(a) A force of 3 N acting horizontally and a force of 4 N acting at the same point on the object at an angle of 60° to the horizontal,
(b) A force of 3 N in a westerly direction and a force of 6 N in a northerly direction.
(c) Forces of 4 N and 5 N with an angle of 65° between them,
(d) Forces of 3 N and 8 N with an angle of 50° between them,
(e) Forces of 5 N and 7 N with an angle of 30° between them,
(f) Forces of 9 N and 10 N with an angle of 40° between them,
(g) Forces of 10 N and 12 N with an angle of 105° between them.

7 In a plane structure a particular point is acted on by forces of 1.2 kN and 2.0 kN in the plane, the angle between the forces being 15°. What is the resultant force?

8 An object of weight 30 N is suspended from a horizontal beam by two chains. The two chains are attached to the same point on the object and are at 30° and 40° to the vertical. Determine the tensions in the two chains.

9 Determine the resultant velocity of each of the following pairs of velocities:

(a) 10 km/h due north and 5 km/h due east,
(b) 6 km/h due east and 4 km/h in a north-westerly direction,
(c) 5 km/h due north and 7 km/h in a direction 60° east of south.

10 A boat is required to cross a river 100 m wide to a point on the bank directly opposite the start point. If the boat can have a velocity of 6 km/h and the river is flowing with a current of 3 km/h, in what direction should the boat be steered?

11 A ship is steering a course due south across a current which is in a westerly direction. The resultant velocity of the ship is 18 km/h in a direction 15° west of south. What is the velocity set by the ship?

10.2.1 Polygon of vectors

The triangle law is used to find the sum of two vectors. Consider now the problem of determining the sum of more than two vectors. Suppose we have vectors **a**, **b**, **c** and **d**. We can use the triangle law to find the sum **e** of vectors **a** and **b**. We can then use the triangle law to find the sum **f** of **e** and **c**. We can then use the triangle law to find the sum **g** of **f** and **d**. Figure 10.13 illustrates the above procedure.

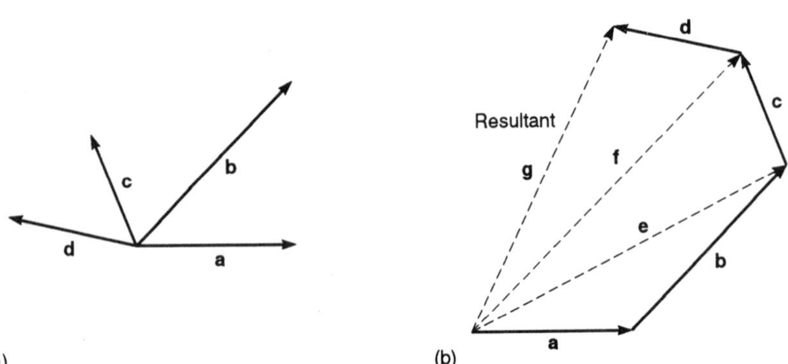

Figure 10.13 *Polygon of vectors*

In fact what we have done is take the vectors in sequence and draw the arrows tail to head. The resultant of all the vectors is the vector required to complete the polygon shape and link the head of the last vector to the tail of the start vector.

Example

Determine the force acting on the eyebolt shown in figure 10.14(a) as a result of the forces applied along the connecting wires.

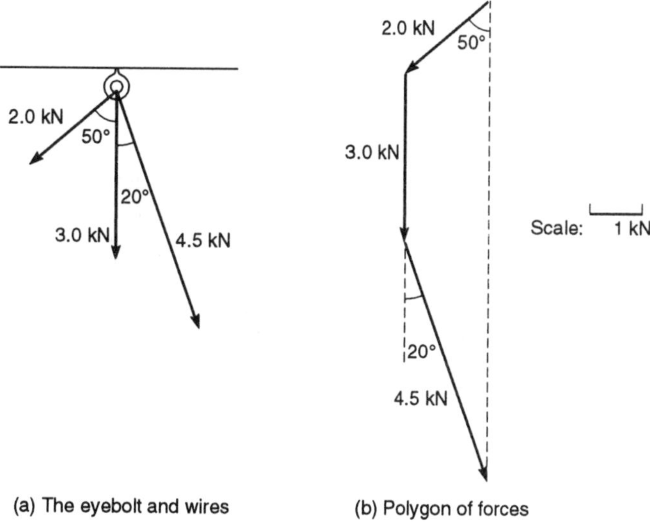

(a) The eyebolt and wires

(b) Polygon of forces

Figure 10.14

Starting with the 2.0 kN force at 50° to the vertical, the vector is drawn to represent it. From the head of that vector we draw a vector to represent the next force in sequence as we proceed anticlockwise round the forces, namely the vertical 3.0 kN force. From the head of this vector we draw a vector to represent the next force in the sequence, namely the 4.5 kN force at 20° to the vertical. The resultant force is the force completing the polygon. We could calculate this by breaking the polygon down into the triangles indicated by figure 10.13 and using the cosine law with each triangle in turn. Alternatively we could draw the polygon to scale. With a scale diagram the force is indicated as being about 8.5 kN vertically.

An alternative technique which will be met later in this chapter is to resolve each of the forces into the horizontal and vertical directions. Then the algebraic sum of the horozontal forces and the sum of the vertical forces are obtained. These two forces can then be added, by vector means, to give the resultant.

Revision

12 Determine the resultant force acting on the gusset plate as a result of the forces shown in figure 10.15.

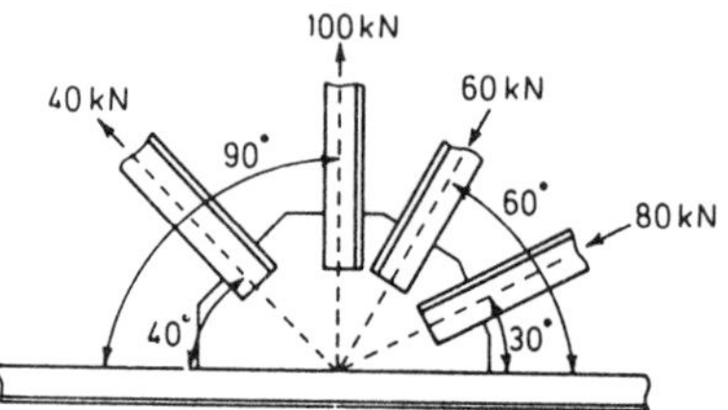

Figure 10.15

13 Three forces act at a point on an object. One of the forces is 6 N horizontally to the left, another 3 N at 70° anticlockwise to the 6 N force, and the third 4 N at 150° anticlockwise to the 6 N force. Determine the resultant force.

10.3 Subtraction of vectors

When we subtract two scalar quantities, say taking a mass of 2 kg from one of 5 kg, we can write 5 – 2 = 3 and so the subtraction gives a mass of 3 kg. With vectors we have to consider both the magnitude and direction of each phasor in carrying out the subtraction.

With vectors we define subtraction of vector **b** from vector **a** as being

$$\mathbf{a} - \mathbf{b} = \mathbf{a} + (-\mathbf{b})$$

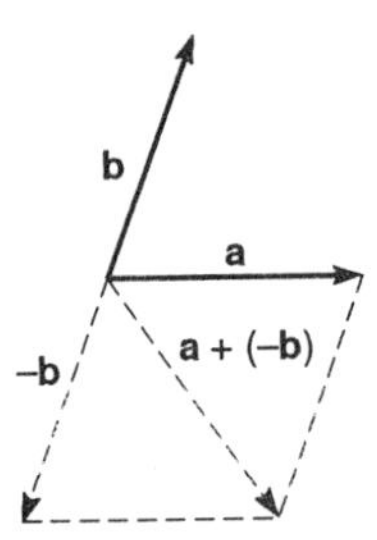

Figure 10.16 *Subtracting vectors*

A negative vector –**b** has the same magnitude but is in the opposite direction to the vector **b**. Thus to subtract **b** from **a**, all we do is reverse the direction of **b** and then add it to **a** by the use of the triangle or parallelogram rule. Figure 10.16 illustrates this process of subtracting vector **b** from vector **a**.

Example

If vector **a** has a magnitude of 3 in an easterly direction and vector **b** a magnitude of 4 in a northerly direction, determine the value of **a** – **b**.

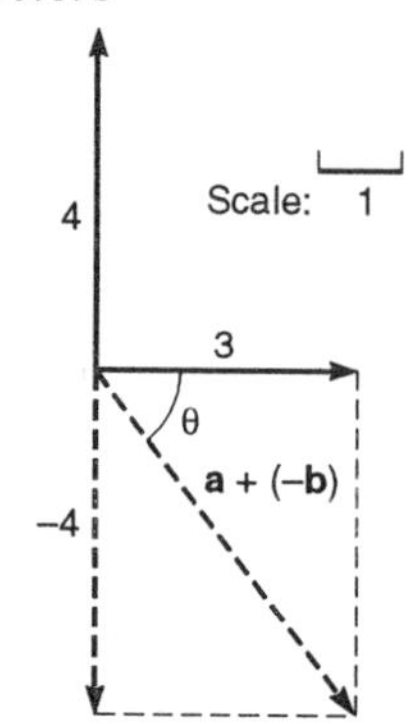

Figure 10.17

Figure 10.17 shows the vectors **a** and **b**. To subtract **b** from **a** we reverse the direction of **b** to give –**b** and then add –**b** and **a**. We thus end up with the parallelogram shown in the figure. The vector **a** + (–**b**) has thus a magnitude given by the Pythagoras theorem as

$$\sqrt{3^2 + (-4)^2} = 5$$

and in a direction which is at an angle θ given by

$$\tan\theta = \frac{-4}{3}$$

Hence $\theta = -53.1°$.

Revision

14 Determine **a** − **b** if:

(a) vector **a** has a magnitude of 4 in a westerly direction and vector **b** a magnitude of 2 in a northerly direction,
(b) vector **a** has a magnitude of 3 in an easterly direction and vector **b** a magnitude of 5 in a direction 60° south of east,
(c) vector **a** has a magnitude of 7 in an easterly direction and vector **b** a magnitude of 4 in a direction 25° south of west,
(d) vector **a** has a magnitude of 4 in a northerly direction and vector **b** a magnitude of 5 in a direction 25° south of west.

10.3.1 Relative velocity

Consider two objects A and B in motion, one having a velocity $\mathbf{v}_A$ and the other a velocity $\mathbf{v}_B$ with respect to the fixed ground. What is the velocity of B as perceived by A? The relative velocity of B with respect to A is the velocity that B would appear to have if A was considered to be the fixed point, rather than the fixed ground, from which the velocity of B was measured. To effectively make A at rest we would have to add to it a velocity of $-\mathbf{v}_A$. We add this same velocity to B. Thus $\mathbf{v}_B + (-\mathbf{v}_A)$ is the velocity of B relative to A.

Example

Aircraft A is flying due west at 400 km/h, while aircraft B is flying due north at 300 km/h. What is the velocity of aircraft B relative to aircraft A?

We need to consider the velocity of aircraft B as perceived by aircraft A if aircraft A is considered to be at rest. Thus we add a velocity of 400 km/h in an easterly direction to that of aircraft A. We then add this same velocity of 400 km/h in an easterly direction to the velocity of aircraft B. Figure 10.18 shows the resulting vector diagram. Using the Pythagoras theorem, then the magnitude of the relative velocity is the length of the diagonal of the parallelogram of vectors, i.e.

$$(\text{relative velocity})^2 = 400^2 + 300^2$$

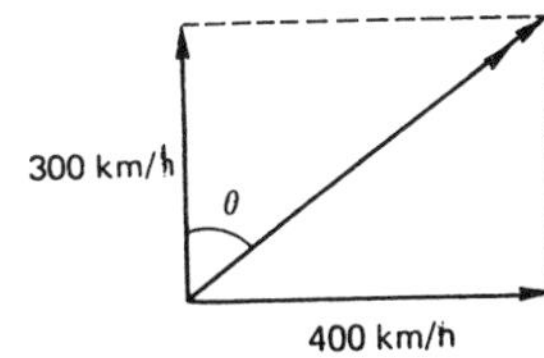

Figure 10.18

Hence the relative velocity has a magnitude of 500 km/h. The direction of this velocity is given by $\tan\theta = 400/300$. Hence $\theta = 53.1°$. Thus the relative velocity is 500 km/h in a direction 53.1° east of north.

Revision

15 A passenger on a train travelling due north at 25 m/s sees a second train on a parallel track moving at 40 m/s in the opposite direction. What is the relative velocity of the second train as perceived by the passenger on the first train?

16 Object A is moving at 5 m/s in a westerly direction. Object B is moving at 12 m/s in a south-westerly direction. What is the velocity of B relative to A?

17 A student is walking along a road at 5 km/h. It is raining and the rain is falling vertically with a velocity of 8 km/h. At what angle will the student perceive the rain to be falling?

18 Object A is moving at 5 m/s due west and object B at 12 m/s in a south-easterly direction. What is the velocity of B relative to A?

10.4 Resolution of vectors

Two vectors can be added together to give a single vector called the resultant. There is a reverse process of taking a single vector **a** and expressing it in terms of two *component vectors* **v** and **h** at right angles to each other. The sum of the components is the original vector, i.e. $\mathbf{a} = \mathbf{v} + \mathbf{h}$. These components are referred to as the *resolved parts* of the vector.

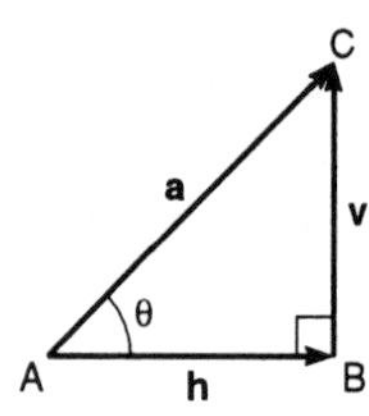

Figure 10.19 *Resolving vectors*

Consider the vector **a** shown in figure 10.19. This vector acts along AC, with triangle ABC being a right-angled triangle. If we take the **v** vector as being along BC and the **h** vector along AB, then for the magnitudes of the vectors we have, for the **v** vector

$$\frac{|\mathbf{v}|}{|\mathbf{a}|} = \sin\theta$$

and so $|\mathbf{v}| = |\mathbf{a}| \sin\theta$. For the **h** vector we have

$$\frac{|\mathbf{h}|}{|\mathbf{a}|} = \cos\theta$$

and so $|\mathbf{h}| = |\mathbf{a}| \cos\theta$.

In figure 10.19 the triangle rule was used to resolve a vector into its two components. An alternative, but absolutely equivalent method, would have been to use the parallelogram rule. Thus, suppose we have a force of 5 N at an angle of 40° to the horizontal. Then using the parallelogram rule we can draw the horizontal and vertical components as the sides of the parallelogram, as shown in figure 10.20. Then we have, using the equation derived above,

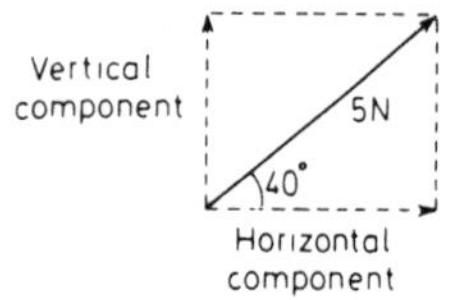

Figure 10.20 *Resolving a force*

horizontal component = 5 cos 40° = 3.8 N

vertical component = 5 sin 40° = 3.2 N

We could therefore replace the 5 N force by the two components of 3.8 N and 3.2 N and the effect would be precisely the same.

Example

Determine the horizontal and vertical components of the 120 kN force acting on the beam shown in figure 10.21.

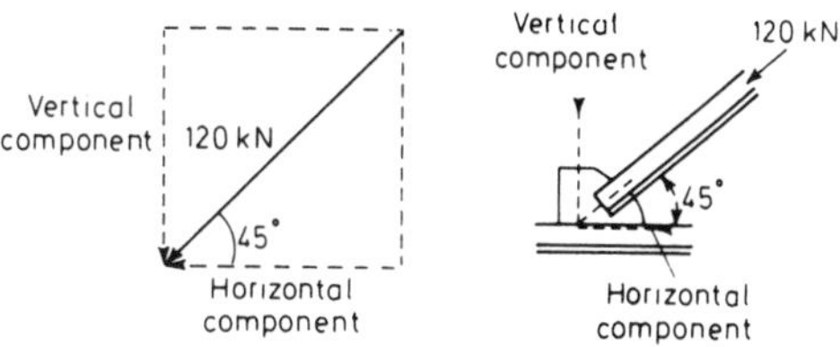

Figure 10.21

Figure 10.21 shows the parallelogram of vectors which can be used to determine the components:

horizontal component = $|\mathbf{F}| \cos \theta = 120 \cos 45° = 84.9$ kN

vertical component = $|\mathbf{F}| \sin \theta = 120 \sin 45° = 84.9$ kN

Example

Determine the resultant force acting on the bracket shown in figure 10.22(a) due to the three forces indicated.

This problem could be solved by drawing the polygon of vectors. However, an alternative method of determining the resultant is to resolve all the forces into their vertical and horizontal components. We can then easily determine the sum of the vertical components and the sum of the vertical components. We then have replaced all the forces by just two components and from these can determine the resultant.

For the 3 kN force we have

horizontal component = $3.0 \cos 60° = 1.5$ kN

vertical component = $3.0 \sin 60° = 2.6$ kN

For the 2.0 kN force we have

horizontal component = $2.0 \cos 30° = 1.7$ kN

vertical component = $2.0 \sin 30° = 1.0$ kN

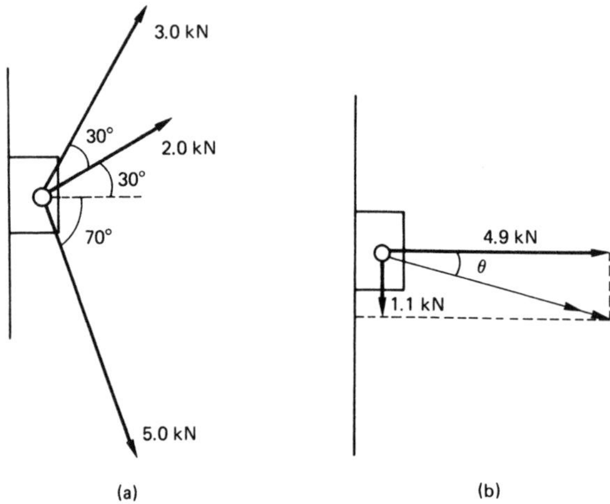

Figure 10.22

For the 5.0 kN force we have

$$\text{horizontal component} = 5.0 \cos 70° = 1.7 \text{ kN}$$

$$\text{vertical component} = -5.0 \sin 70° = -4.7 \text{ kN}$$

The minus sign is because this force is acting downwards and in the opposite direction to the other vertical components which we have taken as being positive. All the horizontal components are in the same direction. Thus

$$\text{sum of horizontal components} = 1.5 + 1.7 + 1.7 = 4.9 \text{ kN}$$

$$\text{sum of vertical components} = 2.6 + 1.0 - 4.7 = -1.1 \text{ kN}$$

Figure 10.22(b) shows how we can use the parallelogram rule to find the resultant with these two components. Since the two components are at right angles to each other, the resultant can be calculated using the Pythagoras theorem. Thus

$$(\text{resultant})^2 = 4.9^2 + 1.1^2$$

Hence the resultant has a magnitude of 5.0 kN. The resultant is at an angle θ downwards from the horizontal given by

$$\tan\theta = \frac{1.1}{4.9}$$

Thus $\theta = 12.7°$.

Example

An object of weight 30 N rests on an incline which is at 35° to the horizontal. What are the components of the weight acting at right angles to the incline and along the incline?

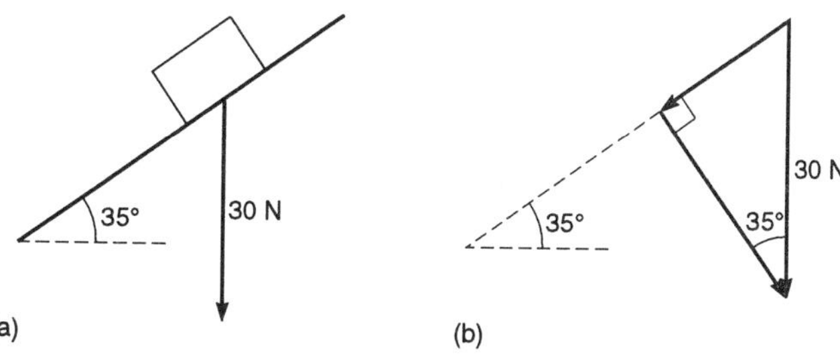

Figure 10.23

Figure 10.23(a) describes the situation and figure 10.23(b) how the triangle rule can be used to determine the components. Thus

component down incline = 30 sin 35° = 17.2 N

component at right angles to incline = 30 cos 35° = 24.6 N

Example

A projectile has an initial velocity of 10 m/s at an angle of 60° to the horizontal. What are the horizontal and vertical components of this velocity?

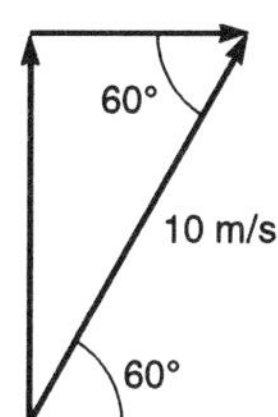

Figure 10.24

Figure 10.24 shows how the triangle rule can be used to determine the velocity components. Thus

vertical component = 10 sin 60° = 8.7 m/s

horizontal component = 10 cos 60° = 5.0 m/s

Revision

19 Determine the horizontal and vertical components of the following forces:

(a) 10 N at 40° to the horizontal,
(b) 15 kN at 70° to the horizontal,

(c) 12 N at 20° to the horizontal,
(d) 30 kN at 80° to the horizontal.

20 An object of weight 30 N rests on an incline which is at 20° to the horizontal. What are the components of the weight at right angles to the incline and parallel to the incline?

21 For each of the following systems of forces as a result of considering the components of each force, determine the resultant force:

(a) 2 N in an easterly direction, 3 N at 60° west of north and 2 N due south,
(b) 4 N in a north-easterly direction, 3 N due west and 5 N at 30° south of east,
(c) 2.8 N in a north-easterly direction, 4 N at 60° south of west and 6 N at 30° south of east.

22 An object has a velocity of 40 m/s in a north-easterly direction. What are the components of this velocity in the north and south directions?

23 A rocket starts its motion from the surface of the earth with a velocity of 400 m/s at an angle of 70° to the horizontal. What are the components of the velocity in the vertical and horizontal directions?

24 An object has a displacement of 10 m in a north-easterly direction. What are the components of this displacement in the easterly and northerly directions?

Problems

1 For the vectors **a** of magnitude 2 in a direction due east, **b** of magnitude 3 in a direction due north and **c** of magnitude 4 in a north-westerly direction, determine (a) **a** + **b**, (b) **b** + **c**, (c) **a** + **c**, (d) **a** – **b**, (e) **b** – **c**, (f) **a** – **c**.

2 Determine the resultants for the following forces acting at the same point on an object:

(a) 2 kN in a northerly direction and 4 kN in a westerly direction,
(b) 10 N in a vertical direction and 20 N in a horizontal direction,
(c) 40 N and 50 N if the angle between the forces is 45°,
(d) 20 N vertically downwards and 10 N in a downward direction at 50° to the vertical,
(e) 10 N and 12 N if the angle between them is 105°,
(f) 20 kN and 12 kN if the angle between them is 120°.

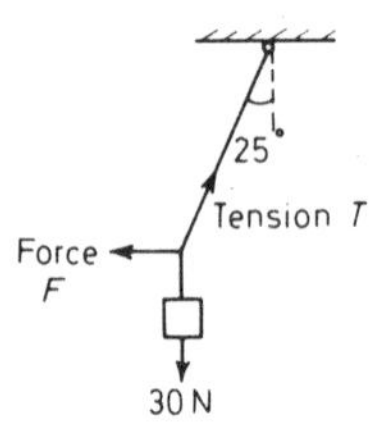

Figure 10.25

3 An object of weight 30 N is suspended by a string from the ceiling. What horizontal force F must be applied to the object if the string is to become deflected and make an angle of 25° with the vertical (figure 10.25)?

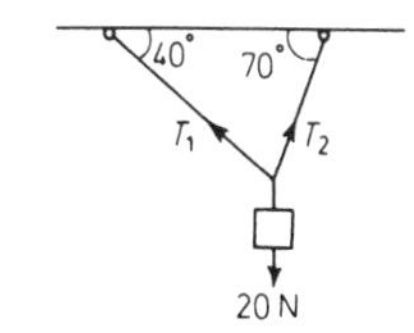

Figure 10.26

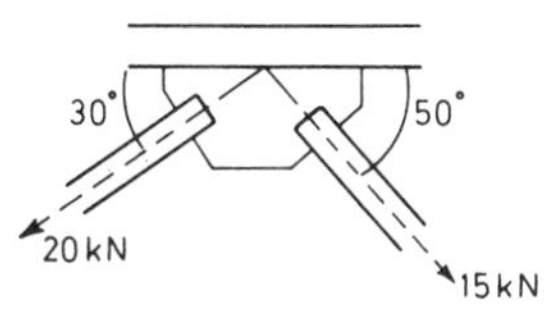

Figure 10.27

4 An object of weight 20 N is supported from the ceiling by two cables inclined at 40° and 70° to the ceiling (figure 10.26). Determine the tensions in the cables.

5 An object of weight 5 N hangs on a vertical string. At what angle to the vertical will the string be when the object is held aside by a force of 3 N in a direction which is 20° above the horizontal?

6 Determine the resultant force acting on an object when two forces of 20 kN and 40 kN are applied to the same point on the object and the angle between the lines of action of the forces is 90°.

7 Three girders in the same plane meet at a point. If there are tensile forces of 70 kN, 80 kN and 90 kN in the girders, what angles will the girders have to be at if there is equilibrium?

8 What is the resultant force acting on the gusset plate shown in figure 10.27 as a result of the forces shown?

9 Determine the resultant velocity of an object if it is subject to the following velocities:

(a) 4 km/h in a northerly direction and 3 km/h in an easterly direction,
(b) 10 m/s vertically downwards and 4 m/s horizontally,
(c) 20 m/s and 10 m/s if the angle between the velocities is 60°,
(d) 10 km/h in a westerly direction and 20 km/h in a north-easterly direction,
(e) 20 m/s in a southerly direction and 10 m/s in a northerly direction.

10 An object is thrown vertically upwards with a velocity of 10 m/s. If there is a horizontal wind blowing of 5 m/s, what will be the resultant velocity of the object?

11 A pilot sets a course due south with an air speed of 300 km/h. If a 100 km/h wind blows from the south-west, what is the actual velocity of the plane?

12 A cyclist travels 5 km in an easterly direction followed by 7 km in a northerly direction. How far, and in what direction, is the cyclist from the start position?

13 A person walks 6 km in a south-westerly direction and then 4 km in a westerly direction. How far, and in what direction, is the walker from the start position?

14 Determine the resultant force acting on an object if it is acted on by four forces acting in the same plane of 1 N in a westerly direction, 3 N in a south-westerly direction, 6 N in a north-easterly direction and 5 N in a northerly direction.

15 Ship A is moving in a north-easterly direction at 15 km/h. Ship B is moving due west at 8 km/h. To ship A, with what velocity does ship B appear to be moving?

16 Object A moves with a velocity of 4 m/s due north and object B moves with a velocity of 3 m/s due east. What is the velocity of B relative to A?

17 An object of weight 10 N rests on an incline which is at 30° to the horizontal. What are the components of this weight in a direction at right angles to the incline and parallel to the incline?

18 Three forces act in the same plane on the same point on an object. If the forces are 4 N in a direction due north, 7 N in a south-easterly direction and 4 N in a direction 60° south of west, what is the resultant force?

19 A cable exerts a force of 15 kN on a bracket. If the cable is at an angle of 35° to the horizontal, what are the horizontal and vertical components of the force?

20 Forces of 10 N, 12 N and 20 N act in the same plane on an object in the directions west, 30° west of north, and north respectively. Determine the resultant force.

21 Forces of 1 N, 2 N, 3 N, 4 N and 5 N act in the same plane on an object in the directions north, north-east, east, 60° west of south, and due west respectively. Determine the resultant force.

11 Phasors

11.1 Introduction

The term *direct* voltage or current is used when the voltage or current is always in the same direction. The term *alternating* voltage or current is used when the direction alternates, continually changing with time. Alternating waveforms oscillate from positive to negative values in a regular, periodic manner. One complete sequence of such an oscillation is called a *cycle* (figure 11.1). The time T taken for one complete cycle is called the *periodic time* and the number of cycles occurring per second is called the *frequency* f. Thus $f = 1/T$. The unit of frequency is the hertz (Hz), 1 Hz being 1 cycle per second.

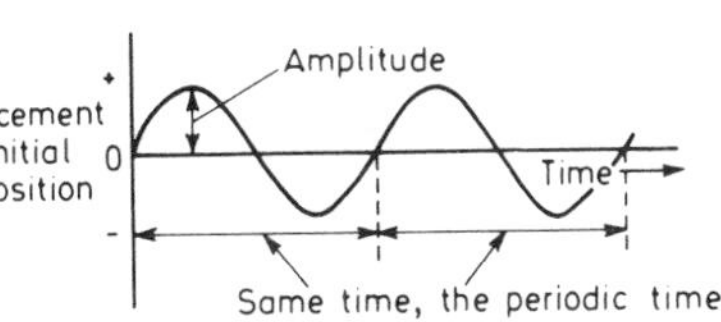

Figure 11.1 *An alternating waveform*

Alternating voltages and currents can take many forms. A particularly important form is one which is in the form of a sine graph, the form shown in figure 11.1. This is because electrical power generation in the entire world is virtually all in the form of such sinusoidal voltages. Consequently the mains electrical supply to houses, offices and factories is sinusoidal. This chapter is about sinusoidal voltages and currents and how phasors can be used to simplify the analysis of electrical circuits when such voltages and currents occur.

11.2 Phasors

Consider the line OA in figure 11.2 rotating in an anticlockwise direction about O with a constant angular velocity ω (note that the unit of angular velocity is radians/second). A constant angular velocity means that the line rotates through equal angles in equal intervals of time. The line starts from the horizontal position and rotates through an angle θ in a time t.

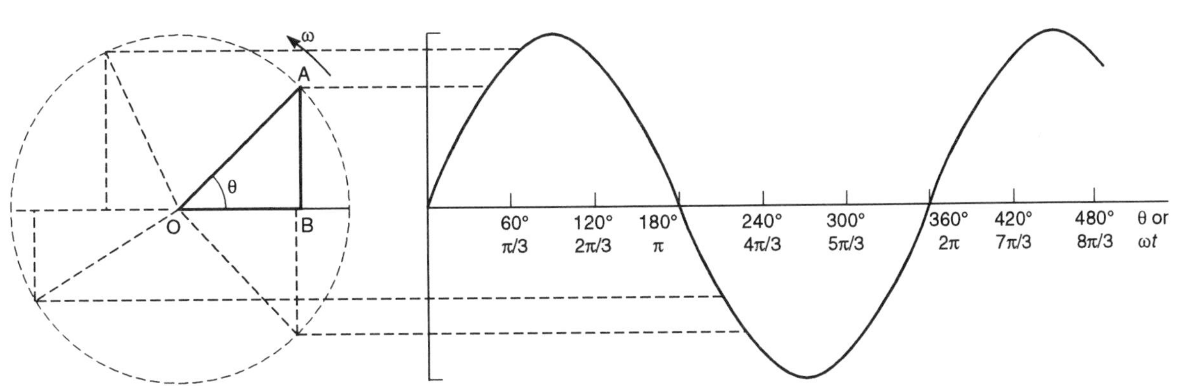

Figure 11.2 *Deriving a sinusoidal waveform*

Since AB/OA = sin θ we can write

$$AB = OA \sin \theta$$

where AB is the vertical height of the line at some instant of time, OA being its length. The maximum value of AB will be OA and occur when $\theta = 90°$. But an angular velocity ω means that in a time t the angle θ covered is ωt. Thus the vertical projection AB of the rotating line will vary with time and is described by the equation

$$AB = OA \sin \omega t$$

If we represent an alternating current i by the perpendicular height AB then its maximum value I_m is represented by OA and we can thus write

$$i = I_m \sin \omega t$$

This is the equation describing the sinusoidal waveform and how its current i at any instant varies with time t. In a similar way we can write for a sinusoidal alternating voltage

$$v = V_m \sin \omega t$$

Thus we can think of an alternating current and voltage in terms of a model in which the instantaneous value of the current or voltage is represented by the vertical projection of a line rotating in an anticlockwise direction with a constant angular velocity. The term *phasor*, being an abbreviation of the term phase vector, is used for such rotating lines. The length of the phasor can represent the maximum value of the sinusoidal waveform or the root-mean-square value (see chapter 16). This is because the maximum value is proportional to the root-mean-square value. The root-mean-square value is the maximum value divided by $\sqrt{2}$. It is often more convenient to use the root-mean-square values but it must be remembered that in an equation such as $v = V_m \sin \omega t$, the V_m must be the maximum value. The line representing a phasor is drawn with an arrowhead at the end that rotates and is drawn in its position at time $t = 0$, i.e. the phasor represents a frozen view of the rotating line at one instant of time of $t = 0$. Thus for the phasor to represent the current or voltage described by figure 11.2 we have the phasor shown in figure 11.3.

Figure 11.3 *Phasor*

It is usual to give angular velocities in units of radians per second, an angular rotation through 360° being a rotation through 2π radians. Since the periodic time T is the time taken for one cycle of a waveform, then T is the time taken for OA to complete one revolution, i.e. 2π radians. Thus

$$T = \frac{2\pi}{\omega}$$

The frequency f is $1/T$ and so $\omega = 2\pi f$. Because ω is just 2π times the frequency, it is often called the *angular frequency*. We can thus write the above equations as

$$i = I_m \sin 2\pi ft$$

and

$$v = V_m \sin 2\pi ft$$

Example

A sinusoidal alternating current has a frequency of 50 Hz and a maximum value of 4.0 A. What is the instantaneous value of the current (a) 1 ms, (b) 2 ms after it is at zero current?

The sinusoidal alternating current has the equation

$$i = I_m \sin 2\pi ft = 4.0 \sin 2\pi \times 50t = 4.0 \sin 314t$$

(a) When $t = 1$ ms we have

$$i = 4.0 \sin 314 \times 1 \times 10^{-3} = 4.0 \sin 0.314$$

The 0.314 is in units of radians, not degrees. This can be worked out using a calculator operating in the radian mode, the key sequence being change mode to rad, press the keys for 0.314, then for sin, the × key, the key for 4 and then the = key. Thus $i = 1.24$ A.
(b) When $t = 2$ ms we have

$$i = 4.0 \sin 314 \times 2 \times 10^{-3} = 4.0 \sin 0.628$$

The 0.314 is in units of radians, not degrees. This can be worked out using a calculator operating in the radian mode.Thus $i = 2.35$ A.

Example

A sinusoidal alternating current is represented by $i = 10 \sin 500t$, where i is in mA. What is (a) the size of the maximum current, (b) the angular frequency, (c) the frequency and (d) the current after 1 ms from when it is zero?

(a) The maximum current is 10 mA.
(b) The angular frequency is 500 rad/s.
(c) The frequency is given by the equation $\omega = 2\pi f$ and so

$$f = \frac{\omega}{2\pi} = \frac{500}{2\pi} = 79.6 \text{ Hz}$$

(d) After 0.01 s we have

$$i = 10 \sin 500 \times 0.01 = 10 \sin 0.50$$

This angle is in radians. Thus $i = 4.79$ mA

Revision

1 A sinusoidal voltage has a maximum value of 10 V and a frequency of 50 Hz. (a) Write an equation describing how the voltage varies with time. (b) Determine the voltages after times from $t = 0$ of (i) 0.002 s, (ii) 0.006 s and (iii) 0.012s.

2 A sinusoidal current has a maximum value of 50 mA and a frequency of 2 kHz. (a) Write an equation describing how the current varies with time. (b) Determine the currents after times from $t = 0$ of (i) 0.4 ms, (ii) 0.8 ms, (iii) 1.6 ms.

3 For a sinusoidal voltage described by $v = 10 \sin 1000t$ volts, what will be (a) the value of the voltage at time $t = 0$, (b) the maximum value of the voltage, (c) the voltage after 0.2 ms?

4 Complete the following table for the voltage $v = 1 \sin 100t$ volts.

t in ms	0	2	4	6	8	10	12	14	16
v in V									

11.2.1 Phase differences

In figure 11.2 the rotating line OA was shown as starting from the horizontal position at time $t = 0$ with the phasor being as shown in figure 11.3. But we could have an alternating voltage or current starting from some value other than 0 at $t = 0$. Figure 11.4 shows such a situation. At the time $t = 0$ the line OA is already at some angle ϕ.

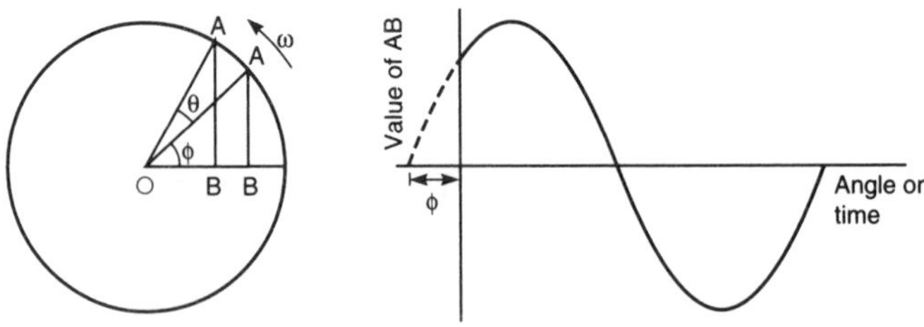

Figure 11.4 *Waveforms not zero at t = 0*

As the line OA rotates with an angular velocity ω then in a time t the angle swept out is ωt and thus at time t the angle with respect to the horizontal is $\omega t + \phi$. Thus we have

$$i = I_m \sin(\omega t + \phi) \text{ or } v = V_m \sin(\omega t + \phi)$$

Figure 11.5 *Phasor*

The phasor for such alternating currents or voltages is represented by a phasor (figure 11.5) at an angle ϕ to some reference line, this line being generally taken as being the horizontal. The angle ϕ is termed the *phase angle*.

In discussing alternating current circuits we often have to consider the relationship between an alternating current through a component and the alternating voltage across it. If we take the alternating voltage as the reference and consider it to be represented by OA for the voltage being horizontal at time $t = 0$, then the current may have some value at that time and so be represented by another line OA at some angle ϕ at $t = 0$. There is said to be a *phase difference* of ϕ between the current and the voltage. If ϕ has a positive value then the current is said to be *leading* the voltage, if a negative value then *lagging* the voltage (figure 11.6).

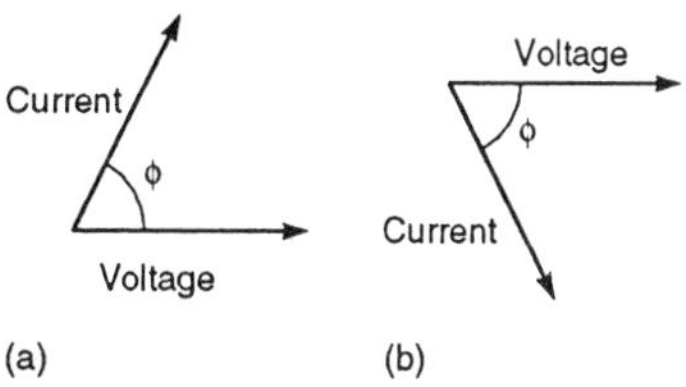

Figure 11.6 *(a) Current leading voltage, (b) current lagging voltage*

As with vectors, in books the usual convention is to write a phasor in a bold font as **I** or **V**. When we refer to the magnitude of the phasor we can write |**I**| or |**V**|. However, the magnitude of a phasor is just the maximum value, or sometimes the root-mean-square value, and so is often written as I_m or V_m, or in the case of root-mean-square values just I or V. To specify a phasor all we need to do is specify its magnitude and whether it has a phase angle. Thus, for example, we can specify a current phasor as $20\angle 30°$ mA (r.m.s.). The magnitude, and hence in this case the root-mean-square value, is thus 20 mA and the phase angle is 30°. The phasor is thus an arrow with a length proportional to 20 mA at an angle of 30° to the horizontal.

Example

The instantaneous value i of a sinusoidal alternating current is described by the equation $i = 2 \sin(\omega t + \pi/4)$ amps. Determine the instantaneous current when (a) $\omega t = 0$, (b) $\omega t = \pi/4$ rad, (c) $\omega t = \pi/2$ rad.

(a) When $\omega t = 0$ then

$$i = 2 \sin \pi/4 = 1.41 \text{ A}$$

(b) When $\omega t = \pi/4$ then

$$i = 2 \sin(\pi/4 + \pi/4) = 2 \sin \pi/2 = 2 \text{ A}$$

(c) When $\omega t = \pi/2$ then

$$i = 2 \sin(\pi/2 + \pi/4) = 2 \sin 3\pi/4 = 1.41 \text{ A}$$

Example

A sinusoidal voltage has a maximum value of 10 V and a frequency of 100 Hz. If the voltage has a phase angle of 30°, what will be the instantaneous voltage at times of (a) $t = 0$, (b) $t = 0.5$ ms.

The equation for the sinusoidal voltage will be

$$v = V_m \sin (2\pi ft + \phi)$$

The term $2\pi ft$, i.e. ωt, is in radians. Thus, for consistency, we should express ϕ in radians. An angle of 30° is $\pi/6$ radians. Thus

$$v = 10 \sin (2\pi \times 100t + \pi/6) \text{ volts}$$

It should be noted that it is quite common in electrical/electronic engineering to mix the units of radians and degrees in such expressions. Thus you might see

$$v = 10 \sin (2\pi \times 100t + 30°) \text{ volts}$$

However, when carrying out calculations involving the terms in the bracket there must be consistency of the units.

(a) When $t = 0$ then

$$v = 10 \sin \pi/6 = 5 \text{ V}$$

(b) When $t = 0.5$ ms then

$$v = 10 \sin (2\pi \times 100 \times 0.5 \times 10^{-3} + \pi/6) \text{ volts}$$

and so $v = 10 \sin 0.838 = 7.43$ V.

Revision

5 Complete the following table for the sinusoidal voltage described by the equation $v = 5 \sin (\omega t + \pi/2)$ volts:

ωt rad	0	$\pi/3$	$\pi/2$	$2\pi/3$	π	$4\pi/3$	$3\pi/2$	$5\pi/3$	2π
v volts									

6 A sinusoidal voltage has a maximum value of 1 V and a frequency of 1 kHz. If the voltage has a phase angle of 60°, what will be the instantaneous voltage at times of (a) $t = 0$, (b) $t = 0.5$ ms?

7 A sinusoidal alternating current has an instantaneous value i at a time t, in seconds, given by $i = 100 \sin (200\pi t - 0.25)$ mA. Determine (a) the maximum current, (b) the frequency, (c) the phase angle.

8 A sinusoidal alternating voltage has an instantaneous value v at a time t, in seconds, given by $v = 12 \sin (100\pi t + 0.5)$ volts. Determine (a) the maximum voltage, (b) the frequency, (c) the phase angle.

9 Draw two phasors to represent the following:

(a) A sinusoidal voltage with a root-mean-square value 2 V leading another voltage with a root-mean-square value of 4 V and the same frequency by 45°.
(b) A sinusoidal current of root-mean-square value of 12 mA lagging another current with a root-mean-square value of 6 mA and the same frequency by 60°.
(c) A sinusoidal voltage with a root-mean-square value 10 V lagging another voltage with a root-mean-square value of 5 V and the same frequency by 30°.

11.3 Adding phasors

Suppose we want to add the voltages across two components in series. If they are alternating voltages we must take account of the possibility that the two voltages may not be in phase, despite having the same frequency since they are supplied by the same source. This means that if we consider the phasors, they will rotate with the same angular velocity but may have different lengths and start with a phase angle between them. Consider one of the voltages to have a magnitude of V_1 and zero phase angle (figure 11.7(a)). The other voltage we will consider as having a magnitude of V_2 and a phase difference of ϕ from the first voltage (figure 11.7 (b)). We can obtain the sum of the two by adding the two graphs, point-by-point, to obtain the result shown in figure 11.7(c). Thus at the instant of time indicated in the figures, the two voltages are v_1 and v_2. Hence the total voltage is $v = v_1 + v_2$. We can repeat this for each instant of time and hence end up with the graph shown in figure 11.7(c).

The following table illustrates this for a number of times when we add the voltages $v_1 = 1 \sin 1000t$ and $v_2 = 1.5 \sin(1000t + \pi/3)$:

t in ms	0	0.5	1.0	1.5	2.0
$v_1 = 1 \sin 1000t$	0	0.48	0.84	1.00	0.91
$v_2 = 1.5 \sin(1000t + \pi/3)$	1.31	1.50	1.34	0.84	0.14
$v = v_1 + v_2$	1.31	1.98	2.18	1.84	1.05

However, exactly the same result is obtained by adding the two phasors by means of the *parallelogram rule* or *triangle rule* of vectors (see section 10.2). If we place the tails of the arrows representing the two phasors together and complete a parallelogram, then the diagonal of that parallelogram drawn from the junction of the two tails represents the sum of the two phasors. Figure 11.7(c) shows such a parallelogram and the resulting phasor with magnitude V.

If the phase angle between the two phasors of sizes V_1 and V_2 is 90°, as in figure 11.8, then the resultant can be calculated by the use of the Pythagoras theorem as having a size V of

$$V^2 = V_1^2 + V_2^2$$

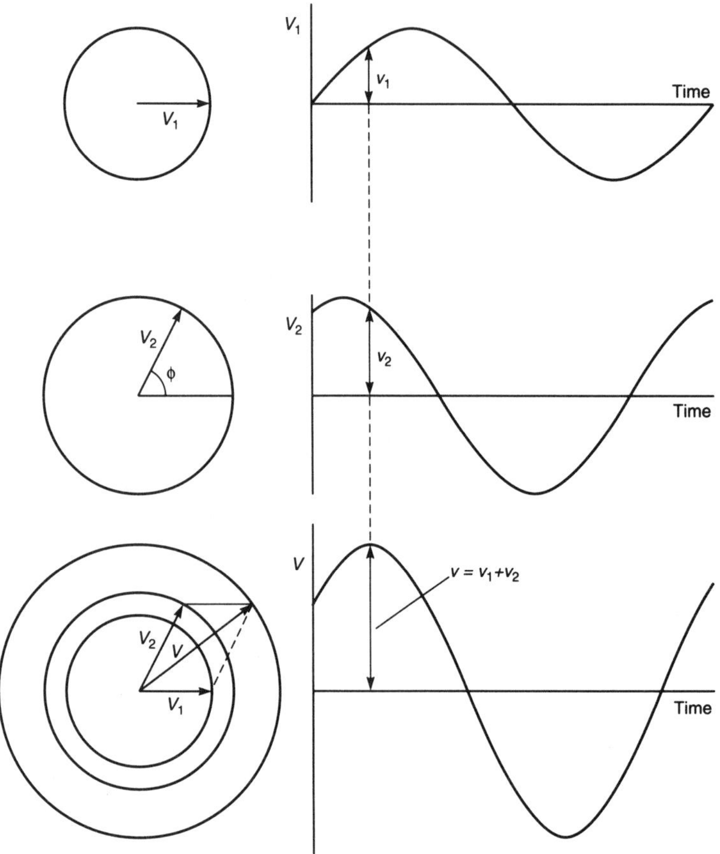

Figure 11.7 *Adding two sinusoidal signals of the same frequency*

and is at a phase angle ϕ relative to the phasor for V_2 of

$$\tan\phi = \frac{V_2}{V_1}$$

Figure 11.8 *Adding two phasors with a phase difference of 90°*

If the phase angle is some angle other than 90° then we might use the cosine rule (see section 9.3) to determine the resultant with the sine rule (see section 9.2) or cosine rule to determine the phase angle. Alternatively we might resolve the phasors into their horizontal and vertical components (see section 10.4), sum the components in each direction, and then use the

Pythagoras theorem to determine the resultant. The following examples illustrate these methods.

Example

Two sinusoidal alternating voltages are described by the equations of $v_1 = 10\sqrt{2}\ \sin\omega t$ volts and $v_2 = 15\sqrt{2}\ \sin(\omega t + \pi/2)$ volts. Determine the sum of these voltages.

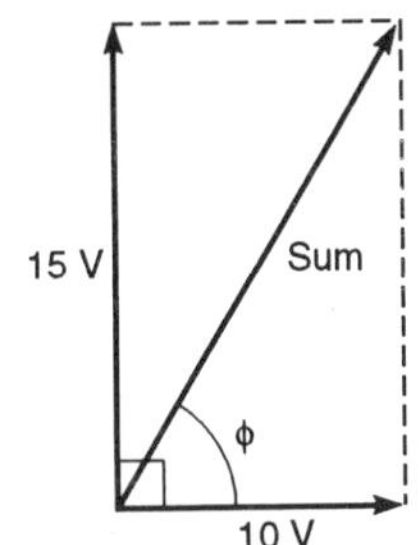

Figure 11.9

Figure 11.9 shows the phasor diagram for the two voltages. The magnitudes of the phasors have been drawn as the root-mean-square values and so are of lengths proportional to 10 V and 15 V. The example could have been worked through with the maximum values and the result would have been the same. The angle between the phasors is $\pi/2$, i.e. 90°. We could determine the sum from a scale drawing or by calculation using the Pythagoras theorem. Thus

$$(\text{sum})^2 = 10^2 + 15^2$$

Hence the magnitude of the sum of the two voltages is 18.0 V. The phase angle is given by

$$\tan\phi = \frac{15}{10}$$

Hence ϕ = 56.3° or 0.983 rad. Thus the sum is an alternating voltage described by a phasor of root-mean-square magnitude 18.0 V and phase angle 56.3° (or 0.983 rad). This alternating voltage is thus described by $v = 18.0\sqrt{2}\ \sin(\omega t + 0.983)$ volts.

Example

Two sinusoidal alternating currents are described by the equations of $i_1 = 50\sqrt{2}\ \sin\omega t$ mA and $i_2 = 100\sqrt{2}\ \sin(\omega t + \pi/3)$ mA. Determine the total current when these currents both flow through the same branch of a circuit.

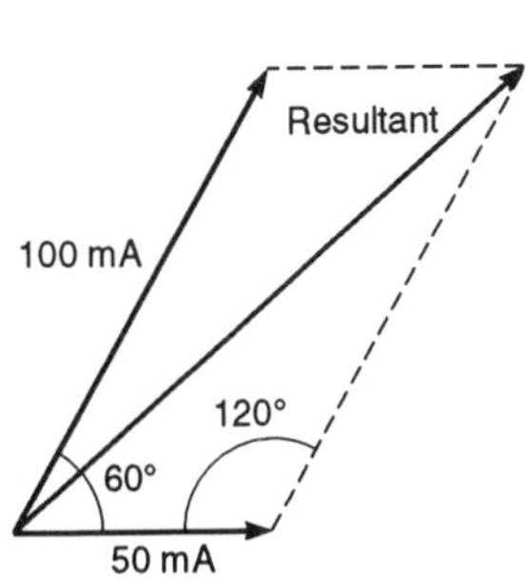

Figure 11.10

Figure 11.10 shows the phasor diagram for the two currents. The magnitudes of the phasors have been drawn as the root-mean-square values and so are of lengths proportional to 50 mA and 100 mA. The angle between the phasors is $\pi/3$ rad, i.e. 60°. We could determine the resultant from a scale drawing. Alternatively we can calculate it by means of the cosine law. Thus

$$(\text{resultant})^2 = 100^2 + 50^2 - 2 \times 100 \times 50 \cos 120°$$

Thus the resultant has a magnitude, i.e. a root-mean-square value, of 132.3 mA. The phase angle ϕ of the resultant can be determined by the use of the sine rule. Thus

$$\frac{132.3}{\sin 120°} = \frac{100}{\sin\phi}$$

Hence $\phi = 40.9°$ or 0.714 rad. The total current is thus a phasor with a root-mean-square magnitude of 132.3 mA and a phase angle of 40.9° (or 0.714 rad). This is a current of $i = 132.3\sqrt{2}\,\sin(\omega t + 0.714)$ mA.

Alternatively we could have resolved the phasors into their horizontal and vertical components. Thus for the $100\angle 60°$ mA phasor we have

$$\text{horizontal component} = 100 \cos 60°$$

$$\text{vertical component} = 100 \sin 60°$$

For the $50\angle 0°$ mA phasor we have

$$\text{horizontal component} = 50$$

$$\text{vertical component} = 0$$

Thus the sum of the horizontal components is

$$100 \cos 60° + 50 = 100 \text{ mA}$$

and the sum of the vertical components is

$$100 \sin 60° + 0 = 86.6 \text{ mA}$$

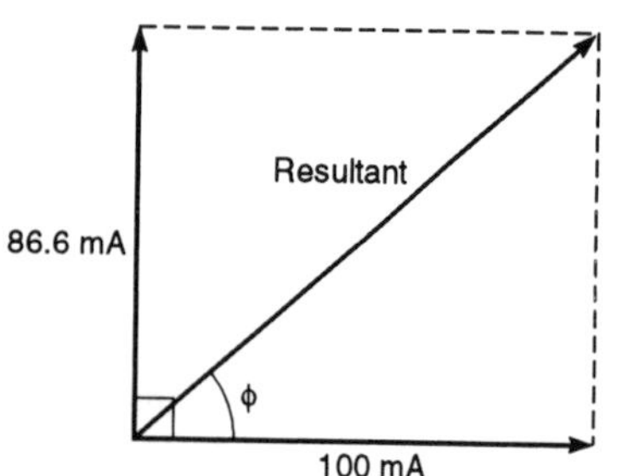

Figure 11.11

The resultant of these components (figure 11.11) is given by the Pythagoras theorem as

$$(\text{resultant})^2 = 100^2 + 86.6^2$$

and hence the resultant has a magnitude of 132.3 mA. The phase angle is given by

$$\tan\phi = \frac{86.6}{100}$$

and so $\phi = 40.9°$ or 0.714 rad. The result is thus as obtained earlier by the use of the cosine and sine rules.

Example

The voltage across one component in a circuit is $10 \sin \omega t$ volts and that across another component in series with it $20 \sin(\omega t - \pi/2)$ volts. What is the total voltage across the two components?

The voltage across two components in series is the sum of the voltages across the two components. Figure 11.12 shows the phasor diagram.

For convenience the magnitudes of the phasors have been drawn as the maximum values. The phase difference of $-\pi/2$ means that the voltage across the second component is drawn as lagging that across the first by 90°. Using the Pythagoras theorem, the sum is thus given by

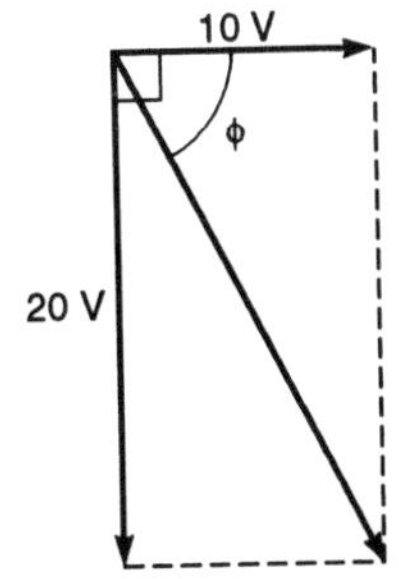

Figure 11.12

$$(\text{sum})^2 = 10^2 + 20^2$$

Hence the sum phasor has a magnitude of 22.4 V. The phase angle is given by

$$\tan\phi = \frac{-20}{10}$$

and so $\phi = -63.4°$ or -1.11 rad. Thus the voltage is described by the expression $22.4 \sin(\omega t - 1.11)$ V.

Revision

10 The currents in two parallel branches of a circuit are $10 \sin \omega t$ milliamps and $20 \sin(\omega t + \pi/2)$ milliamps. What is the total current entering the parallel arrangement?

11 The voltage across a component in a circuit is $5.0 \sin \omega t$ volts and across another component in series with it $2.0 \sin(\omega t + \pi/6)$ volts. Determine the total voltage across both components.

12 The sinusoidal alternating voltage across a component in a circuit is $50\sqrt{2}\, \sin(\omega t + 40°)$ volts and across another component in series with it $100\sqrt{2}\, \sin(\omega t - 30°)$ volts. What is the total voltage across the two components?

13 The currents in two parallel branches of a circuit are $4\sqrt{2}\, \sin \omega t$ amps and $6\sqrt{2}\, \sin(\omega t - \pi/3)$ amps. What is the total current entering the parallel arrangement?

11.4 Subtracting phasors

Consider a situation where we know the circuit alternating current entering a junction of two parallel components and the current through one of them. What is the current through the other component? What we need to do is to subtract, at each instant of time, the value of the current through one component from the circuit current. Figure 11.13 illustrates this. Figure 11.13(a) shows the circuit current and figure 11.13(b) the current through one of the components. Figure 11.13(c) shows the result of subtracting, at each instant of time, the values given by (b) from those given by (a).

Suppose we have the circuit current of figure 11.13(a) as $1 \sin 1000t$ and the current through the first component as $1.5 \sin(1000t + \pi/3)$, as in figure 11.13(b). Then the following table illustrates some of the results of a subtraction:

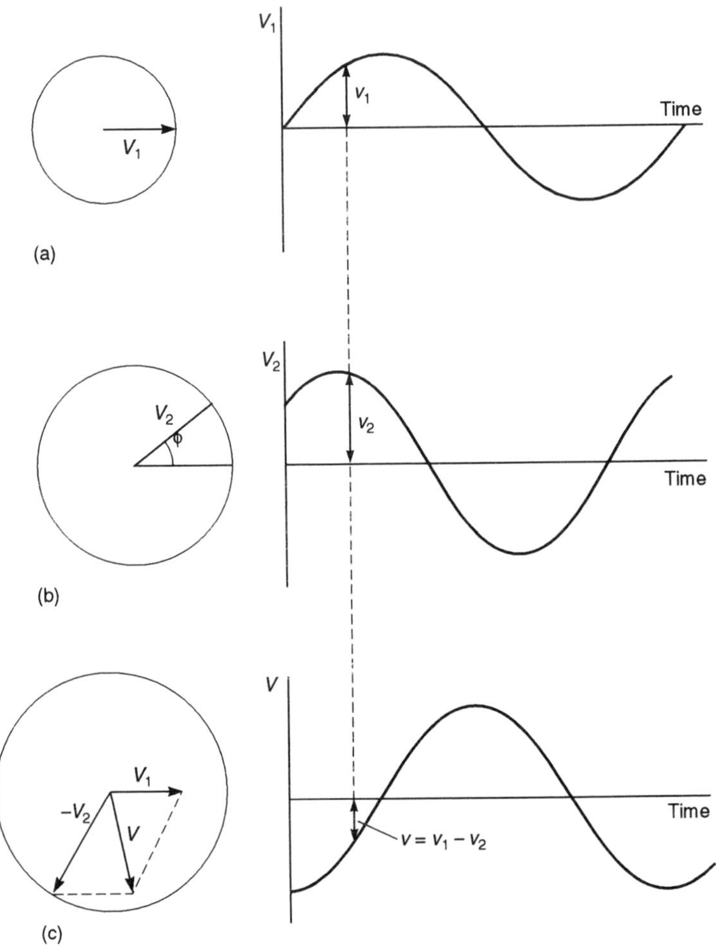

Figure 11.13 *Subtracting two sinusoidal signals of the same frequency*

t in ms	0	0.5	1.0	1.5	2.0
$v = 1 \sin 1000t$	0	0.48	0.84	1.00	0.91
$v_1 = 1.5 \sin (1000t + \pi/3)$	1.31	1.50	1.34	0.84	0.14
$v_2 = v - v_1$	−1.31	−1.02	−0.50	0.16	0.77

However, there is a simpler way of subtracting two sinusoidal alternating quantities and that is by subtracting the phasors. Figure 11.13(c) illustrates this. Subtraction of phasors is carried out in the same way as the subtraction of vectors (see section 10.3). Thus to subtract a phasor $\mathbf{I}_2$ from a phasor $\mathbf{I}_1$ we add the phasor $-\mathbf{I}_2$ to phasor $\mathbf{I}_1$. The phasor $-\mathbf{I}_2$ is just the phasor $\mathbf{I}_2$ with its direction reversed.

Example

The current flowing into a junction in a circuit is $i = 10 \sin \omega t$ milliamps. If the circuit branches at the junction and there are two currents leaving the junction and one is $i_1 = 6 \sin (\omega t + \pi/4)$ milliamps, what is the other current i_2?

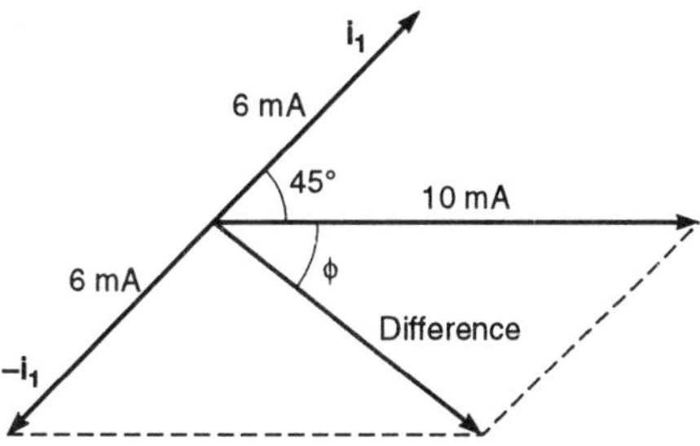

Figure 11.14

We can tackle this example by subtracting phasors. Figure 11.14 shows the phasors for the currents i and i_1. To subtract i_1 from i we add the phasor for $-i_1$ to that for i. The phasor for $-i_1$ has the same magnitude as that for i_1 but is in the opposite direction. Thus, using the cosine rule

$$(\text{difference})^2 = 10^2 + 6^2 - 2 \times 10 \times 6 \cos 45°$$

Hence the difference has a magnitude of 7.15 mA. The phase angle ϕ can be obtained by the use of the sine rule. Thus

$$\frac{7.15}{\sin 45°} = \frac{6}{\sin \phi}$$

Thus $\phi = 36.4°$ or 0.635 rad. Hence the current i_2 is given by the equation $i_2 = 7.15 \sin (\omega t - 0.635)$ mA.

Alternatively we could have obtained the solution by resolving the phasors for i and $-i_1$ into the horizontal and vertical directions. The sum of these phasor components in the horizontal direction is

$$\text{horizontal component} = 10 - 6 \cos 45° = 5.76 \text{ mA}$$

The sum of the vertical components is

$$\text{vertical component} = -6 \sin 45° = -4.24 \text{ mA}$$

The minus sign has been included to indicate that this component is directed downwards and not upwards. The magnitude of the difference phasor is thus obtained by the use of the Pythagoras theorem as

$$(\text{difference})^2 = 5.76^2 + (-4.24)^2$$

Hence the difference has a magnitude of 7.15 mA. The phase angle of this difference phasor is given by

$$\tan\phi = \frac{-4.24}{5.76}$$

Hence the phase angle is 36.4° or 0.635 rad. The result is thus as obtained by the cosine and sine rules.

Revision

14 If the voltage across a pair of series components is 4 sin ωt volts and that across one of the components is 2 sin $(\omega t + \pi/4)$ volts, what is the voltage across the other component?

15 The current through a component is equal to the current 5 sin ωt amps in the circuit minus the current 4 sin $(\omega t - \pi/2)$ amps through a parallel component. Determine the current through the component.

16 The current through a component is equal to the current 120 sin ωt milliamps in the circuit minus the current 80 sin $(\omega t - \pi/4)$ milliamps through a parallel component. Determine the current through the component.

Problems

1 Complete the following tables:

(a)

t in ms	0	2	4	6	8	10
$v = 2 \sin 500t$ in V						
$i = 10 \sin 100t$ in mA						
$v = 5 \sin 1000t$ in V						

(b)

ωt rad	0	$\pi/3$	$\pi/2$	$2\pi/3$	π
$v = 10 \sin \omega t$ in V					
$v = 10 \sin(\omega t + \pi/2)$ in V					
$v = 10 \sin(\omega t + \pi/3)$ in V					

2 A sinusoidal alternating current has an angular frequency of 100 rad/s and a maximum value of 10 A. What is the current 5 ms after it has a zero value?

3 A sinusoidal alternating voltage has a frequency of 2 MHz and a maximum value of 10 V. What is the instantaneous voltage 0.1 μs after it is at a zero value?

4 A sinusoidal alternating voltage in mV is represented by the equation $v = 50 \sin (500t + 20°)$. What is (a) the angular frequency, (b) the frequency, (c) the value of the voltage at $t = 0$?

5 Draw the phasors to represent the following sinusoidal alternating quantities when each has the same frequency:

(a) Two currents with root-mean-square values of 3 A and 5 A, with the second current leading the first by a phase difference of 60°.
(b) Two voltages with root-mean-square values of 1 V and 2 V, with the second voltage lagging the first by a phase difference of 30°.

6 A sinusoidal alternating current has an instantaneous value i at a time t, in seconds, given by $i = 20 \sin (500\pi t - 0.5)$ mA. Determine (a) the maximum current, (b) the frequency, (c) the phase angle.

7 A sinusoidal alternating voltage has an instantaneous value v at a time t, in seconds, given by $v = 5 \sin (1000\pi t + 0.2)$ volts. Determine (a) the maximum voltage, (b) the frequency, (c) the phase angle.

8 The voltage across a component in a circuit is $4.0 \sin \omega t$ volts and across another component in series with it $2.0 \sin (\omega t + \pi/2)$ volts. Determine the total voltage across both components.

9 The voltage across a component in a circuit is $5.0 \sin \omega t$ volts and across another component in series with it $2.0 \sin (\omega t - \pi/6)$ volts. Determine the total voltage across both components.

10 The voltage across a component in a circuit is $4.0 \sin (\omega t + \pi/3)$ volts and across another component in series with it $3.0 \sin (\omega t - \pi/4)$ volts. Determine the total voltage across both components.

11 The currents in two parallel branches of a circuit are $20 \sin \omega t$ milliamps and $30 \sin (\omega t + \pi/3)$ milliamps. What is the total current entering the parallel arrangement?

12 The currents in two parallel branches of a circuit are $30 \sin \omega t$ milliamps and $40 \sin (\omega t + \pi/2)$ milliamps. What is the total current entering the parallel arrangement?

13 The voltage across a component in a circuit is $4.0 \sin (\omega t + \pi/2)$ volts and across another component in series with it $4.0 \sin \omega t$ volts. Determine the total voltage across both components.

14 The voltage across a component in a circuit is $20 \sin \omega t$ volts and across another component in series with it $15 \sin (\omega t - \pi/2)$ volts. Determine the total voltage across both components.

15 The current through a component is equal to the current $20 \sin \omega t$ milliamps in the circuit minus the current $10 \sin (\omega t + \pi/4)$ milliamps through a parallel component. Determine the current through the component.

16 The current through a component is equal to the current $25 \sin \omega t$ milliamps in the circuit minus the current $16 \sin (\omega t - \pi/4)$ milliamps through a parallel component. Determine the current through the component.

17 If the voltage across a pair of series components is $4 \sin \omega t$ volts and that across one of the components is $2 \sin (\omega t - \pi/2)$ volts, what is the voltage across the other component?

18 If the voltage across a pair of series components is $10 \sin \omega t$ volts and that across one of the components is $5 \sin (\omega t + \pi/3)$ volts, what is the voltage across the other component?

19 Determine for the alternating currents $i_1 = 10 \sin (\omega t + \pi/6)$ milliamps and $i_2 = 8 \sin (\omega t - \pi/3)$ milliamps (a) the sum of the two currents and (b) the difference when the second current is subtracted from the first.

20 A circuit contains two parallel branches. The circuit current is $10 \sin \omega t$ amps and the current in one branch is $10 \sin (\omega t - \pi/3)$ amps. What is the current in the other branch?

21 A circuit contains two parallel branches. The root-mean-square current in one branch is 1.2 A in the reference direction. The current in the other branch has a root-mean-square magnitude of 2.0 A and lags the current in the first branch by 30°. What is the circuit current entering the parallel branches?

Section C
Graphs

The aims of this section are to enable the reader to:

- Plot graphs, choosing suitable cordinates and scales.
- Plot and use linear, parabolic, sinusoidal and exponential graphs.
- Determine gradients, intercepts and areas from graphs.
- Interpret graphs.
- Plot polar graphs
- Reduce non-linear laws to linear form.
- Use and interpret graphs in an engineering context.

This section is about the construction and interpretation of the types of graphs that are encountered in engineering problems. There is thus a consideration of linear, parabolic, exponential, logarithmic and sinusoidal graphs.

Chapter 12 parallels the development of the mathematics in Section A: Algebra, illuminating some of the aspects considered there. It is the basic chapter on which the rest of the section is built. Basic algebraic techniques are assumed. Chapter 13 assumes that chapter 12, and much of the algebra in section A, has been covered. Chapter 14 assumes a basic knowledge of trigonometry and that the basics of chapter 12 have been covered. Chapter 15 assumes that chapter 12 has been covered and links closely with chapter 6. Chapter 16 assumes that chapter 12 has been covered and links closely with chapter 11.

12 Graphs

12.1 Introduction

Graphs are an indispensable element in engineering. They enable trends in experimental data to be more easily seen, and determined, than is possible by just looking at the numbers. Also, equations can be pictorially displayed by means of graphs and the relationships described by the equations more easily comprehended. This chapter is a review of the basic techniques of Cartesian graph drawing and the interpretation of such graphs. Functions which give linear and non-linear graphs are considered.

Chapter 13 is a more detailed consideration of linear graphs and how relationships can be organised to give linear relationships. Chapter 14 is a consideration of polar graphs.

12.1.1 Cartesian graphs

It is possible to specify the position of a point along a line from some zero point on that line by specifying its distance from the point. In order to specify which side of the zero point the distance is, distances measured to the right of the zero point are given as positive and distances to the left negative (figure 12.1). Thus, with such a line, a distance specified as +2 is located 2 units to the right of the zero point.

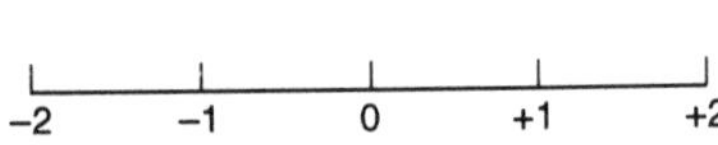

Figure 12.1 *Number line*

If we want to specify the position of a point on a plane then we can use two such number lines at right angles to each other and intersecting at their zero points. The two lines are then called *coordinate axes* and their point of intersection the *origin*. The horizontal axis is called the *x-axis* and the vertical axis the *y-axis* (figure 12.2). The positive half of the x-axis is to the right of the origin and the negative half to the left. The positive half of the y-axis is upwards from the origin and the negative half downwards.

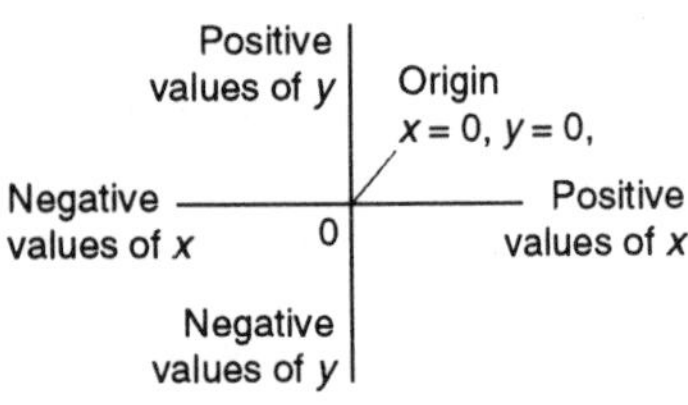

Figure 12.2 *Cartesian coordinates*

To specify a point on the plane then we can specify its horizontal displacement and its vertical displacement from the origin, i.e. the intersection of the vertical and horizontal lines through the point with the x- and y-axes. If these intersections are x_A and y_A for some point A then the point is said to have the *coordinates* of (x_A, y_A). Such coordinates of points are called the *Cartesian coordinates*. The first number in such a pair of numbers is always the *x-coordinate* (or *abscissa)* and the second the *y-coordinate* (or *ordinate*).

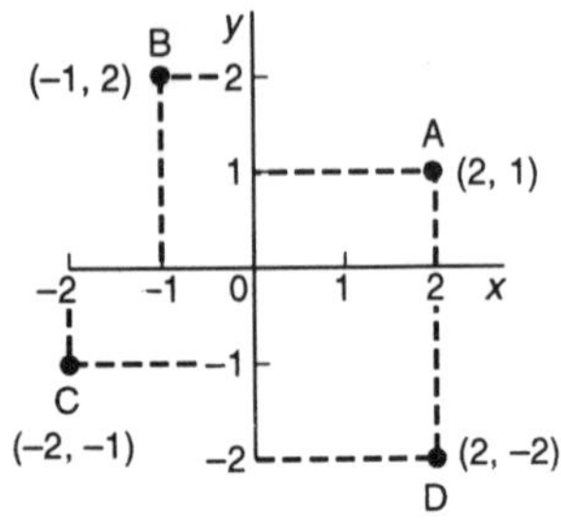

Figure 12.3 *Examples of coordinates*

Figure 12.3 illustrates this for a number of points. Thus point A is located 2 units in the x-direction from the origin and 1 unit in the y-direction and so has the coordinates of (2, 1). Point B is located –1 units in the x-direction and +2 units in the y-direction and so has the coordinates (–1, 2). Point C is located –2 units in the x-direction and –1 units in the y-direction and so has the coordinates (–2, –1). Point D is located +2 units in the x-direction and –2 units in the y direction and so has the coordinates of (2, –2). The sequence of the numbers in the specified coordinate pair is important. Thus (2, 1) and (1, 2) specify two quite different points. The first has an *x-coordinate* of 2 and a *y-coordinate* of 1 while the second has an *x-coordinate* of 1 and a *y-coordinate* of 2.

Each of the axes in figure 12.3 has a *scale*. These are the numbers which indicate how numbers are assigned to distances measured out along the axes. The scales used in figure 12.3 are *linear scales* in that the number is proportional to the distance from the origin. Chapter 13 shows another type of scale called the *logarithmic scale*.

In selecting the scales to be used for the axes of a graph and in plotting the graph, several points should be taken into account.

1 The scales should be chosen to that the points to be plotted on the graph occupy the full range of the axes used for the graph. There is no point in having a graph with scales from 0 to 100 if all the data points have values between 0 and 10.
2 The scales should not start at zero if starting at zero produces an accumulation of points within a small area of the graph. Thus if all the points have values between 80 and 100 then a scale from 0 to 100 means the points are concentrated in just the end zone of the scale. It is better, in this situation, to have a scale running from 80 to 100.
3 Scales should be chosen so that the location of points between scale marks is made easy. Thus with graph paper subdivided into squares of 10 small squares, it is easy to locate a point of 0.2 if one large square corresponds to 1 but much more difficult if one large square corresponds to, say, 3.
4 The axes should be labelled with the quantities they represent and their units.
5 The data points should be clearly marked, e.g. a large dot or a cross.
6 With experimental data there will be some errors associated with the values being plotted. This will result in some scatter of the points and so the best line should be drawn. This is a smooth line for which there is the same amount of scatter on one side of the line as the other.

Example

The following data is for the voltage across a resistor at different electrical currents. Plot the data as a graph.

Voltage in volts	0	0.5	1.0	1.5	2.0	2.5
Current in amps	0	0.12	0.18	0.30	0.42	0.48

If we take the voltage as being represented by distances measured from the origin along the y-axis then a suitable scale would be one from 0 to 2.5. For the current, represented by distances measured along the x-axis, then a suitable scale would be one from 0 to 0.5. The points to be plotted are (0, 0), (0.1, 0.5), (0.2, 1.0), (0.3, 1.5), (0.4, 2.0) and (0.5, 2.5). Figure 12.4 shows the graph. The best line through the data points is a straight line.

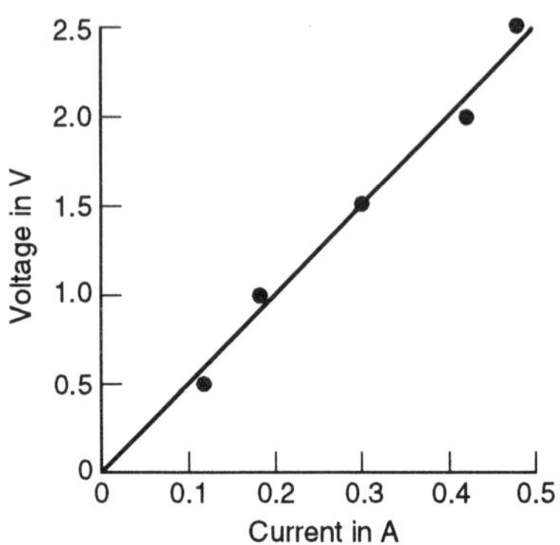

Figure 12.4

Example

The following data is for the length of a metal rod at different temperatures. Plot the data as a graph.

Length in m	20.000	20.004	20.008	20.012	20.016
Temperature in °C	20	40	60	80	100

Starting the scale for the length at 0 would squash all the data into the region of the scale devoted to 20. Thus it is better to start this scale at 20 and have the scale going from 20 to 20.016. For the scale for temperature, we could have the scale starting at 20 and going to 100. Figure 12.5 shows the resulting graph. All the points lie on a straight line.

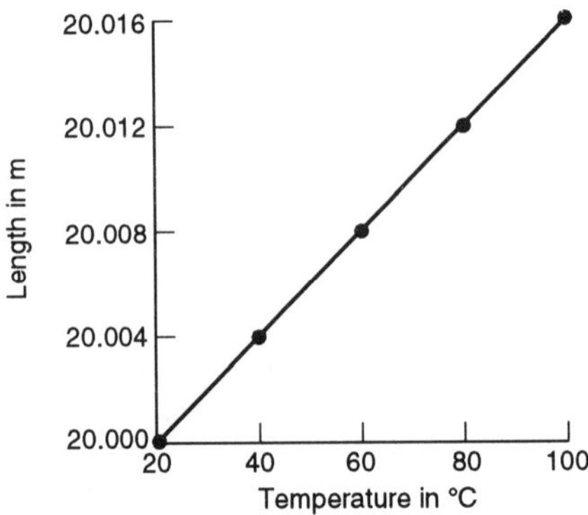

Figure 12.5

Revision

1 The following data is for the force applied to stretch a spring and the extension that results. Plot the data as a graph.

Force in N	0	0.2	0.4	0.6	0.8	1.0
Extension in mm	0	5	10	15	20	25

2 The following data is for the distance travelled and the time at which the distance is measured. Plot the data as a graph.

Distance in m	10	15	20	25	30	35
Time in s	10	12	14	16	18	20

12.2 Graphs of equations

Consider the equation $y = x + 2$. We can plot a graph of this equation determining the coordinates of a few points, plotting them and then connecting the points with a smooth curve. Thus suppose we take values of x of −2, −1, 0, +1, and + 2. The corresponding y values are:

x	−2	−1	0	+1	+2
y	0	+1	+2	+3	+4

Figure 12.6 shows the points and the resulting graph. It is a straight line.

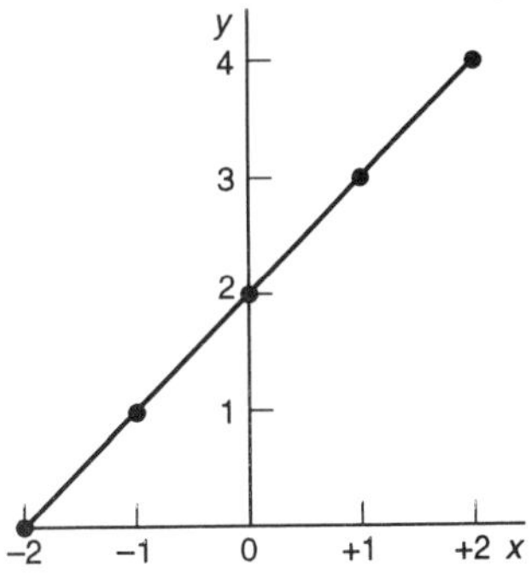

Figure 12.6 $y = x + 2$

Consider the equation $y = x^2 - 2$. We can plot this graph by determining the coordinates of a few points, plotting them and then connecting the points with a smooth curve. Thus suppose we take values of x of −2, −1, 0, +1, and + 2. The corresponding y values are:

x	−2	−1	0	+1	+2
y	2	−1	−2	−1	+2

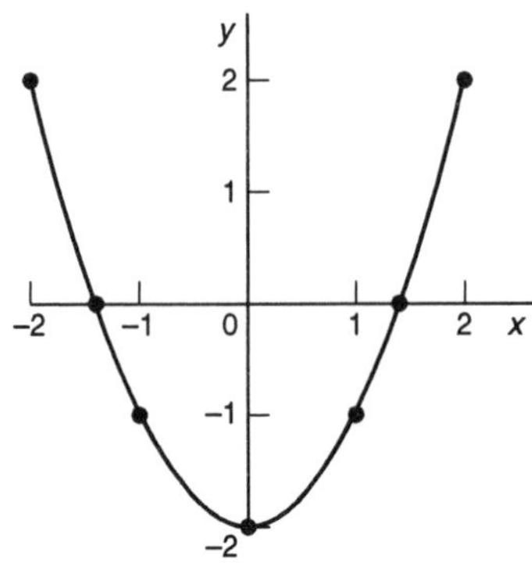

Figure 12.7 $y = x^2 - 2$

Figure 12.7 shows the points and the resulting graph. In plotting the graph it helps if we determine where the graph line should cut the x-axis, i.e. the values of x when $y = 0$. With $y = 0$ we have $x^2 = 2$ and so $x = \pm\sqrt{2}$.

The graph of any function can be obtained by simply plotting enough points and drawing a smooth curve through the points. However, in many cases, the amount of labour involved can often be reduced by determining key features.

1 Is the graph a straight-line graph? A straight-line graph has only y or x to the power 1 and there are no products of x and y or expressions such as trigonometric ratios or exponentials. The equation $y = x + 2$ gives a straight line graph (see figure 12.6). Section 12.3 discusses such graphs in more detail.
2 Is the graph symmetrical about one of the axes, or both of the axes? If the graph is symmetrical about the y-axis then it contains only even powers of x. If it is symmetrical about the x-axis then it contains only even powers of y. Thus for the graph in figure 12.7 of $y = x^2 - 2$, we have only even powers of x. Thus the graph is symmetrical about the y-axis.
3 Is the graph symmetrical about the origin? The graph is symmetrical about the origin if a change in the sign of x causes a change in the sign of y without changing its numerical value. The graph of $y = 1/x$ is symmetrical about the origin because if we change the sign of x we change the sign of y without changing its numerical value. Figure 12.8 shows such a graph.

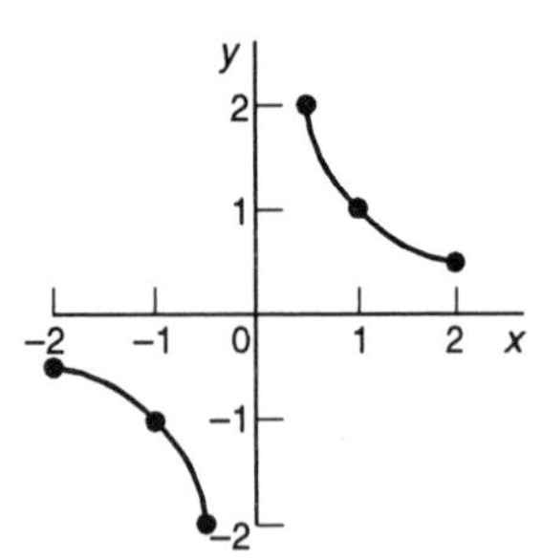

Figure 12.8 $y = 1/x$

4 Where does the graph line cut the axes? The value of where a graph cuts the x-axis is given by putting $y = 0$, the value of where it cuts the y-axis by putting $x = 0$. In the discussion of figure 12.7 the points where the x-axis was cut were determined.
5 Are there any values of x which make y have an infinite value, or values of y which make x have an infinite value? For $y = 1/x$, when $x = 0$ then $y = 1/0$ which is infinite. Figure 12.8 shows how the graph goes off to infinity as x approaches zero. When $y = 0$ then $x = 1/0$ which is infinite and so the graph goes off to infinity as y approaches zero.

6 Does the graph have any maximum or minimum values? To determine such points we need to use calculus and this is discussed in section D.

Example

Plot the graph of $y = 2x^2$.

We have x to the power 2. Thus the graph will not be a straight-line graph. The equation contains only even powers of x and is symmetrical about the y-axis. Thus we need only find values of y for positive values of x since we know that the graph for negative values of x will be just a mirror image of that for the positive values. When $x = 0$ then $y = 0$. If we now take some values of x then we can find the corresponding values of y.

x	0	1	2	3
y	0	2	8	18

Figure 12.9 shows the resulting graph. Graphs of equations of the form $y = ax^2 + bx + c$, where a, b and c are some constant (a must not be zero but b and c can), give cup-shaped graphs called *parabolas*.

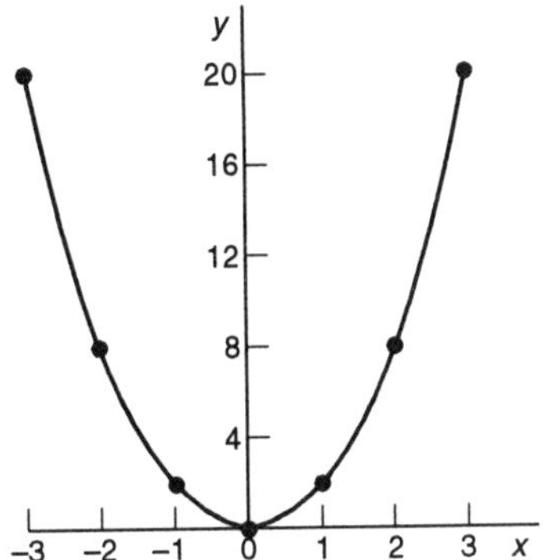

Figure 12.9 $y = 2x^2$

Revision

3 Plot graphs of the functions (a) $y = 5x$, (b) $y = 2x + 3$, (c) $y = x^2$, and (d) $y = x^2 + 4$

12.2.1 Graphs of trigonometric ratios

See section 8.4.2 for a discussion of the graphs of trigonometric ratios and section 11.2 for a discussion of how such waveforms can be generated. Chapter 16 continues that discussion.

Consider the graph of $y = \sin x$. The values of x that will give $\sin x = 0$ and hence $y = 0$ are the values for which $x = 0, \pi, 2\pi, 3\pi$, etc. (i.e. 0°, 180°, 360°, 540°, etc.). The values of x that will give $\sin x = +1$ and hence $y = +1$ are $x = \pi/2, 5\pi/2$, etc. (i.e. 90°, 450°, etc.). The values of x that will give $\sin x = -1$ and hence $y = -1$ are $x = 3\pi/2, 7\pi/2$, etc. (i.e. 270°, 630°, etc.). We thus have:

x	0	$\pi/2$	π	$3\pi/2$	2π	$5\pi/2$	3π	$7\pi/2$	etc.
$y = \sin x$	0	+1	0	−1	0	+1	0	−1	

Figure 12.10(a) shows the graph.

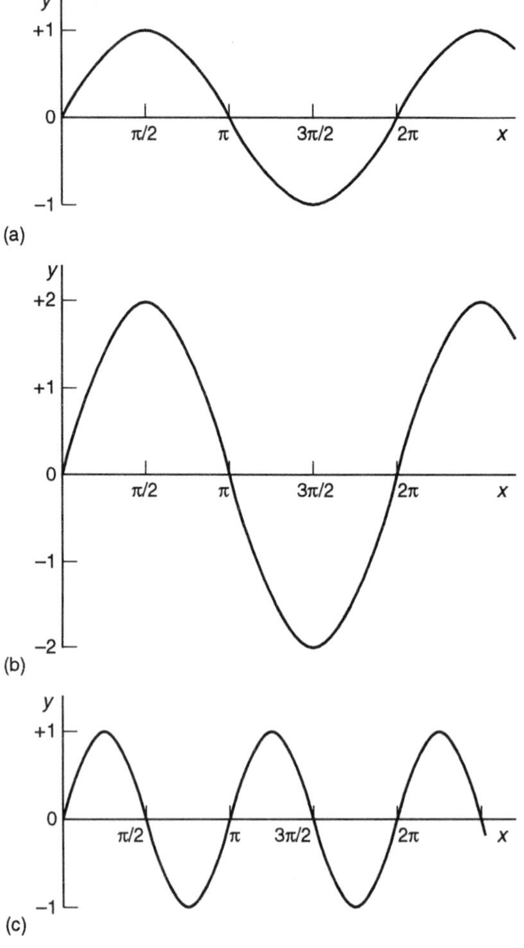

Figure 12.10 *(a)* $y = \sin x$, *(b)* $y = 2 \sin x$, *(c)* $y = \sin 2x$

Consider the graph of $y = 2 \sin x$. This just multiplies the values of $\sin x$ by 2. Thus we have:

x	0	$\pi/2$	π	$3\pi/2$	2π	$5\pi/2$	3π	$7\pi/2$	etc.
$y = \sin x$	0	+1	0	−1	0	+1	0	−1	
$y = 2 \sin x$	0	+2	0	−2	0	+2	0	−2	

Figure 12.10(b) shows the graph. Comparing this with the graph in figure 12.10(a) for $y = \sin x$, then the 2 has just multiplied the amplitude of the $\sin x$ graph by 2.

Consider the graph of $y = \sin 2x$. For $y = \sin 2x$, the values of x that will give $y = 0$ are the values for which $2x = 0, \pi, 2\pi$, etc. (i.e. 0°, 180°, 360°, etc.). We have $\sin 2x = +1$ when $2x = \pi/2, 5\pi/2$, etc. (i.e. 90°, 450°, etc.). We have $\sin 2x = -1$ when $2x = 3\pi/2, 7\pi/2$, etc. (i.e. 270°, 630°, etc.). Thus we have:

x	0	$\pi/4$	$\pi/2$	$3\pi/4$	π	$5\pi/4$	$3\pi/2$	etc.
$2x$	0	$\pi/2$	π	$3\pi/2$	2π	$5\pi/2$	3π	
$y = \sin 2x$	0	+1	0	−1	0	+1	0	

Figure 12.10(c) shows the graph. Comparing this with the graph in figure 12.10(a) for $y = \sin x$, then the 2 in this case has doubled the frequency, i.e. the number of cycles for a given value of x.

Consider the graph of $y = \sin (x + \pi/4)$. The values of x that will give $\sin (x + \pi/4) = 0$ and hence $y = 0$ are the values for which $x + \pi/4 = 0, \pi, 2\pi, 3\pi$, etc. (i.e. 0°, 180°, 360°, 540°, etc.), i.e. $x = -\pi/4, 3\pi/4, 7\pi/4, 11\pi/4$, etc. The values of x that will give $\sin (x + \pi/4) = +1$ and hence $y = +1$ are $x + \pi/4 = \pi/2, 5\pi/2$, etc. (i.e. 90°, 450°, etc.), i.e. $x = \pi/4, 9\pi/4$, etc. The values of x that will give $\sin (x + \pi/4) = -1$ and hence $y = -1$ are when $x + \pi/4 = 3\pi/2, 7\pi/2$, etc. (i.e. 270°, 630°, etc.), i.e. $x = 5\pi/4, 13\pi/4$, etc. We thus have:

x	0	$\pi/4$	$\pi/2$	$3\pi/4$	π	$5\pi/4$	$3\pi/2$	etc.
$x + \pi/4$	$\pi/4$	$\pi/2$	$3\pi/4$	π	$5\pi/4$	$3\pi/2$	$7\pi/4$	
$\sin (x + \pi/4)$	+0.7	+1	0.7	0	−0.7	−1	−0.7	

Figure 12.11(a) shows the graph.

Figure 12.11(b) shows the graph for the equation $y = \sin (x - \pi/4)$. The following table shows the data which can be derived in a similar way to that above.

x	0	$\pi/4$	$\pi/2$	$3\pi/4$	π	$5\pi/4$	$3\pi/2$	etc.
$x - \pi/4$	$-\pi/4$	0	$\pi/4$	$\pi/2$	$3\pi/4$	π	$5\pi/4$	
$\sin (x - \pi/4)$	−0.7	0	+0.7	+1	+0.7	0	−0.7	

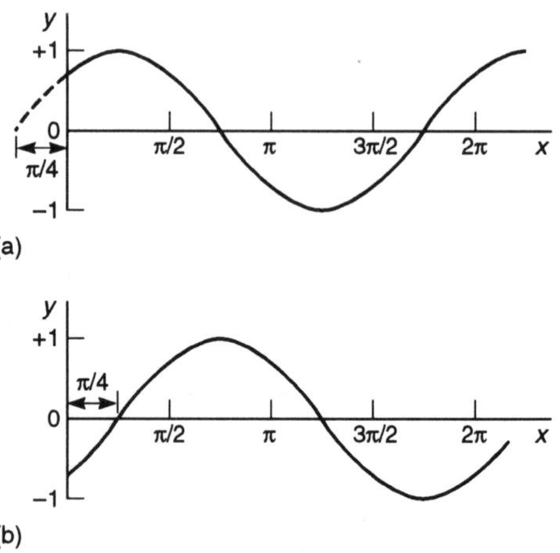

Figure 12.11 *(a) y = sin (x + π/4), (b) y = sin (x − π/4)*

Example

Plot the graph of $y = 2 \sin 2x$.

For $y = 2 \sin 2x$, the values of x that will give $y = 0$ are the values for which $2x = 0, \pi, 2\pi$, etc. (i.e. 0°, 180°, 360°, etc.). Since the sine of an angle oscillates in value between +1 and −1 then the value of y, in this case, will oscillate between +2 and −2. We have $\sin 2x = +1$ when $2x = \pi/2, 5\pi/2$, etc. (i.e. 90°, 450°, etc.). We have $\sin 2x = -1$ when $2x = 3\pi/2, 7\pi/2$, etc. (i.e. 270°, 630°, etc.). Thus we have:

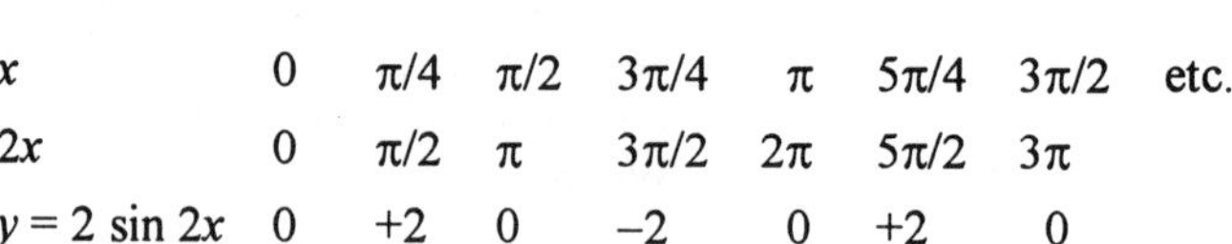

x	0	$\pi/4$	$\pi/2$	$3\pi/4$	π	$5\pi/4$	$3\pi/2$	etc.
$2x$	0	$\pi/2$	π	$3\pi/2$	2π	$5\pi/2$	3π	
$y = 2 \sin 2x$	0	+2	0	−2	0	+2	0	

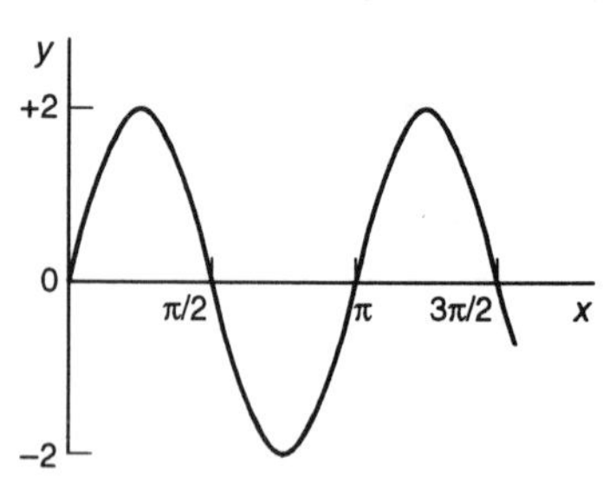

Figure 12.12 *y = 2 sin 2x*

Figure 12.12 shows the graph.

Revision

4 Plot graphs of the functions:

(a) $y = 2 \sin x$, (b) $y = 2 + \sin x$, (c) $y = \cos x$, (d) $y = \cos 2x$, (e) $y = \cos \frac{1}{2}x$

12.2.2 Graphs of exponentials

See chapter 4 and figures 4.1, 4.2 and 4.3 for a discussion and graphs of exponentials.

Revision

5 Plot graphs of the functions:

(a) $y = e^x$, (b) $y = 2\,e^x$, (c) $y = e^{2x}$, (d) $y = e^{-x}$, (e) $y = e^{-2x}$, (f) $y = 1 - e^{-x}$

12.2.3 Adding and subtracting functions

Consider the equation $y = x + 2$. The value of y at a particular value of x is obtained by adding the 2 to the value of x. We can thus consider the graph of this equation to be formed as a result of adding the graph of $y = x$ to that of $y = 2$. Figure 12.13 shows such an addition. Thus, for example, we can add CD and BD to give AD and so the point A on the line for $y = x + 2$.

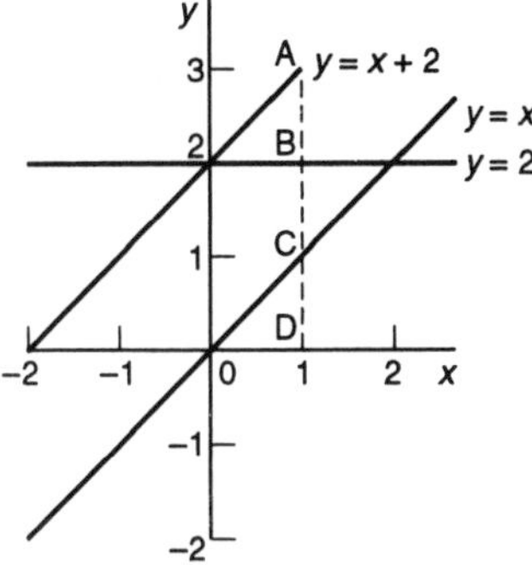

Figure 12.13 $y = x + 2$

Consider the equation $y = x^2 - 2$. We can obtain this graph by adding the graphs for $y = x^2$ and $y = -2$, or by subtracting the graph for $y = 2$ from that for $y = x^2$. Figure 12.14 illustrates this subtraction. We have BC minus AC giving CD and so the point D on the line for $y = x^2 - 2$.

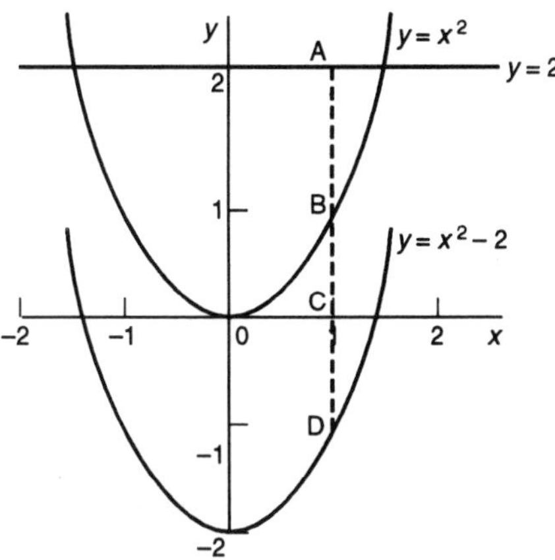

Figure 12.14 $y = x^2 - 2$

The graph for the equation $y = 2 + \sin x$ (see revision problem 4(b)) can be obtained by adding the graphs for $y = 2$ and $y = \sin x$, the graph for $y = \sin x + \sin 2x$ by adding the graphs for $y = \sin x$ and $y = \sin 2x$.

Example

The distance s travelled by an object when starting at time $t = 0$ with an initial velocity u and having a uniform acceleration a is given by the equation $s = ut + \frac{1}{2}at^2$. Plot a graph showing how the distance varies with time for the first 4 s if $u = 10$ m/s and $a = 4$ m/s^2.

We can consider the required graph to be the sum of the graphs of $s = ut$ and $s = \frac{1}{2}at^2$. Thus, with the values given, we have:

t in s	0	1	2	3	4
$s = 10t$ in m	0	10	20	30	40

t in s	0	1	2	3	4
$s = 2t^2$ in m	0	2	8	18	32

Figure 12.15 shows the two graphs and their sum.

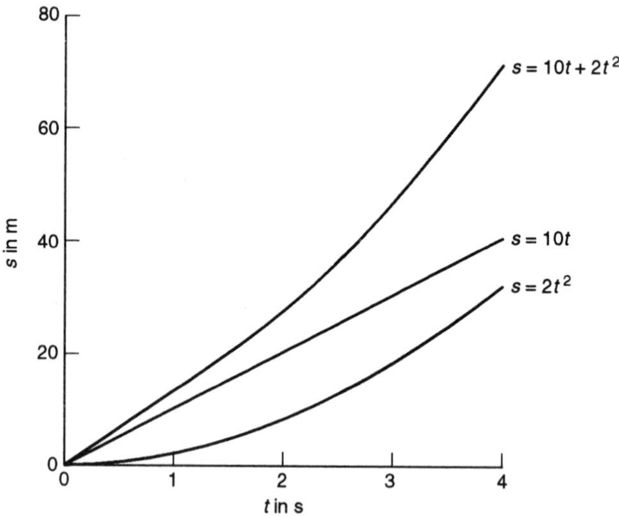

Figure 12.15 $s = 10t + 2t^2$

Revision

6 Plot the graphs of the following equations:

(a) $y = 2x + 1$, (b) $y = x^2 + 2x$, (c) $y = \sin x + \sin 2x$, (d) $y = 2 + e^x$

12.3 Solving equations graphically

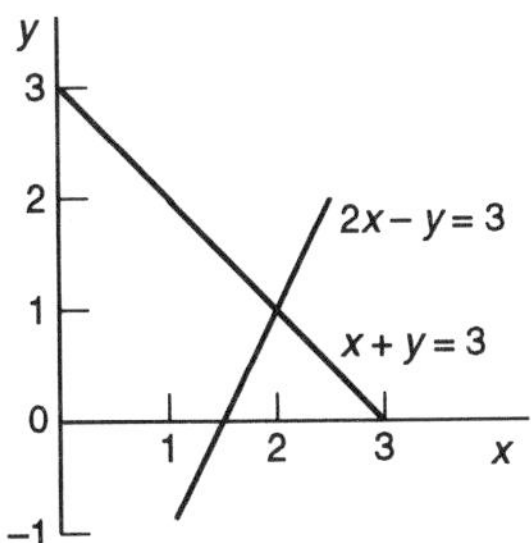

Figure 12.16 *Graphs of simultaneous equations*

Consider the following pair of simultaneous equations

$$x + y = 3 \text{ and } 2x - y = 3$$

Solving the equations means finding the values of x and y which simultaneously fit both equations. In graphical terms, this means finding the values of x and y at the points of intersections of their graphs. These are the only points at which the same values of x and y occur for the two graphs. Figure 12.16 shows the graphs of the above equations. There is just one point of intersection, this being when $x = 2$ and $y = 1$. These are then the solutions. We can check that they are by substituting them in the equations. Thus $2 + 1 = 3$ and $2 \times 2 - 1 = 3$.

As a further illustration of this technique of solving simultaneous equations, consider the solution of

$$x - y + 2 = 0 \text{ and } x^2 - y = 0$$

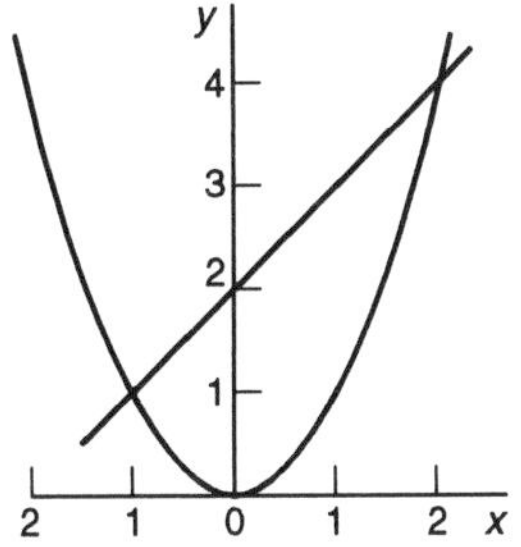

Figure 12.17 *Graphs of simultaneous equations*

Figure 12.17 shows the graphs of the two equations. There are two points of intersection, these being at $x = 2$ and $y = 4$ and at $x = -1$ and $y = 1$. We can check that these are the solutions by substituting them in the equations. Thus $2 - 4 + 2 = 0$ and $2^2 - 4 = 0$, $-1 - 1 + 2 = 0$ and $(-1)^2 - 1 = 0$.

Revision

7 Solve graphically the following pairs of simultaneous equations:

(a) $2x + y = 4$ and $x + y = 3$, (b) $x + y = 1$ and $2x - y = 5$,
(c) $y - x^2 = 0$ and $2y + 3x = 3$, (d) $y - 4x^2 = 0$ and $y - x = 1$

12.3.1 Solving equations

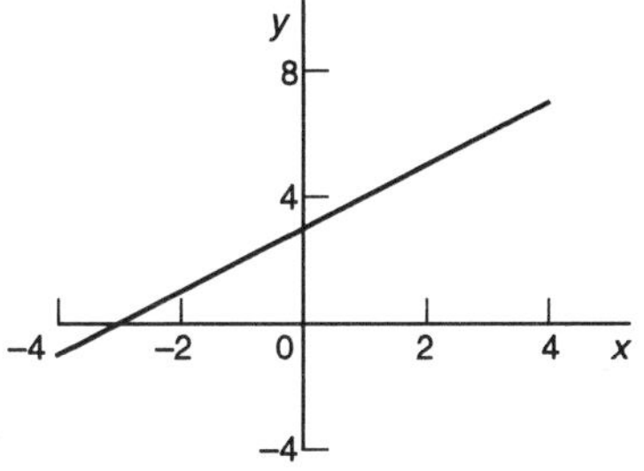

Figure 12.18 *Solving $x + 3 = 0$*

Suppose we need to solve the equation $x + 3 = 0$. This is fairly straightforward and the answer is easily obtained by algebraic manipulation of the equation (see chapter 1) as $x = -3$. We could, however, have obtained the result graphically. What we are looking for is where the graph of $y = x + 3$ cuts the x-axis. Then we have the value of $y = 0$. Figure 12.18 shows the graph. What we are essentially doing is finding where the graph lines for $y = x + 3$ and $y = 0$ intersect, i.e. solving the simultaneous equations

This technique can be used for quadratic, or more complex, equations. Consider the graphical solution of the equation $x^2 + 6x + 5 = 0$. We can do this by plotting the graph of $y = x^2 + 6x + 5$. Taking coordinate values we obtain:

x	-6	-5	-4	-3	-2	-1	0
$y = x^2 + 6x + 5$	$+5$	0	-3	-4	-3	0	$+5$

then the graph has the form shown in figure 12.19. The intercepts with the x-axis occur at -1 and -5. Thus the solutions are $x = -1$ and $x = -5$. What

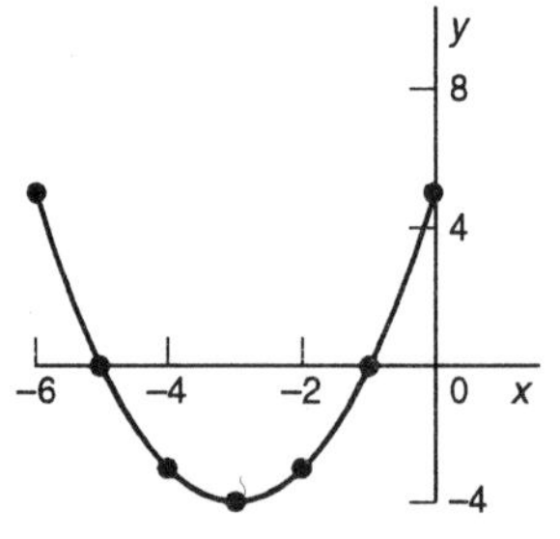

Figure 12.19 $y = x^2 + 5x + 6$

we are essentially doing is finding where the graph lines for $y = x^2 + 6x + 5$ and $y = 0$ intersect, i.e. solving the simultaneous equations.

We could have tackled the solution of this equation in a different way. With the above solution what we have done with $x^2 + 6x + 5 = 0$ is let $y = 0$ and so $x^2 + 6x + 5 = y$ and so have this pair of simultaneous equations to solve. We could, however, have let $y = 6x + 5$ and so $x^2 + y = 0$. We now have a different pair of simultaneous equations to solve. Alternatively we could have let $y = 6x$ and so $x^2 + y + 5 = 0$ and then have to solve this pair of simultaneous equations. The results of such simultaneous equations, i.e. their intercepts, and hence the solutions of the equation $x^2 + 6x + 5 = 0$, are at $x = -1$ and $x = -5$.

Example

Solve the equation $x^3 - 4x^2 - 2x + 3 = 0$

For the equation $y = x^3 - 4x^2 - 2x + 3$ then we can obtain the following co-ordinate points:

x	−2	−1	0	1	2	3	4	5
y	−17	0	3	−2	−9	−12	−5	18

Figure 12.20 shows the resulting graph. The intercepts with the x-axis occur at about −1, +0.75 and +4.2. Thus the solutions are $x = -1$, $x = 0.75$ and $x = 4.2$.

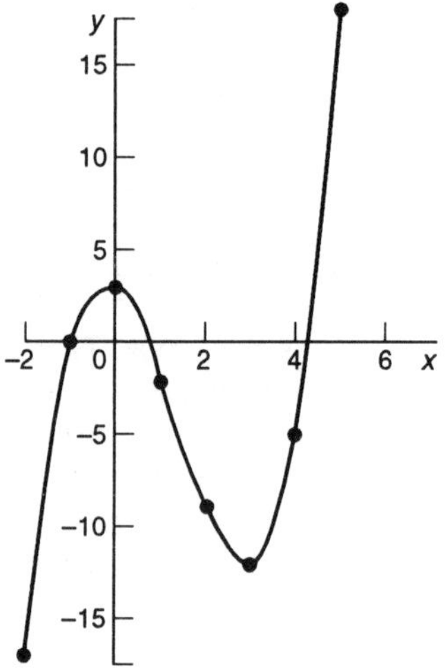

Figure 12.20

Example

Solve the equation $x^2 + 4x - 3 = 0$.

Suppose we let $y = x^2$; we then have $y + 4x - 3 = 0$ and a pair of simultaneous equations to solve. Figure 12.21 shows the graphs of these two equations. They intersect at $x = 0.6$ and $x = -4.6$. These are thus the required solutions.

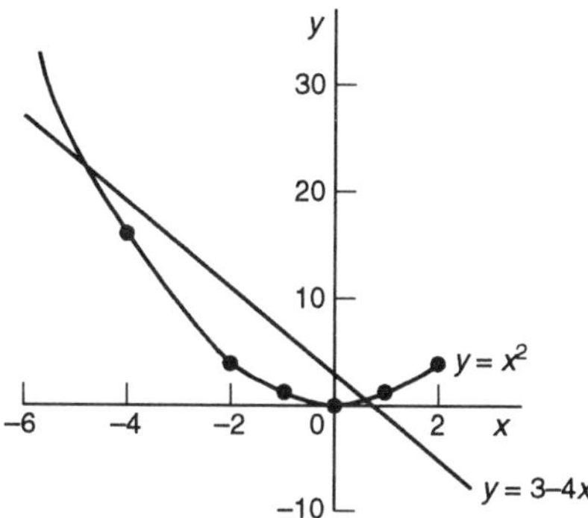

Figure 12.21

Revision

8 Solve graphically the following equations:

(a) $x - 5 = 0$, (b) $x^2 + 3x + 2 = 0$, (c) $x^2 - 4x + 2 = 0$,
(d) $x^3 - x^2 - 5x = 0$, (e) $x^3 - 1 = 0$, (f) $x^3 - 2x^2 - 5x + 6 = 0$

12.3.2 Solutions of quadratic equations

The quadratic equation has the general form $ax^2 + bx + c = 0$ (see chapter 3), where a, b and c are constants. We can solve such equations graphically by plotting $ax^2 + bx + c = y$ and determining the intercepts of the graph with the $y = 0$ axis. Such graphs are parabolic in shape.

Figure 12.22(a) shows the graph of $y = x^2$, i.e. a is 1 but b and c are zero. This graph touches the x-axis at just one point, namely $x = 0$. Thus there is just one value for the roots of this equation. Figure 12.22(b) shows the graph of $y = 2x^2$, i.e. a is 2 but b and c are zero. This graph still touches the x-axis at just one point, namely $x = 0$. However, changing the value of a has changed the steepness of the curve. Figure 12.22(c) shows the graph of $y = x^2 - 2$, i.e. a is 1 and c is -2 but b is 0. Now there are two roots, namely $x = +\sqrt{2}$ and $x = -\sqrt{2}$. Figure 12.22(d) shows the graph of $y = x^2 + 2$, i.e. $a = 1$ and c is $+2$ but b is 0. There are now no real roots. Changing the value of c shifts the curve up or down the y-axis. If the curve is shifted above the x-axis then there are no real roots. Figure 12.22(e) shows the graph of the equation $y = x^2 - 4x + 3$, i.e. a is 1, b is -4 and c is $+3$. The effect of the b term is to shift the graph down the y-axis and that of the c term to shift the graph sideways along the x-axis. There are two roots. Since $x^2 - 4x + 3$ has the factors $(x - 1)(x - 3)$ the intercepts with the x-axis

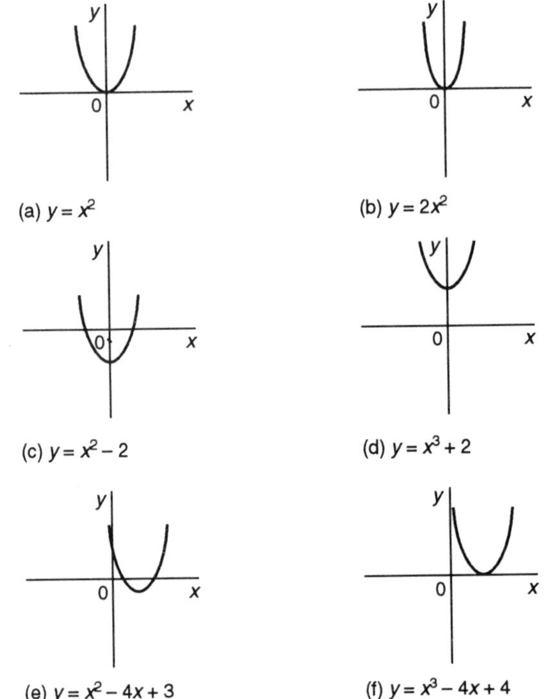

Figure 12.22 *Graphs of quadratic equations*

are at $x = +1$ and $x = +3$. Figure 12.22(f) shows the graph of the equation $y = x^2 - 4x + 4$, i.e. a is 1, b is –4 and c is +4. This graph has only one value for its roots. This is because the quadratic has the factors $(x - 2)(x - 2)$.

The graph of $y = ax^2 + bx + c$ cuts the x-axis at two points, i.e. has two real roots, if $b^2 > 4ac$ (see section 3.4). If $b^2 = 4ac$ then the graph just touches the x-axis at just one point. If $b^2 < 4ac$ then the curve will never cut the x-axis and there are no real roots.

Example

At what points, if any, will the graph of $y = x^2 + 5x + 6$ cut the x-axis?

The quadratic equation $x^2 + 5x + 6 = 0$ can be factorised to give $(x + 3)(x + 2) = 0$ and so there will be intercepts at $x = -2$ and $x = -3$.

Example

How will a graph of $y = x^2 + 5$ differ from one of $y = x^2$?

The $y = x^2$ is the basic parabolic shape (as in figure 12.22(a)). Including a c term of +5 shifts the entire graph up the y-axis by 5 units.

Revision

9 What will be the intercepts, if any, with the x-axis of graphs of the following equations:

(a) $y = 3x^2$, (b) $y = x^2 - 4$, (c) $y = x^2 + 2x + 1$, (d) $y = x^2 + 3x + 2$,

(e) $y = x^2 + 3x + 4$

10 How will the graph of the equation $y = x^2 - 2x + 3$ differ from that of $y = x^2 - 2x + 1$?

12.4 Gradients

Consider the line in figure 12.23. From point A to point B on the line there is a vertical rise of BC in a horizontal distance of AC. We say that the line between the points A and B has a *gradient* (slope) of BC/AC. Since BC is $y_2 - y_1$ and AC is $x_2 - x_1$ then

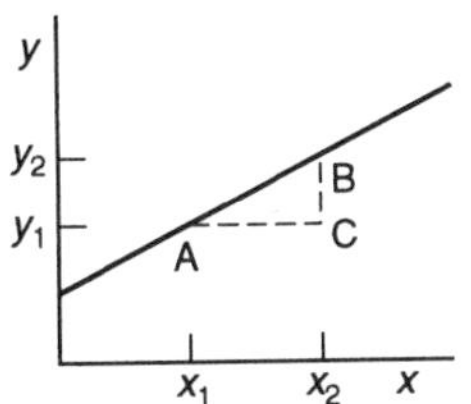

Figure 12.23 *Gradient*

$$\text{gradient} = \frac{y_2 - y_1}{x_2 - x_1}$$

With a straight line it does not matter what points we consider along the line: the gradient is the same at all positions and so a constant. Note that a horizontal line, i.e. one for which there is no change in y when x increases, has a zero gradient. A line has a positive gradient if when there is an increase in x there is an increase in y. It has a negative gradient if when there is an increase in x there is a decrease in y. Figure 12.24 illustrates such lines.

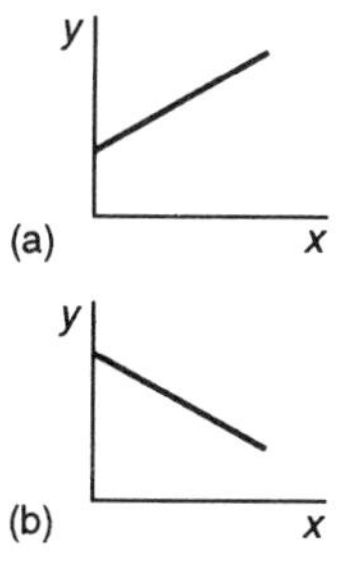

Figure 12.24 *(a) Positive, (b) negative gradient*

For a straight-line graph, as in figure 12.22, if we let the gradient be m, then

$$m = \frac{y_2 - y_1}{x_2 - x_1}$$

Hence

$$y_2 - y_1 = m(x_2 - x_1)$$

We can consider any pair of points on a straight line. Thus suppose we consider A to be where the graph of the straight line intercepts with the y axis (figure 12.25) and has the value c, i.e. $y_1 = c$ when $x_1 = 0$. If we take the point B to have the coordinates (x, y), then we can write the above equation as

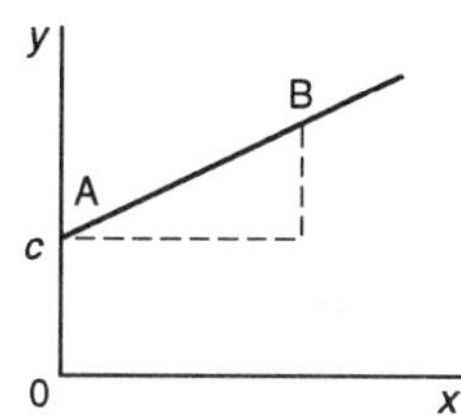

Figure 12.25 *Straight-line graph*

$$y - c = m(x - 0)$$

Thus, when rearranged, we have the equation

$$y = mx + c$$

This is the *general equation for a straight line*. All equations of this form give straight-line graphs with a gradient of m and an intercept with the y axis of c.

The gradient measures the rate of change of y compared with x. This is an important concept which is returned to in section D and the discussion of differentiation.

Example

Determine the gradient of the line joining the pair of points (5, 3) and (2, 4).

Using the equation given above for the gradient

$$\text{gradient} = \frac{y_2 - y_1}{x_2 - x_1} = \frac{4-3}{2-5} = -\frac{1}{3}$$

The minus sign indicates that the value of y is decreasing as x increases.

Example

Determine the gradient of the line given by the equation $y = 2x - 3$.

We could draw the graph of this equation and determine, by taking points on the straight line, the gradient. We could however just consider the coordinates of two points on such a line. Thus, suppose we determine the gradient between the points on the line for which $x = 2$ and $x = 3$. These points have $y = 2 \times 2 - 3 = 1$ and $y = 2 \times 3 - 3 = 3$. The gradient is thus

$$\text{gradient} = \frac{y_2 - y_1}{x_2 - x_1} = \frac{3-1}{3-2} = 2$$

Note that this is the coefficient of x in the original equation. For an equation of the form $y = mx + c$ the gradient is always m.

Example

For the equation $y = 3x - 4$, what is the gradient of the graph and the value of its intercept with the y-axis?

This equation is of the form $y = mx + c$ and thus the gradient is 3 and the intercept with the y-axis is -4.

Example

Determine the equation of the straight-line graph which has a gradient of 5 and an intercept with the y-axis of +2.

The equation is of the form $y = mx + c$ and thus is $y = 5x + 2$.

Revision

11 Determine the gradients of the lines joining the following pairs of points:

(a) (3, 1) and (4, 2), (b) (5, 1) and (10, 2), (c) (−3, 2) and (1, 3), (d) (5, 3) and (0, 0), (e) (0, 2) and (1, 0)

12 Determine the gradients of the lines given by the following equations and their intercepts with the y-axis:

(a) $y = 5x + 3$, (b) $y = x - 3$, (c) $2y = 4x + 1$, (d) $y = 2 - 2x$

13 Determine the equations of the straight-line graphs which have:

(a) a gradient of 2 and an intercept with the y-axis of 3,
(b) a gradient of 4 and an intercept with the y-axis of −3,
(c) a gradient of −2 and an intercept with the y-axis of 1,
(d) a gradient of 5 and an intercept with the y-axis of 0.

14 The velocity v of an accelerating object after a time t is given by the equation $v = 4 + 3t$. What is the gradient and the intercept with the y-axis for the straight-line graph of v plotted against t?

12.4.1 Gradient of a curve

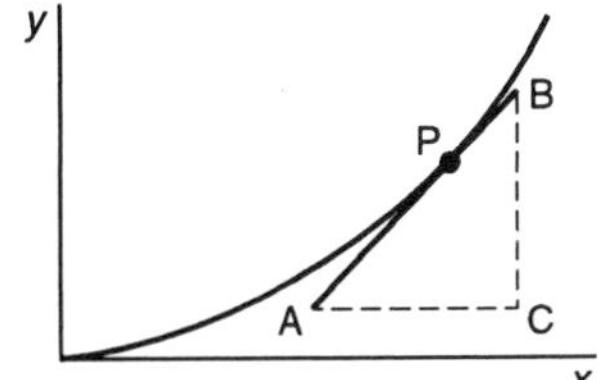

Figure 12.26 *Gradient of a curve*

The gradient of a straight line is the same for all points on the line. However, the gradient of a curve is not a constant and varies from point to point along the curve. The gradient of a particular point on a curve is defined as being the gradient of the tangent to the curve at that point. Figure 12.26 illustrates this, the gradient at point P being BC/AC.

Example

Plot the graph of the equation $y = x^2$ for values of x from 0 to 3 and estimate the value of the gradient at $x = 2$.

Figure 12.27 shows the graph. A ruler placed against the curve at the point $x = 2$ can be used to draw the tangent to the curve at that point. This leads to an estimate of the gradient as $8/2 = 4$.

Revision

15 Plot graphs of the following equations between the values of x indicated and estimate the gradients of the graphs at the indicated points:

(a) $y = x^2 + 1$, x between 0 and 3, gradient at $x = 2$,
(b) $y = x^2 + x$, x between 0 and 3, gradient at $x = 2$,
(c) $y = x^3 + 1$, x between 0 and 3, gradient at $x = 2$.

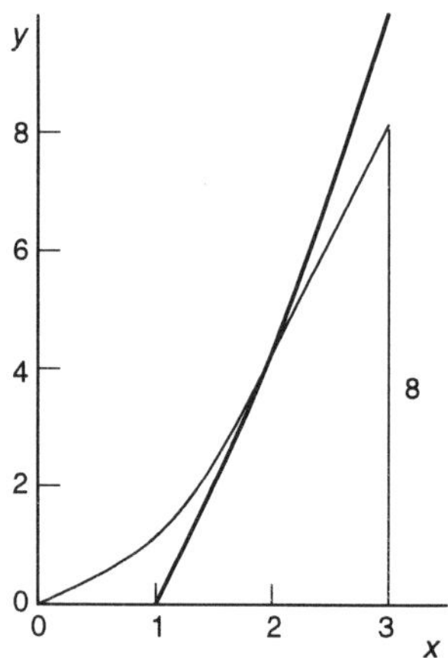

Figure 12.27

12.5 Areas under graphs

It is often necessary to estimate the area enclosed between a graph line and the axis of a graph. For example, in estimating the work done when a spring is stretched through a distance x by a force F, the area under the graph line between extensions 0 and x is the work done. With a straight-line graph, the area can be calculated. Where the graph is non-linear, a simple method that can be used is to count the squares under the graph and then multiply the number of squares by the value corresponding to their area.

Example

Estimate the area under the graph of $y = x^2$ between $x = 0$ and $x = 2$.

Figure 12.28 shows the graph. An estimate of the area under the graph between $x = 0$ and $x = 2$, the shaded area in the graph, is about one and a quarter times graph squares. The square has a vertical height of 2 and a horizontal length of 1 and so an area of 2 square units. Thus the area is about 2.5 square units.

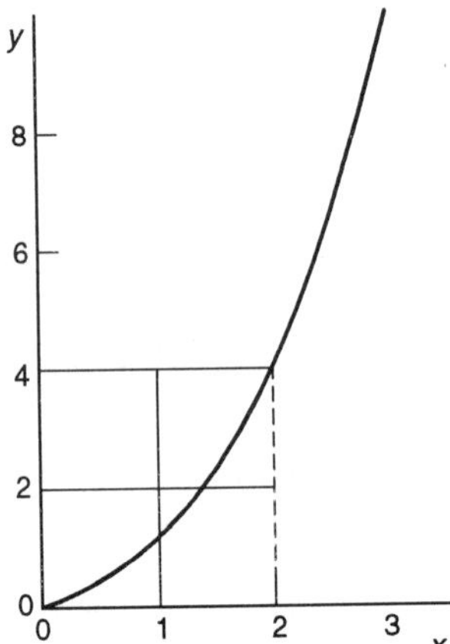

Figure 12.28

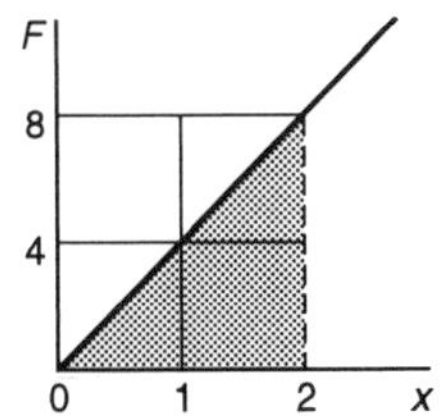

Figure 12.29

Example

Estimate the area under the graph of $F = 4x$ between $x = 0$ and $x = 2$.

Figure 12.29 shows the graph. It is a straight-line graph through the origin. Thus the area is that of a triangle and so is half the base × height, i.e. $\frac{1}{2} \times 2 \times 8 = 8$ square units.

Revision

16 Estimate the areas under the following graphs:

(a) $y = 2x$ between $x = 0$ and $x = 2$,
(b) $y = x + 2$ between $x = 0$ and $x = 1$,
(c) $y = x^2$ between $x = 0$ and $x = 1$,
(d) $y = 2x^2 + 3$ between $x = 1$ and $x = 2$.

12.5.1 The mid-ordinate rule

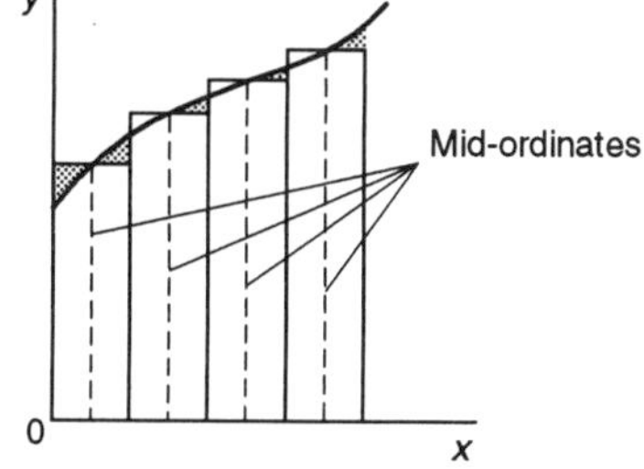

Figure 12.30 *Mid-ordinate rule*

Another way of determining the area under a curve is the *mid-ordinate rule*. The area under the graph is divided into a number of equal width vertical strips, as shown in figure 12.30. The area of a strip is obtained by assuming that multiplying the mid-ordinate, i.e. the ordinate at the middle of the strip, by the width of the strip gives a reasonable approximation. This assumes that the shaded areas are near enough equal.

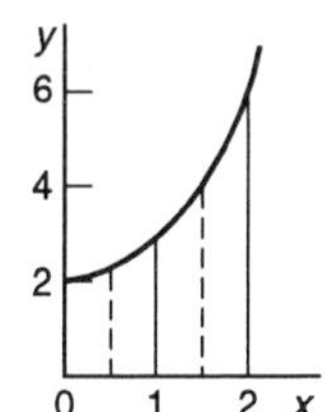

Figure 12.31

Example

Estimate, by the use of the mid-ordinate rule, the area under the graph of $y = x^2 + 2$ between $x = 0$ and $x = 2$.

Figure 12.31 shows the graph with the area divided into just two strips. The mid-ordinate for the first strip occurs at $x = 0.5$. This ordinate has a height, given by the equation $y = x^2 + 2$, of $0.5^2 + 2 = 2.25$. Thus, since the strip has a width of 1, the area of the first strip is $2.25 \times 1 = 2.25$ square units. For the second strip, the mid-ordinate occurs at $x = 1.5$. The height of this ordinate is thus $1.5^2 + 2 = 4.25$. Thus, since the width of the strip is 1, the area is $4.25 \times 1 = 4.25$ square units. The estimate of the total area is thus $2.25 + 4.25 = 6.50$ square units.

We could improve the accuracy of the estimate by using more strips. Thus with strips of width 0.5, and hence four strips, we have:

	Strip 1	Strip 2	Strip 3	Strip 4
Mid-ordinate $x =$	0.25	0.75	1.25	1.75
Height of ordinate	2.06	2.56	3.56	5.06
Area of strip	1.03	1.28	1.78	2.53

The total area is thus 1.03 + 1.28 + 1.78 + 2.53 = 6.62 square units.

Increasing the number of strips gives yet more accuracy. With a very large number of strips the area is 6.67 square units.

Revision

17 Use the mid-ordinate rule to obtain estimates of the areas under the graphs of the following equations:

(a) $y = 2x^2$ between $x = 0$ and $x = 2$,
(b) $y = x^2 + 3$ between $x = 0$ and $x = 2$,
(c) $y = x^3$ between $x = 1$ and $x = 3$.

18 Use the mid-ordinate rule to obtain an estimate of the area under the graph between $x = 0$ and $x = 6$ for the graph which is plotted from the following data:

y	10	18	22	18	10	10	14
x	0	1	2	3	4	5	6

19 Use the mid-ordinate rule to obtain an estimate of the area under the graph between $t = 0$ and $t = 60$ s for the graph plotted from the velocity–time data given below (note that the area gives the distance travelled in 60 s, see chapter 15):

v in m/s	0	14	20	24	27	28	29
t in s	0	10	20	30	40	50	60

Problems

1 Plot graphs of the following equations:

(a) $y = 3x$, (b) $y = x - 1$, (c) $y = x^2 + 1$, (d) $y = x^2 - 1$, (e) $y = 2\cos 2x$,

(f) $y = \cos x + 1$, (g) $y = 2\sin \frac{1}{2}x$, (h) $y = 4 + e^{-x}$

2 Solve, by graphical means, the following equations:

(a) $x + 6 = 0$, (b) $x^2 - 3x + 2 = 0$, (c) $x^2 + 4x + 3 = 0$,

(d) $2x^2 + 3x - 3 = 0$, (e) $x^3 + 2x^2 - 3x = 0$, (f) $x^3 - 3x + 1 = 0$,

(g) $2x^3 - 5x^2 - 10x + 3 = 0$, (h) $3\,e^{2x} + 20 = 0$, (i) $\sin \frac{1}{2}x - e^{-x} = 0$

3 Solve, by graphical means, the following simultaneous equations:

(a) $y + 2x = 3$ and $2y - x = 1$, (b) $y = x^2$ and $y - 2x = 3$,

(c) $y + 3x = 5$ and $2y - 3x = 1$, (d) $y = 3x^2$ and $y - 8 = 0$,

(e) $y = x^2$ and $y = x + 5$, (f) $y = x^2$ and $y = 2x + 1$

4 What will be the intercepts, if any, with the x-axis of graphs of the following equations:

(a) $y = 5x^2$, (b) $y = x^2 - 9$, (c) $y = x^2 + 3x - 4$, (d) $y = x^2 + 5x + 2$,

(e) $y = x^2 + 4x + 4$, (f) $y = x^2 + 2$, (g) $y = x^2 + x - 6$

5 Plot graphs, using the first few positive-value points, of the following relationships:

(a) The power P dissipated is a function of the current I passing through a resistor and is given by $P = 2I^2$.
(b) The time T taken for one oscillation of a pendulum is related to the length L of the pendulum by $T = 2\sqrt{L}$.
(c) The velocity v of a moving object is related to the time t for which it is accelerated by $v = 2 + 3t$.
(d) The force F required to stretch a spring to an extension of x is given by $F = 10x$.
(e) The resistance R of a resistor depends on the temperature θ and is given by $R = 10 + 0.01\theta$.
(f) The voltage v across a capacitor varies with time t and is given by $v = 10\ e^{-0.2t}$.

6 Determine the equations of the straight-line graphs which have:

(a) a gradient of 1 and an intercept with the y-axis of 2,
(b) a gradient of 5 and an intercept with the y-axis of -1,
(c) a gradient of -3 and an intercept with the y-axis of 2,
(d) a gradient of -5 and an intercept with the y-axis of -2,
(e) a gradient of 0 and an intercept with the y-axis of 3,
(f) a gradient of 2 and an intercept with the y-axis of -5.

7 Plot graphs of the following equations between the values of x indicated and estimate the gradients of the graphs at the indicated points:

(a) $y = x^2 + 3x$, x between 0 and 3, gradient at $x = 2$,
(b) $y = x^2 - 2x$, x between -1 and $+1$, gradient at $x = 0$,
(c) $y = x^3$, x between 0 and 3, gradient at $x = 1$,
(d) $y = 2x^2 + 1$, x between 0 and 3, gradient at $x = 2$,
(e) $y = x^2 + 2x$, x between 0 and 3, gradient at $x = 2$.

8 Estimate the areas under the graphs of the following equations by using the mid-ordinate rule:

(a) $y = x^3 + 3$ between $x = 0$ and $x = 2$, with 4 strips,

(b) $y = \sin x$ between $x = 0$ and $x = \pi/2$, with 5 strips,
(c) $y = 3\ e^{-x}$ between $x = 0$ and $x = 4$, with 4 strips,
(d) $y = x^3$ between $x = 0$ and $x = 2$, you select the number of strips,
(e) $y = x^2 + 1$ between $x = 0$ and $x = 3$, you select the number of strips,
(f) $y = 2x + 1$ between $x = 1$ and $x = 5$, you select the number of strips.

9 The following are points used to plot a graph:

y	1.00	1.20	1.36	1.50	1.68	1.78	1.86	1.90
x	0	2	4	6	8	10	12	14

Use the mid-ordinate rule to estimate the area under the graph between $x = 0$ and $x = 14$.

10 The following are points used to plot a graph:

y	1	2	5	10	17	26
x	0	1	2	3	4	5

Use the mid-ordinate rule to estimate the area under the graph between $x = 0$ and $x = 5$.

11 Use the mid-ordinate rule to estimate the area under the graph of $y = 20 + x - x^2$ between $x = 0$ and $x = 5$. Use five strips.

12 The following are points used to plot a graph:

y	0	1.43	2.06	2.44	2.69	2.84	2.94	3.00
x	0	1	2	3	4	5	6	7

Use the mid-ordinate rule to estimate the area under the graph between $x = 0$ and $x = 7$. Use seven strips.

13 Linear graphs

13.1 Introduction

If two variables x and y give a straight-line graph then the equation relating them is of the form $y = mx + c$, where m is the gradient and c the intercept of the graph line with the y-axis (see section 12.4). Thus $y = 2x + 3$ is the equation of a straight-line graph with a gradient of 2 and an intercept with the y-axis of +3. The equation $y = -2x + 3$ describes a straight-line graph with a gradient of −2 and an intercept with the y-axis of +3 (see figure 12.24 for graphs showing the difference between positive and negative gradients).

This chapter is about how, for a set of data points involving two variables which have a linear relationship, we can establish the linear law relating them. It also is about how we can reduce non-linear laws to linear forms and hence determine them.

13.2 Finding linear laws

If when we are given, or determine from an experiment, a set of corresponding values of two variables and plot them on a graph to give a straight line graph, then we know that the variables must be connected by a law of the form $y = mx + c$. To determine the law, all we have to do is find the values of m and c.

For example, suppose we have the following data for two related variables y and x:

y	−3	−1	1	3
x	−1	0	1	2

We will consider two alternative methods of determining m and c.

Figure 13.1 shows the above data plotted as a graph. The gradient of the graph can be found by drawing lines such as AB and BC in order to find by how much the line has risen in some horizontal distance. AB/BC = 4/2 and so $m = +2$. The line intercepts the y-axis at −1 and so $c = -1$. Thus the law is $y = 2x - 1$.

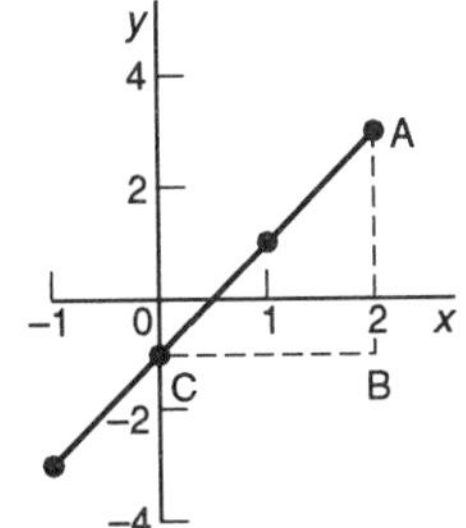

Figure 13.1 *Linear graph*

An alternative method is to choose two points, from those given if they can be assumed to fall on the straight-line graph or take values from the straight-line graph, and substitute their values in the equation $y = mx + c$. Thus, if we take the values given of $y = -3$ when $x = -1$ we obtain

$$-3 = -1m + c$$

If we take another set of values, say $y = 3$ when $x = 2$, then we can also write

$$3 = 2m + c$$

We have a pair of simultaneous equations which have to be solved to give m and c:

$$\begin{array}{lrl} & -\ m + c & = -3 \\ \text{minus} & 2m + c & = \ \ 3 \\ \hline & -3m & = -6 \end{array}$$

Thus $m = 2$. Substituting this in the first equation gives $-2 + c = -3$ and hence $c = -1$. The equation is thus $y = 2x - 1$.

Example

The following is the data obtained from experimental measurements of the load lifted by a machine and the effort expended. Determine if the relationship between the effort E and the load W is linear and if so the relationship.

E in N	18	27	32	43	51
W in N	40	80	120	160	200

Figure 13.2 shows the graph. Within the limits of experimental error the results appear to indicate a straight-line relationship. The gradient is AB/BC = 41/200 or about 0.21. The intercept with the E axis is at 10. Thus the relationship is $E = 0.21W + 10$.

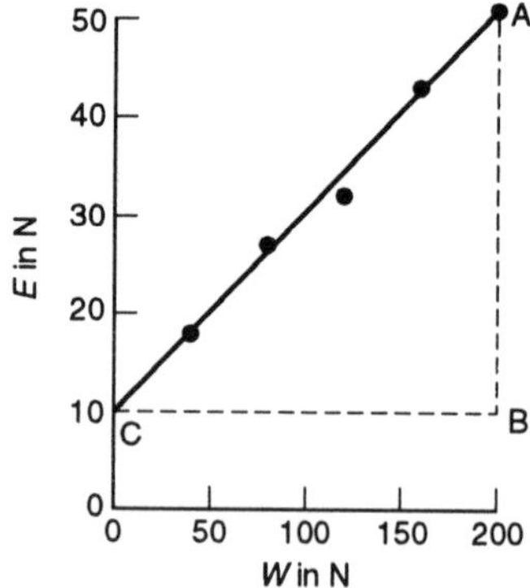

Figure 13.2

Revision

1 Determine, assuming linear, the relationships between the following measured variables:

(a) The volume V and the temperature θ of a gas if we have:

V in m^3	19.7	20.5	21.6	22.7	23.1	23.5
θ in °C	20	30	40	50	60	70

(b) The rate of rotation n of the shaft of a motor and the applied voltage V if we have:

n in rev/s	3	10	16	20	26	33
V in volts	40	80	120	160	200	240

(c) The specific latent heat L of steam and the temperature θ if we have:

L in kJ	2340	2300	2280	2250	2220	2160
θ in °C	70	80	90	100	110	120

(d) The tensile strength σ of an alloy at temperature θ if we have:

σ in kPa	83.7	80.2	77.7	75.2	72.7
θ in °C	100	200	300	400	500

13.3 Reduction of laws to linear forms

When a graph is a straight line then the relationship between the two variables can easily be stated as being of the form $y = mx + c$. With a curved graph we can, however, have many possible forms of relationship. Thus we might, for example, have $y = ax^2 + b$, $y = \frac{a}{x} + b$, $y = ax^b$, $y = a\,e^{bx}$, etc. To determine the form of the law in such cases the technique is adopted of changing the variables to put them into a linear form.

The following illustrate how we can do this for a number of common forms of curved graphs. The principle used does, however, apply to other forms.

13.3.1 $y = ax^2 + b$

Consider the equation $y = ax^2 + b$, where a and b are constants. Let $X = x^2$. We can now write $y = aX + b$. This is a linear relationship between y and X and will give a straight-line graph with a gradient of a and an intercept on the y-axis of b. We thus obtain a straight-line graph by plotting y against x^2.

Example

It is believed that the relationship between y and x for the following data is of the form $y = ax^2 + b$. Determine the values of a and b.

y	2.5	4.0	6.5	10.0	14.5
x	1	2	3	4	5

Let $X = x^2$ to give $y = aX + b$.

y	2.5	4.0	6.5	10.0	14.5
$X = x^2$	1	4	9	16	25

Figure 13.3 shows the graph of y against X. The graph has a gradient of AB/BC = 12.5/25 = 0.5 and an intercept with the y-axis of 2. Thus the relationship is $y = 0.5X + 2$ and so we have $y = 0.5x^2 + 2$.

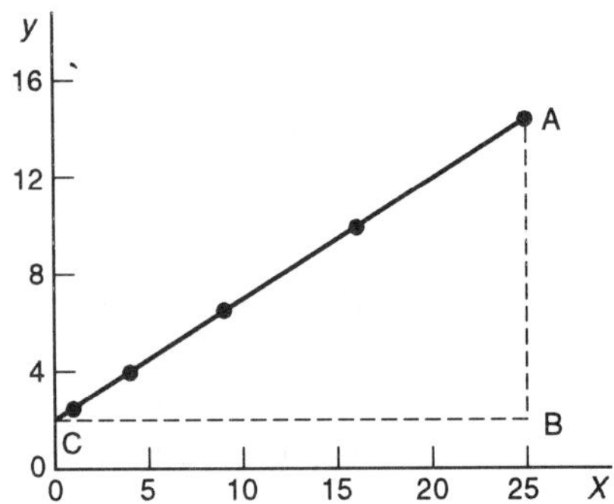

Figure 13.3

Revision

2 It is believed that the relationships between y and x for the following data are of the form $y = ax^2 + b$. Determine the values of a and b.

(a) y	−10	−7	2	17	44
x	0	1	2	3	4

(b) y	2	3	6	11	18
x	0	1	2	3	4

(c) y	5	3	−3	−13	−26
x	0	1	2	3	4

13.3.2 $y = (a/x) + b$

Consider the equation $y = a/x + b$, where a and b are constants. Let $X = 1/x$. We can now write $y = aX + b$. This is a linear relationship between y and X and will give a straight-line graph with a gradient of a and an intercept on the y-axis of b. We thus obtain the straight-line graph by plotting y against $1/x$.

Example

It is believed that the relationship between y and x for the following data is of the form $y = (a/x) + b$. Determine the values of a and b.

y	5	4	3.7	3.5	3.4
x	1	2	3	4	5

Let $X = 1/x$ to give $y = aX + b$.

y	5	4	3.7	3.5	3.4
$X = 1/x$	1	0.50	0.33	0.25	0.20

Figure 13.4 shows the graph of y against X. The graph has a gradient of AB/BC = (5 − 3)/1.0 = 2 and an intercept with the y-axis of 3. Thus the equation for the graph is $y = 2X + 3$ and so we have $y = (2/x) + 3$.

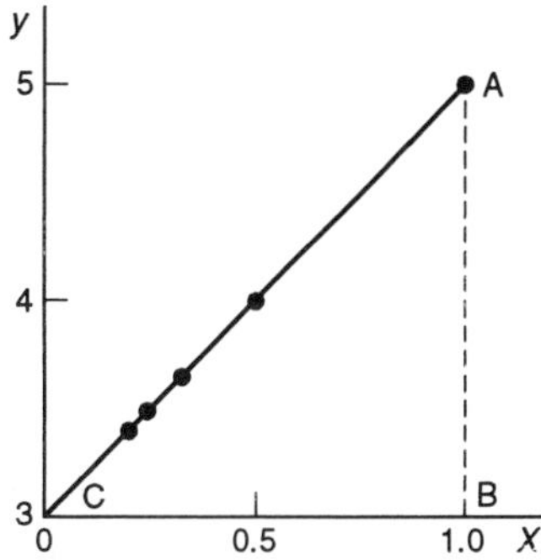

Figure 13.4

Revision

3 It is believed that the relationships between y and x for the following data are of the form $y = (a/x) + b$. Determine the values of a and b.

(a)

y	2	1.5	1.33	1.25	1.20
x	1	2	3	4	5

(b)

y	1	0	−0.33	−0.50	−0.60
x	1	2	3	4	5

(c)

y	1	1.50	1.67	1.75	1.80
x	1	2	3	4	5

13.3.3 $y = ax^2 + bx$

Consider the equation $y = ax^2 + bx$, where a and b are constants. We can rearrange this equation to give

$$\frac{y}{x} = ax + b$$

Let $Y = y/x$. We can now write $Y = ax + b$. This is a linear relationship between Y and x and will give a straight-line graph with a gradient of a and an intercept on the y-axis of b. We thus obtain the straight-line graph by plotting y/x against x.

Example

It is believed that the relationship between y and x for the following data is of the form $y = ax^2 + bx$. Determine the values of a and b.

y	3	8	15	24	35
x	1	2	3	4	5

Let $Y = y/x$ to give $Y = ax + b$. Then we have

$Y = y/x$	3	4	5	6	7
x	1	2	3	4	5

Figure 13.5 shows the resulting graph of Y against x. The gradient of the graph is AB/BC = (7 − 2)/5 = 1. The intercept with the Y axis is 2. Hence the equation of the graph is $Y = 1x + 2$ and so the required relationship is $y = x^2 + 2x$.

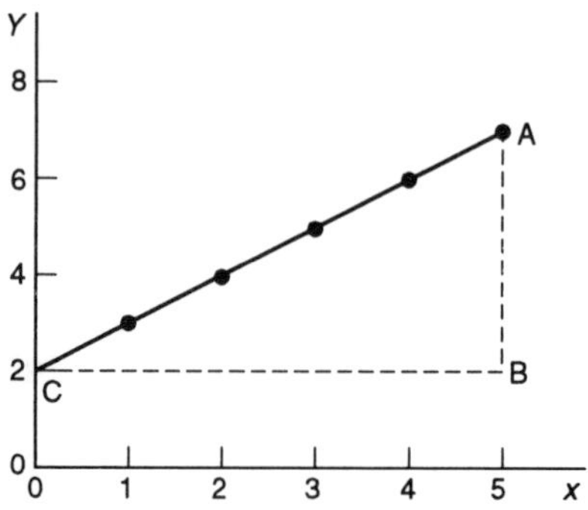

Figure 13.5

Revision

4 It is believed that the relationships between y and x for the following data are of the form $y = ax^2 + bx$. Determine the values of a and b.

(a) y	2	10	24	44	70
x	1	2	3	4	5

(b) y	3	8	15	24	35
x	1	2	3	4	5

(c) y	0	−2	−6	−12	−20
x	1	2	3	4	5

13.3.4 $y = ax^b$

Consider the equation $y = ax^b$, where a and b are constants. We can put this equation in a linear form by taking logarithms, usually to base 10. Thus

$$\lg y = \lg (ax^b) = \lg x^b + \lg a = b \lg x + \lg a$$

If we let $Y = \lg y$ and $X = \lg X$ then we have $Y = bX + c$, where $c = \lg a$. This is a linear relationship between Y and X and will give a straight-line graph with a gradient of b and an intercept on the y-axis of c. We thus obtain the straight-line graph by plotting $\lg y$ against $\lg x$.

Example

It is believed that the relationship between y and x for the following data is of the form $y = ax^b$. Determine the values of a and b.

y	3.0	4.2	5.2	6.0	6.7
x	1	2	3	4	5

Let $Y = \lg y$ and $X = \lg X$. Then we have

$Y = \lg y$	0.48	0.62	0.72	0.78	0.83
$X = \lg x$	0	0.30	0.48	0.60	0.70

Figure 13.6 shows the graph of Y plotted against X. The graph has a gradient of AB/BC = (0.83 − 0.48)/0.70 = 0.5 and an intercept on the Y-axis of 0.48. Thus the graph has the equation $Y = 0.5X + 0.48$. Thus, since b is the gradient 0.5 and the intercept $c = 0.48 = \lg a$, then the required equation is $y = 3.0x^{0.5}$ or $y = 3\sqrt{x}$.

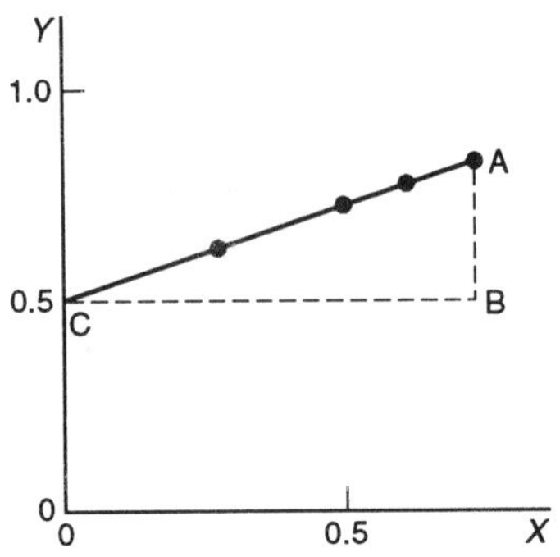

Figure 13.6

Revision

5 It is believed that the relationships between y and x for the following data are of the form $y = ax^b$. Determine the values of a and b.

(a)

y	1	8	27	64	125
x	1	2	3	4	5

(b)

y	2	0.50	0.22	0.13	0.08
x	1	2	3	4	5

(c)

y	−2	−2.83	−3.46	−4.00	−4.47
x	1	2	3	4	5

13.3.5 $y = a\,e^{bx}$

Consider the equation $y = a\,e^{bx}$, where a and b are constants. We can put this equation in a linear form by taking logarithms to base e. Thus

$$\ln y = \ln (a\,e^{bx}) = \ln e^{bx} + \ln a = bx + \ln a$$

If we let $Y = \ln y$ then we have $Y = bx + c$, where $c = \ln a$. This is a linear relationship between Y and x and will give a straight-line graph with a gradient of b and an intercept on the y-axis of c. We thus obtain the straight-line graph by plotting $\ln y$ against x.

Example

It is believed that the relationship between y and x for the following data is of the form $y = a\,e^{bx}$. Determine the values of a and b.

y	5.53	6.11	6.75	7.46	8.24
x	1	2	3	4	5

Let $Y = \ln y$.

$Y = \ln y$	1.71	1.81	1.91	2.01	2.11
x	1	2	3	4	5

Figure 13.7 shows the graph of Y plotted against x. The graph has a gradient AB/BC = (2.11 – 1.71)/(5 – 1) = 0.1 and an intercept with the Y-axis of 1.61. Thus the graph has the equation $Y = 0.1x + 1.61$. Hence, we have the relationship $\ln y = 0.1x + 1.61$ and thus the required equation is $y = e^{1.61}\, e^{0.1x} = 5\, e^{0.1x}$.

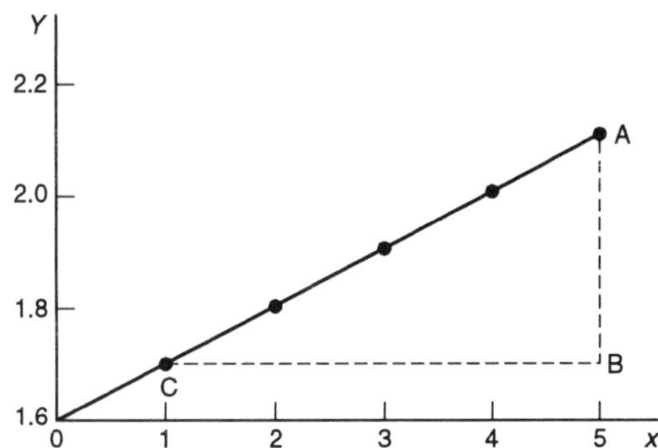

Figure 13.7

Revision

6 It is believed that the relationships between y and x for the following data are of the form $y = a\, e^{bx}$. Show that this is the case and determine the values of a and b.

(a)	y	0.67	0.91	1.23	1.66	2.24
	x	1	2	3	4	5
(b)	y	2.46	2.01	1.65	1.35	1.10
	x	1	2	3	4	5
(c)	y	–2.21	–2.44	–2.70	–2.98	–3.30
	x	1	2	3	4	5

13.4 Log graph paper

When we have relationships involving powers it is often the case that logarithms of the data are taken in order to give a straight-line graph. Thus, as illustrated in figure 13.8, the data is transformed into log (data) and plotted on the linear axis of the graph. This involves calculating the logarithm of each data point before it can be plotted. A more convenient way is, however, to transform the axis of the graph into a *logarithmic axis* as illustrated in figure 13.8. This then enables the data values to be directly plotted, the work of calculating the logarithms having been done by the conversion of the axes. The scale markings on a logarithmic axis do not have equal divisions.

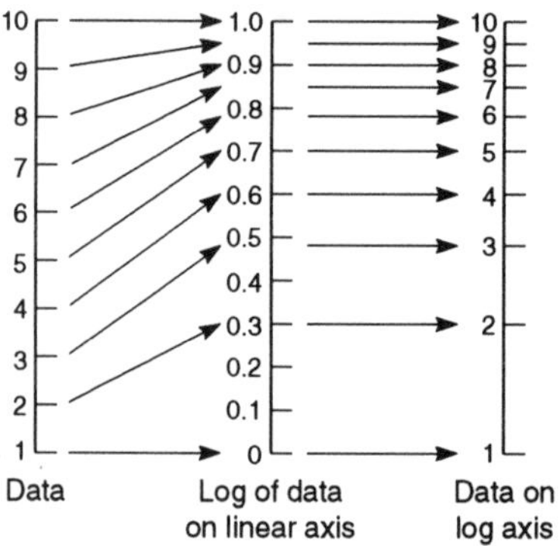

Figure 13.8 *Log axis*

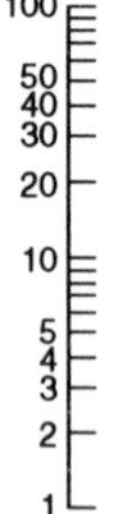

Figure 13.9 *Two cycles*

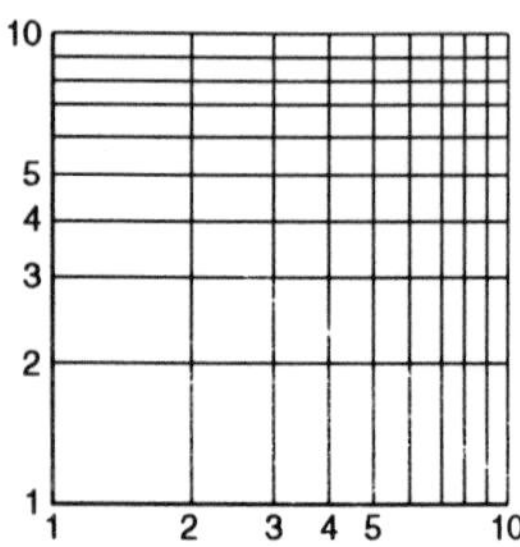

Figure 13.10 *Log-log paper*

The logarithm, to base 10, of 1 is zero and of 10 is 1. The logarithm of 100 is 2. Thus, on a logarithmic scale, the distance between 1 and 10 is the same as between 10 and 100. Each of these distances is referred to as a *cycle*. Figure 13.9 thus shows what an axis would look like with two cycles. Within each cycle we have the same pattern of markings.

Consider the equation $y = 3x^2$. We can convert this into a linear equation by taking logarithms. Thus $\lg y = 2 \lg x + 3$. With conventional, linear, graph paper we would then plot values of $\lg y$ on the Y-axis and $\lg x$ on the X-axis. We can, however, have logarithmic axes for both the axes. Such graph paper is then referred to as *log-log paper*. Figure 13.10 shows the form such graph paper takes. We can determine how many cycles are needed by considering the range of values to be plotted. Thus suppose we have x-values ranging from 1 to 50. Then $\lg x$ ranges from $\lg 1 = 0$ to $\lg 50 = 1.7$. Thus for the x-axis we require two cycles, i.e. from 0 to 1 and 1 to 10. For the y-values we have the range extending from 3×1^2 to 3×50^2. Thus $\lg y$ ranges from $\lg 3 = 0.5$ to $\lg 7500 = 3.9$. Thus for the y-axis we require two cycles, i.e. 0 to 1 and 1 to 10. The graph paper required is thus 2 cycle by 2 cycle.

Consider the equation $y = 3\,e^{2x}$. We can convert this into a linear equation by taking logarithms to base e. Thus $\ln y = 2x + \ln 3$. With conventional, linear, graph paper we would then plot values of $\ln y$ on the Y-axis and x on the x-axis. We can, however, have a logarithmic axis for the

y-axis and a linear axis for the x-axis. Such graph paper is referred to as *log-linear paper*. Figure 13.11 shows the form such graph paper takes. The same log-linear paper can be used for logarithms to base e as to base 10. Such logarithms are related by

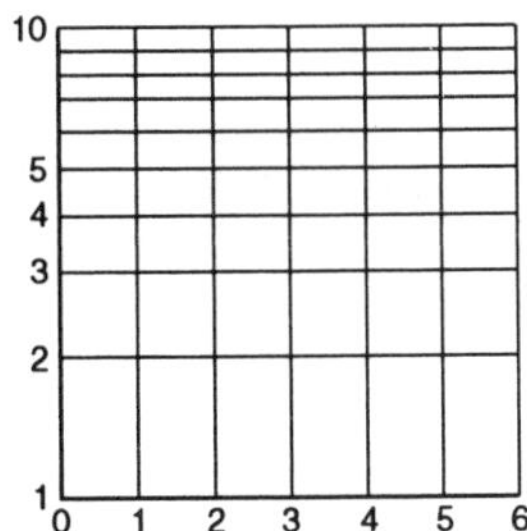

Figure 13.11 *Log-linear paper*

$$\ln x = 2.3026 \lg x$$

Note that the above equation arises because if we take $x = 10^y$ then $\lg x = y$. If we replace the 10 by the identical number $e^{2.3026}$ we obtain $x = (e^{2.3026})^y$ and thus $\ln x = 2.3026y = 2.3026 \lg x$. Thus $\ln x$ is just the logarithm to base 10 scale multiplied by a constant. We can therefore have the scale markings for the cycles on the log axis representing the values ln 1, ln 10, ln 100, etc.

Example

It is believed that the relationship between y and x for the following data is of the form $y = ax^b$. Show that this is the case and determine, using log-log graph paper, the values of a and b.

y	3.0	4.2	5.2	6.0	6.7
x	1	2	3	4	5

This is a repeat of the example used in section 13.3.3. Taking logarithms to base 10 gives $\lg y = b \lg x + \lg a$. The y-axis has to range from $\lg 3 = 0.48$ to $\lg 6.7 = 0.83$ and so just one cycle from 0 to 1 is required. The x-axis has to range from $\lg 1 = 0$ to $\lg 5 = 0.69$ and so one cycle from 0 to 1 is required. Thus 1 cycle × 1 cycle graph paper is required. Figure 13.12 shows the above data plotted on log-log graph paper. The graph is a straight line and so the relationship is of this form.

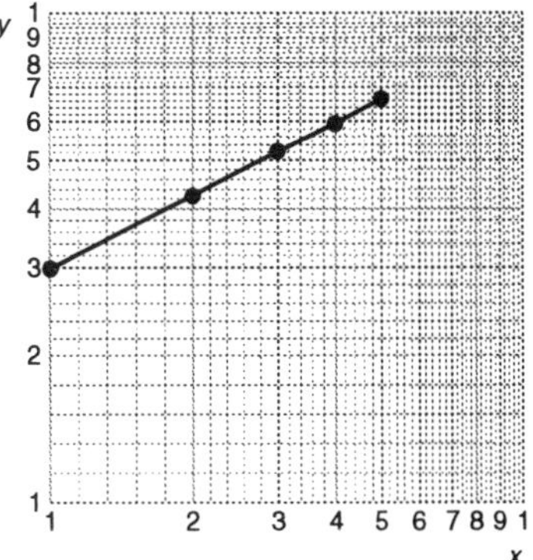

Figure 13.12

The gradient of the graph is

$$\text{gradient} = \frac{\lg 6.7 - \lg 3.0}{\lg 5 - \lg 1} = \frac{0.83 - 0.48}{0.70 - 0} = 0.5$$

When $x = 1$ then $\lg x = 0$ and so $\lg y = \lg a$. Thus the intercept of the log-log graph with the ordinate for $x = 1$ gives $a = 3$. The equation is thus $y = 3x^{0.5}$.

Example

It is believed that the relationship between y and x for the following data is of the form $y = a\,e^{bx}$. Show that this is the case and determine, using log-linear graph paper, the values of a and b.

y	5.53	6.11	6.75	7.46	8.24
x	1	2	3	4	5

This is a repeat of the example in section 13.3.5. Taking logarithms to base e gives $\ln y = bx + \ln a$. We thus require log-linear graph paper. The y-axis, which is the ln axis, has to range from $\ln 5.53 = 1.7$ to $\ln 8.24 = 2.1$ and so just one cycle from 1 to 10 is required. Figure 13.13 shows the resulting graph. The graph is straight line and so the relationship is valid. The gradient is

$$\text{gradient} = \frac{\ln 8.44 - \ln 5.53}{5 - 1} = \frac{2.13 - 1.71}{4} = 0.10$$

The intercept with the y-axis, i.e. $x = 0$ ordinate, is at 5. Thus the required equation is $y = 5\,e^{0.10x}$.

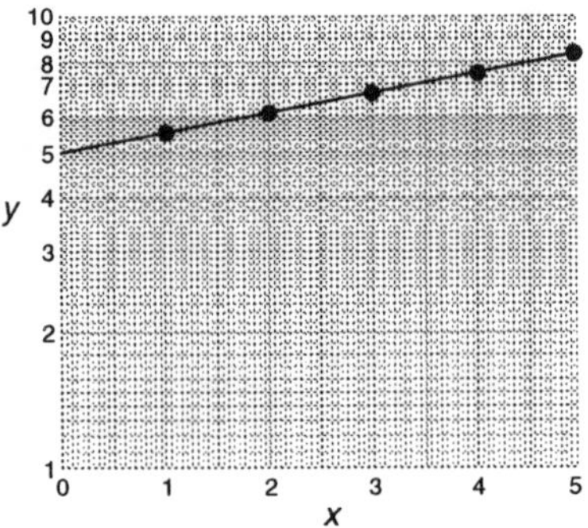

Figure 13.13

Revision

7 Solve revision problems 5 and 6 using log-log or log-linear graph paper.

8 State the type of logarithmic graph paper, including the number of cycles, that should be used when plotting the following relationships.

(a) $\theta = a\,e^{-bt}$ when t varies from 0 to 30 minutes and θ varies from 200°C to 10°C.
(b) $F = av^b$ when v varies from 1 m/s to 6 m/s and the value of F varies from 10 N to 360 N,
(c) $y = ax^b$ when x varies from 2 to 350 and y varies from 14 to 187.

Problems

1 Determine, assuming linear, the relationships between the following variables:

(a) The load L lifted by a machine for the effort E applied.

E in N	9.5	11.8	14.1	16.3	18.5
L in N	10	15	20	25	30

(b) The resistance R of a wire for different lengths L of that wire.

R in Ω	2.1	4.3	6.3	8.3	10.5
L in m	0.5	1.0	1.5	2.0	2.5

(c) The velocity v of an object at different times t.

v in m/s	4.5	5.1	5.5	5.9	6.5
t in s	5	10	15	20	25

(d) The yield stress σ of a steel sample at different temperatures.

σ in MPa	140	130	122	111	101
θ in °C	200	300	400	500	600

2 Determine, on the basis of the data given, the values of a and b in the following relationships:

(a) $y = ax^2 + b$

y	−1	1	7	17	31
x	0	1	2	3	4

(b) $y = ax^2 + b$

y	3	2	−1	−6	−13
x	0	1	2	3	4

(c) $y = (a/x) + b$

y	4	2.5	2	1.75	1.60
x	1	2	3	4	5

(d) $y = (a/x) + b$

y	−6	−5	−4.7	−4.5	−4.4
x	1	2	3	4	5

(e) $y = ax^2 + bx$

y	3	10	21	36	55
x	1	2	3	4	5

(f) $y = ax^2 + bx$

y	−2	−5	−12	−20	−30
x	1	2	3	4	5

(g) $y = ax^b$

y	0.50	0.71	0.87	1.00	1.12
x	1	2	3	4	5

(h) $y = ax^b$

y	0.6	4.8	16.2	38.4	75.0
x	1	2	3	4	5

(i) $y = a\,e^{bx}$

y	905	819	741	670	607
x	1	2	3	4	5

(j) $y = a\,e^{bx}$

y	13.5	18.2	24.6	33.2	44.8
x	1	2	3	4	5

3 The following are examples of relationships which occur in engineering. Determine what form the variables should take when plotted in order to give straight-line graphs and what the values of the gradient and intercept will have.

(a) The period of oscillation T of a pendulum is related to the length L of the pendulum by the equation

$$T = 2\pi\sqrt{\frac{L}{g}}$$

where g is a constant.
(b) The distance s travelled by a uniformly accelerating object after a time t is given by the equation

$$s = ut + \tfrac{1}{2}at^2$$

where u and a are constants.
(c) The e.m.f. e generated by a thermocouple at a temperature θ is given by the equation

$$e = a\theta + b\theta^2$$

where a and b are constants.
(d) The resistance R of a resistor at a temperature θ is given by the equation

$$R = R_0 + R_0\alpha\theta$$

where R_0 and α are constants.
(e) The current i in a circuit when a capacitor is being charged varies with time t and is given by the equation

$$i = I\,e^{t/RC}$$

where I and RC are constants.
(f) The pressure p of a gas and its volume V are related by the equation

$$pV = k$$

where k is a constant.
(g) The deflection y of the free end of a cantilever due to it own weight of w per unit length is related to its length L by the equation

$$y = \frac{wL^4}{8EI}$$

where w, E and I are constants.

4 The resistance R of a lamp is measured at a number of voltages V and the following data obtained. Show that the law relating the resistance to the voltage is of the form $R = (a/V) + b$ and determine the values of a and b.

R in Ω	70	62	59	56	55
V in V	60	100	140	200	240

5 The resistance R of wires of a particular material are measured for a range of wire diameters d and the following results obtained. Show that the relationship is of the form $R = (a/d^2) + b$ and determine the values of a and b.

R in Ω	0.25	0.16	0.10	0.06	0.04
d in mm	0.80	1.00	1.25	1.60	2.00

6 The periodic time T of a simple pendulum is measured for a number of different lengths L and the following results obtained. Show that the relationship is of the form $T = aL^b$ and determine the values of a and b.

T in s	1.41	1.55	1.67	1.79	1.90
L in m	0.50	0.60	0.70	0.80	0.90

7 The volume V of a gas is measured at a number of pressures p and the following results obtained. Show that the relationship is of the form $V = ap^b$ and determine the values of a and b.

V in m^3	13.3	11.4	10.0	8.9	8.0
p in 10^5 Pa	1.2	1.4	1.6	1.8	2.0

8 When a gas is compressed adiabatically the pressure p and temperature T are measured and the following results obtained. Show that the relationship is of the form $T = ap^b$ and determine the values of a and b.

p in 10^5 Pa	1.2	1.5	1.8	2.1	2.4
T in K	526	560	589	615	639

9 The cost C per hour of operating a machine depends on the number of items n produced per hour. The following data has been obtained and is

anticipated to follow a relationship of the form $C = an^3 + b$. Show that this is the case and determine the values of a and b.

C in £	31	38	67	94	155
n	10	20	30	40	50

10 The following are suggested braking distances s for cars travelling at different speeds v. The relationship between s and v is thought to be of the form $s = av^2 + bv$. Show that this is so and determine the values of a and b.

s in m	5	15	30	50	75
v in m/s	5	10	15	20	25

11 The luminosity I of a lamp depends on the voltage V applied to it. The relationship between I and V is thought to be of the form $I = aV^b$. Use the following results to show that this is the case and determine the values of a and b.

I in candela	3.6	6.4	10.0	14.4	19.6
V in volts	60	80	100	120	140

12 The energy E of electrons emitted from a metal when light of different frequency f is incident on it gives the following results. Show that the data fits the equation $E = hf - \phi$ and determine the constants h and ϕ.

E in electron-volts	0.37	1.36	2.10	2.75	3.90
f in 10^{15} Hz	1.14	1.38	1.56	1.72	2.00

13 How many cycles will be needed for log-log graph paper used to plot the graph of $pV^{1.4} = 2000$ if V ranges from 2 to 12?

14 How many cycles will be needed for log-linear graph paper used to plot the graph of $y = 3\,e^x$ if x ranges from 0 to 3?

15 The following data gives the radius R of the orbit of the planets about the sun and the length T of their year. It is considered that the relationship between T and R might be of the form $T = aR^b$. Show that this is so and determine the values of a and b.

	Mercury	Venus	Earth	Mars	Jupiter
R in 10^6 km	57.9	108.2	149.6	227.9	778.3
T in days	88	225	365	687	4329

	Saturn	Uranus	Neptune	Pluto
R in 10^6 km	1 427	2 870	4 497	5 907
T in days	10 753	30 660	60 150	90 470

16 The following data indicates how the voltage v across a component in an electrical circuit varies with time t. It is considered that the relationship between V and t might be of the form $v = V\,e^{-bt}$. Show that this is so and determine the values of V and b.

v in volts	3.75	1.38	0.51	0.19	0.07
t in s	10	20	30	40	50

17 A hot object cools with time. The following data shows how the temperature θ of the object varies with time t. The relationship between q and t is expected to be of the form $\theta = a\,e^{-bx}$. Show that this is so and determine the values of a and b.

θ in °C	536	359	241	162	108
t in min	2	4	6	8	10

18 The rate of flow Q of water over a V-shaped notch weir was measured for different heights h of the water above the point of the V and the following data obtained. The relationship between Q and h is thought to be of the form $Q = ah^b$. Show that this is so and determine the values of a and b.

Q in m^3/s	0.13	0.26	0.46	2.12	1.07
h in m	0.3	0.4	0.5	0.6	0.7

19 The amplitude A of oscillation of a pendulum decreases with time t and gives the following data. Show that the relationship is of the form $A = a\,e^{bt}$ and determine the values of a and b.

A in mm	268	180	120	81	54
t in s	20	40	60	80	100

20 The tension T and T_0 in the two sides of a belt driving a pulley and in contact with the pulley over an angle of θ is given by the equation $T = T_0\,e^{\mu\theta}$. Determine the values of T_0 and μ for the following data:

T in N	69.5	80.8	91.1	109.1	126.7
θ in radians	1.1	1.6	2.0	2.6	3.1

14 Polar coordinates

14.1 Introduction

The Cartesian coordinate system enables the position of a point on a plane to be specified by stating its x and y coordinates, i.e. the displacement of the point relative to two perpendicular axes. There is another way of specifying the position of a point on a plane and that is to use *polar coordinates*. A reference axis is specified as emanating from some fixed point O, often referred to as the *pole*. A line is then drawn from O to the point being specified. The location of the point is then specified by the angle θ that line makes with the reference axis and the distance along that line to the point, i.e. the radius r of a circle with its centre at O and which would pass through the point concerned (figure 14.1). The polar coordinates are then (r, θ). The angle θ is a positive angle when measured anticlockwise from the reference axis and negative when measured in a clockwise direction. The angles can be specified in either degrees or radians. This chapter is a consideration of the polar coordinate method of specifying points on a plane and the relationship between polar and Cartesian coordinates.

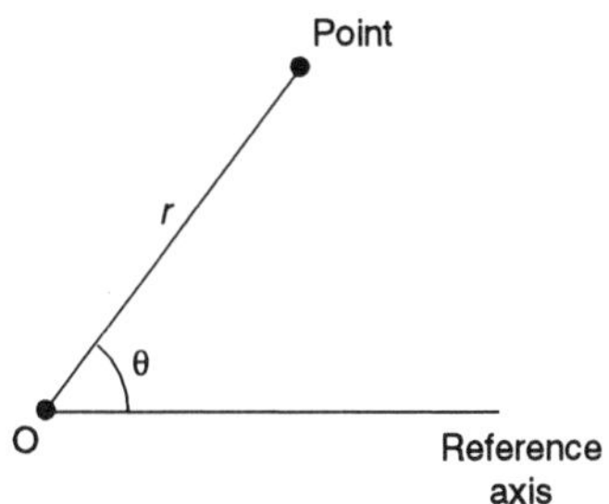

Figure 14.1 *Polar coordinates*

Such a method of specifying a point on a plane is widely used in electrical and electronic engineering for describing phasors (see chapter 11). With phasors the important items are the length of a phasor and its angle to a reference axis. Thus the polar coordinate method of specifying it has great advantages. For example, a current phasor of magnitude 2 A with a phase angle of 30° could be specified in polar coordinates as being (2, 30°) A, this more usually being written as $2\angle 30°$ A.

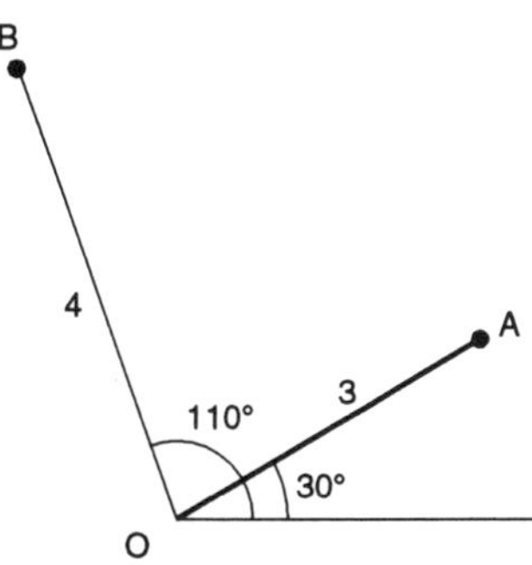

Figure 14.2

Example

Specify the polar coordinates of the points shown in figure 14.2.

Point A has a radius of 3 and an angle of 30° and so is located at the point (3, 30°). Point B has a radius of 4 and an angle of 110° and so is located at the point (4, 110°).

Revision

1 The following points are in polar coordinates. Indicate their positions on a polar graph.

(a) (2, 60°), (b) $(3, \pi/4)$, (c) (2, −45°), (d) $(2.5, 4\pi/3)$, (e) $(2, -\pi/2)$

14.1.1 Conversion from Cartesian to polar coordinates

Consider a point P (figure 14.3) which has Cartesian coordinates (x, y). We will take the polar reference axis to coincide with the x-axis. Then radius r

can be calculated by the use of the Pythagoras theorem. Thus, we have $r^2 = x^2 + y^2$ and so

$$r = \sqrt{x^2 + y^2}$$

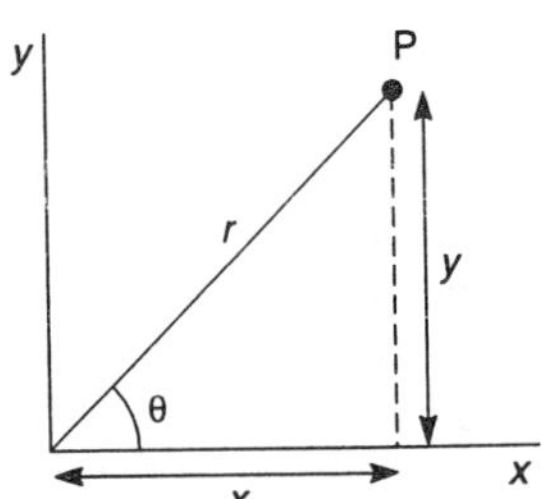

Figure 14.3 *Coordinates*

The angle θ is given by

$$\tan\theta = \frac{y}{x}$$

We can thus write

$$\theta = \tan^{-1}\frac{y}{x}$$

This reads as 'the angle whose tangent is y/x'.

Example

What are the polar coordinates of a point with Cartesian coordinates (3, 2)?

The radius r is given by

$$r = \sqrt{3^2 + 2^2} = 3.6$$

and the angle θ by

$$\theta = \tan^{-1}\frac{2}{3} = 33.7° \text{ or } 0.59 \text{ rad}$$

Thus the polar coordinates are (3.6, 33.7°) or (3.6, 0.59).

Example

What are the polar coordinates of a point with the cartesian coordinates of (−2, 3)?

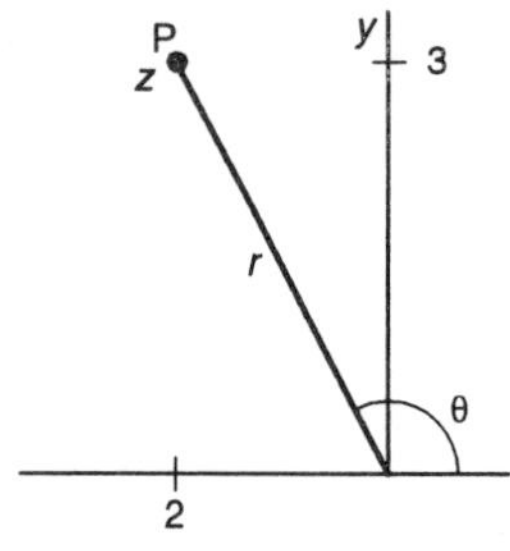

Figure 14.4

Figure 14.4 shows the position of the point. The radius r is given by

$$r = \sqrt{(-2)^2 + 3^2} = 3.6$$

As figure 14.4 indicates, θ is greater than 90° but less than 180°. Thus

$$\theta = \tan^{-1}\frac{3}{-2} = 123.7° \text{ or } 2.16 \text{ rad.}$$

The polar coordinates are thus (3.6, 123.7°) or (3.6, 2.16).

Revision

2 Determine the polar coordinates of the points with the following Cartesian coordinates:

(a) (4, 2), (b) (2, −2), (c) (−2, 4), (d) (−2, −2), (e) (2, 0), (f) (0, 2), (g) (−4, 0), (h) (0, −4)

14.1.2 Conversion from polar to Cartesian coordinates

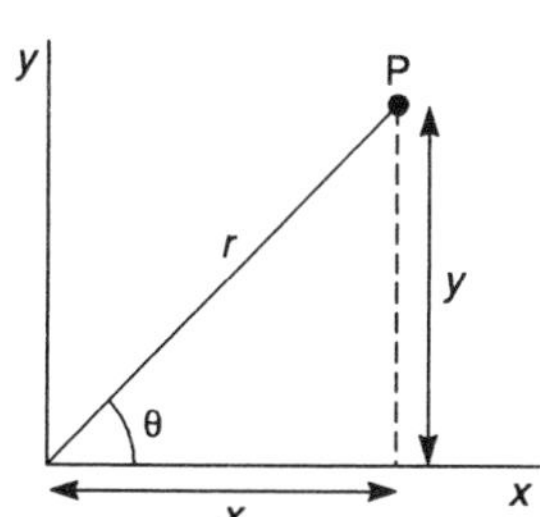

Figure 14.5 *Coordinates*

Consider a point P (figure 14.5) with the polar coordinates (r, θ). We have

$$\cos\theta = \frac{x}{r}$$

and so $x = r \cos \theta$. We can also write

$$\sin\theta = \frac{y}{r}$$

and so $y = r \sin \theta$.

Example

Determine the Cartesian coordinates of the point with the polar co-ordinates (2, 30°).

For the x coordinate we have $x = r \cos \theta = 2 \cos 30° = 1.7$ and for the y coordinate $y = r \sin \theta = 2.0 \sin 30° = 1.0$. Hence the coordinates are (1.7, 1.0).

Revision

3 Determine the Cartesian coordinates of the points with the following polar coordinates:

(a) (4, 45°), (b) (3, 60°), (c) (2, 150°), (d) (3, 290°), (e) (2, π/4), (f) (4, π/6), (g) (3, π).

14.2 Polar graphs

To draw a polar graph we need to know the values of the radius r and angle θ for points that are to fall on the graph line.

Consider the graph of the function which has its coordinates described by the equation

$$r = \text{a constant}$$

r is the same for all values of θ. Such an equation can only describe a circle with its centre at the pole (figure 14.6). Only then will all the points on its graph have the same radius.

radius r

Figure 14.6 r = *a constant*

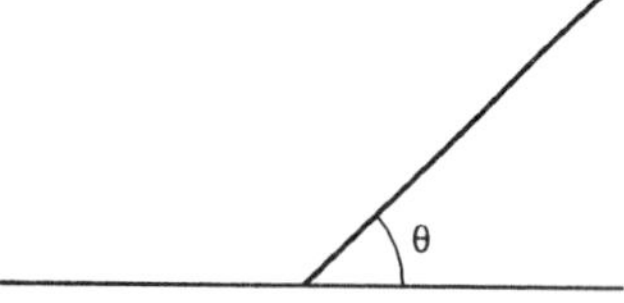

Figure 14.7 $\theta = a$ *constant*

Now consider the graph of the function which has its coordinates described by the equation

θ = a constant

θ is the same for all values of r. This can only describe a line emanating from the pole (figure 14.7) so that all the points on its graph are at the same angle.

Consider the graph of the function for which $r = \sin\theta$. Values of r and θ which fit this equation are:

θ	0°	30°	60°	90°	120°	150°	180°
r	0	0.50	0.87	1.00	0.87	0.50	0

θ	210°	240°	270°	300°	330°	360°
r	−0.50	−0.87	−1.00	−0.87	−0.50	0

Figure 14.8 shows the resulting graph. Plotting the values from 0° to 180° results in a curve which gives a complete circle. When plotting the points from 210° to 360° we are faced with negative values for the radius. A negative value magnitude $|r|$ plots as a distance $|r|$ in the opposite direction out from the pole as indicated by the angle. Thus we plot for the angle 210° a magnitude of 0.50 in the 30° direction. The values of r from 210° to 360° thus end up repeating the circle given by the 0° to 180° values.

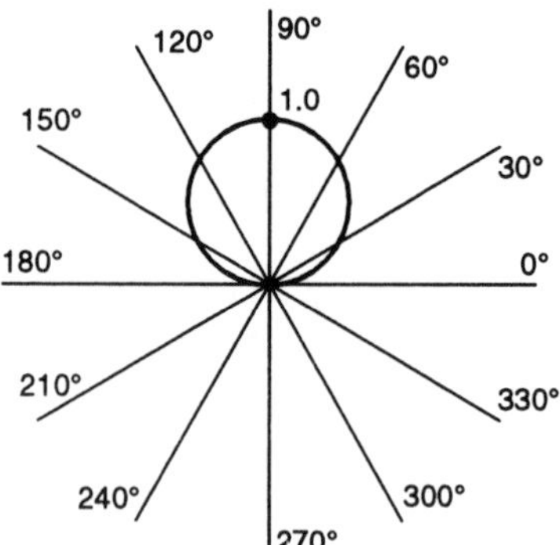

Figure 14.8 $r = \sin\theta$

All functions of the form $r = a \sin\theta$ give circles in the same way. The only effect of the constant a is to change the diameter of the circle, the circle being centred on the y-axis at $y = a/2$ and with a diameter of a. In a similar way, all functions of the form $r = a\cos\theta$ give circles centred on the x-axis at $x = a/2$ and with a diameter a (figure 14.9).

Consider the graph of the function for which $r = 1 + \cos\theta$. Values of r and θ which fit this equation are:

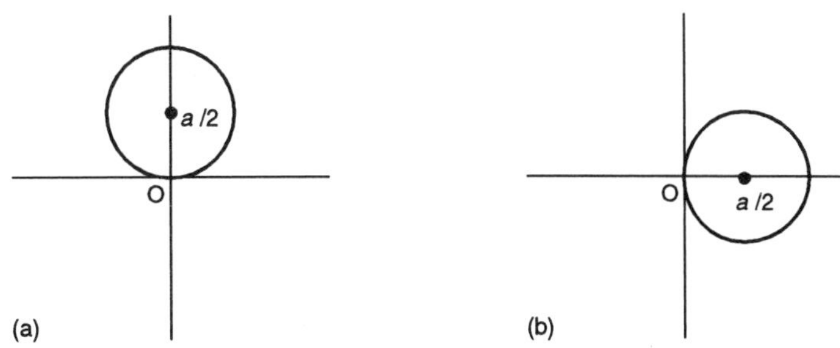

Figure 14.9 *(a)* $r = a \sin \theta$, *(b)* $r = a \cos \theta$

θ	0°	30°	60°	90°	120°	150°	180°
r	2.0	1.9	1.5	1.0	0.5	0.1	0

θ	210°	240°	270°	300°	330°	360°
r	0.1	0.5	1.0	1.5	1.9	2.0

Figure 14.10 shows the resulting graph. It is called a *cardioid* (i.e. a heart shape).

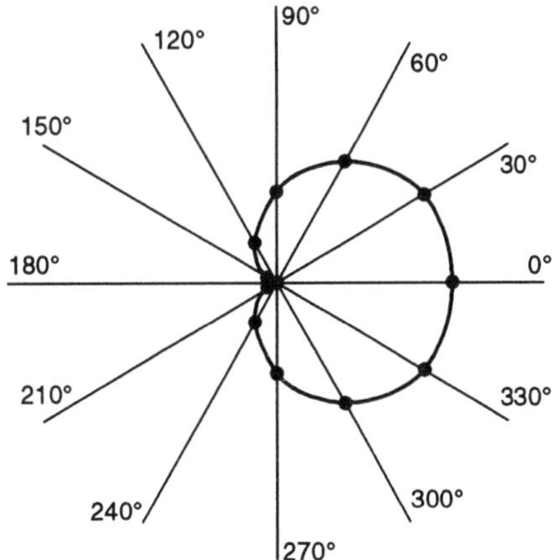

Figure 14.10 $r = 1 + \cos \theta$

Consider the graph of the function for which $r = \theta$. Values of r and θ which fit this equation are:

θ	0	$\pi/6$	$\pi/3$	$\pi/2$	$2\pi/3$	$5\pi/6$	π
r	0	$\pi/6$	$\pi/3$	$\pi/2$	$2\pi/3$	$5\pi/6$	π

θ	$7\pi/6$	$4\pi/3$	$3\pi/2$	$5\pi/3$	$11\pi/6$	2π
r	$7\pi/6$	$4\pi/3$	$3\pi/2$	$5\pi/3$	$11\pi/6$	2π

θ	$13\pi/6$	$7\pi/3$	$5\pi/2$	$8\pi/3$	$15\pi/6$	3π
r	$13\pi/6$	$7\pi/3$	$5\pi/2$	$8\pi/3$	$15\pi/6$	3π

We can continue taking further values. The result is the graph shown in figure 14.11, a spiral.

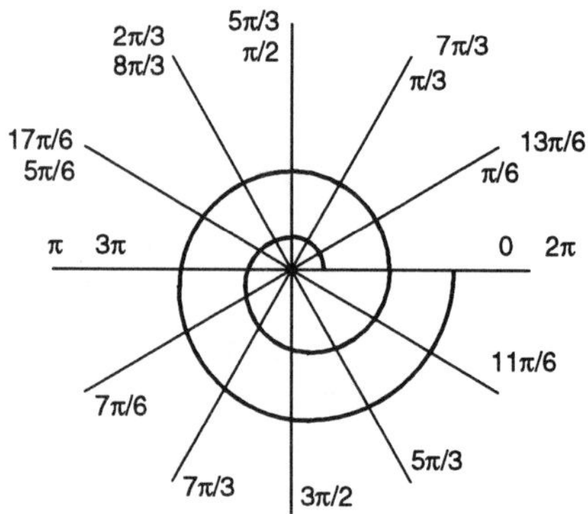

Figure 14.11 $r = \theta$

Revision

4 Plot graphs of the following equations:

(a) $r = 3$, (b) $\theta = \pi/6$, (c) $r = 2(1 + \cos\theta)$, (d) $r = 2(1 - \cos\theta)$

Problems

1 Determine the polar coordinates of the points with the following Cartesian coordinates:

(a) (0, 3), (b) (–2, 0), (c) (–2, 2), (d) (3, 4), (e) (–3, 4), (f) (4, –2), (g) (3, 0)

2 Determine the Cartesian coordinates of the points with the following polar coordinates:

(a) (2, 50°), (b) (3, 160°), (c) (3, 260°), (d) (4, 340°), (e) (2, $\pi/3$), (f) (1, $5\pi/3$), (g) (3, π), (h) (4, $\pi/2$)

3 Plot polar graphs for the following equations:

(a) $r = 4$, (b) $r = \pi/4$, (c) $r = 2 + \cos\theta$, (d) $r = 2\sin 2\theta$

4 A cam has been designed with the following values of radius r at the different angles θ. Sketch the shape of the cam.

θ	0	$\pi/6$	$\pi/3$	$\pi/2$	$2\pi/3$	$5\pi/6$	π
r in mm	10	14	18	22	26	30	34

θ	$7\pi/6$	$4\pi/3$	$3\pi/2$	$5\pi/3$	$11\pi/6$	2π
r in mm	30	26	22	18	14	10

5 Demonstrate by plotting examples of polar graphs that the polar graph of $r = \sin nq$ or $r = \cos nq$ has n loops when n is an odd positive integer and $2n$ loops when n is an even positive integer.

15 Straight-line motion

15.1 Introduction

This chapter illustrates how graphical techniques involving the gradients of graphs and the areas under graphs can be used to solve problems involving motion in a straight line.

15.1.1 Terms

For definitions of terms associated with straight-line motion, see section 6.1.1. Velocity is the rate at which distance measured in a straight line from some point changes with time, hence the term displacement is often used for such distances. Speed is the rate at which distance is covered with time when we are not concerned about the whether the distance is measured in a straight line or not. For simplicity, in this chapter the discussion is restricted to straight-line motion and so the term velocity is always used.

15.2 Distance-time graphs

If the distance moved by an object in a straight line, from some reference point on the line, is measured for different times then a distance–time graph can be plotted. Since velocity is the rate at which distance along a straight line changes with time then for the distance-time graph shown in figure 15.1, where the distance changes from s_1 to s_2 when the time changes from t_1 to t_2, the velocity is

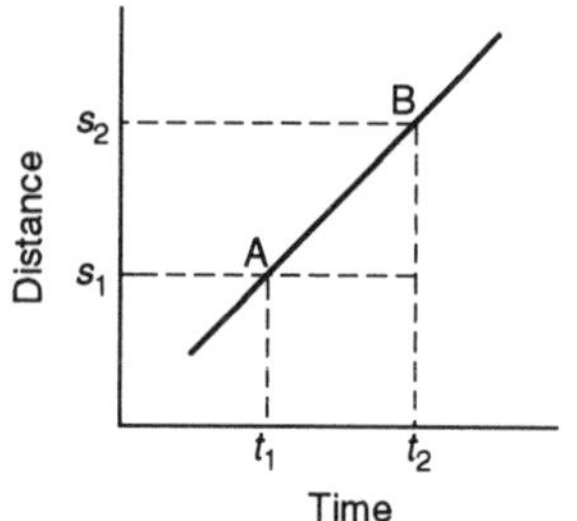

Figure 15.1 *Distance–time graph (straight line)*

$$\text{velocity} = \frac{s_2 - s_1}{t_2 - t_1}$$

But this is just the gradient of the graph. Thus

velocity = gradient of distance–time graph

In figure 15.1 the graph is a straight-line graph. The distance changes by equal amounts in equal intervals of time. There is thus a uniform velocity.

Consider the situation when the graph is not a straight line, as in figure 15.2. Equal distances are not now covered in equal intervals of time and the velocity is no longer uniform. $(s_2 - s_1)$ is the distance travelled in a time of $(t_2 - t_1)$ and thus the average velocity over that time is

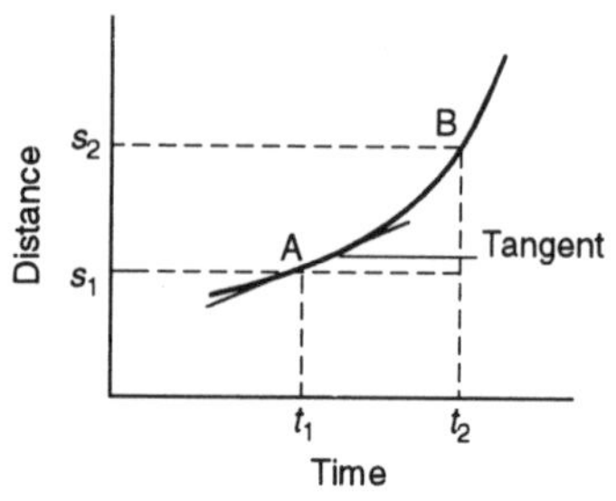

Figure 15.2 *Distance–time graph (curve)*

$$\text{average velocity} = \frac{s_2 - s_1}{t_2 - t_1}$$

The smaller we make the times between A and B then the more the average is taken over a smaller time interval and so the more it approximates to the instantaneous velocity. An infinitesimal small time interval means we have the tangent to the curve. Thus if we want the velocity at an instant of time, say A, then we have to determine the gradient of the tangent to the graph at A. Thus

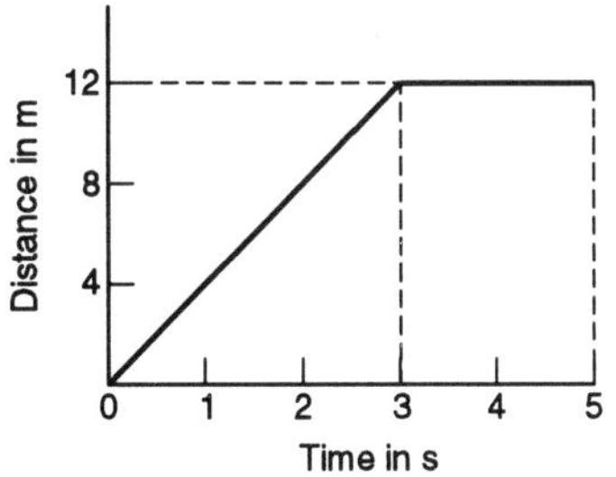

Figure 15.3

instantaneous velocity = gradient of tangent to the distance–time graph at that instant

Example

A car has a constant velocity of 4 m/s for the time from 0 to 3 s and then zero velocity from 3 s to 5 s. Sketch the distance–time graph.

From 0 to 3 s the gradient of the distance-time graph is 4 m/s. From 3 s to 5 s the velocity is zero and so the gradient is zero. The graph is thus as shown in figure 15.3.

Example

Figure 15.4 shows an example of a distance–time graph. Describe how the velocity is changing in the motion from A to F.

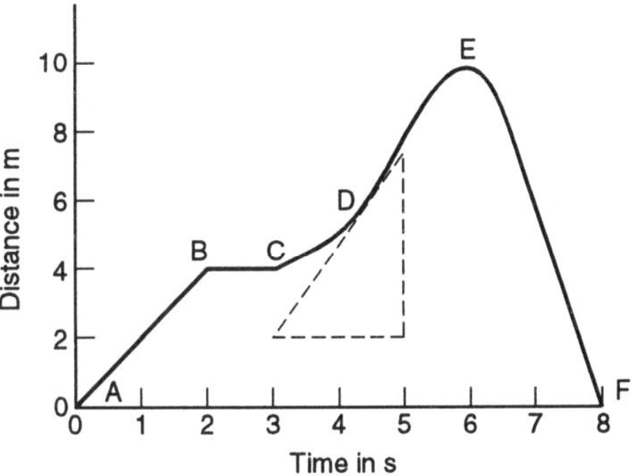

Figure 15.4

From the start position at A to B, the distance increases at a constant rate with time. The gradient is constant over that time and so the velocity is constant and equal to (4 – 0)/(2 – 0) = 2 m/s.

From B to C there is no change in the distance from the reference point and so the velocity is zero, i.e. the object has stopped moving. The gradient of the line between B and C is zero.

From C to E the distance increases with time but in a non-uniform manner. The gradient changes with time. Thus the velocity is not constant during that time.

The velocity at an instant of time is the rate at which the distance is changing at that time and so is the gradient of the graph at that time. Thus to determine the velocity at point D we draw a tangent to the graph curve at that instant. So at point D my estimate of the velocity is about (7.6 – 2.4)/(5 – 3) = 2.6 m/s.

At point E the distance–time graph shows a maximum. The gradient changes from being positive prior to E to negative after E. At point E the gradient is momentarily zero. Thus the velocity changes from being positive prior to E to zero at E and then negative after E. At E the velocity is zero.

From E to F the gradient is negative and so the velocity is negative. A negative velocity means that the object is going in the opposite direction and so is moving back to its starting point, i.e. the distance is becoming smaller rather than increasing. In this case, the object is back at its starting point after a time of 8 s.

Revision

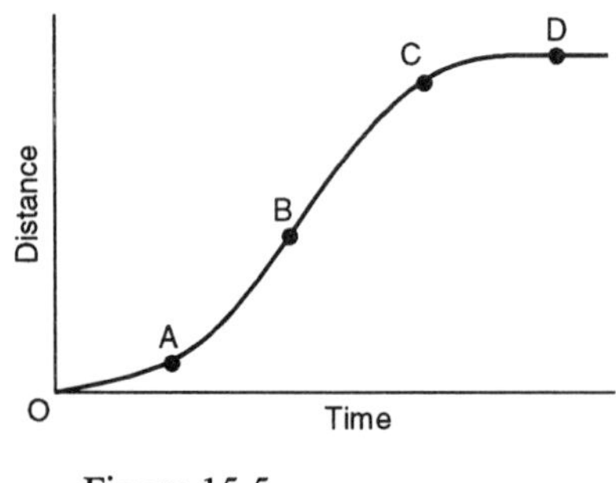

Figure 15.5

For the distance–time graph shown in figure 15.5, state whether at points A, B, C and D the object responsible for it is at rest, moving with a constant velocity or moving with a changing velocity.

For the following distance–time data, plot the distance–time graph and hence estimate the velocities at times of (a) 2 s, (b) 4 s, (c) 6 s and (d) the average velocity over the time 0 to 6 s.

distance in m	0	2	4	6	6	6	7	9
time in s	0	1	2	3	4	5	6	7

3 A stone falls off the edge of a cliff. The following data shows how the distance the stone has fallen changes with time. Plot the distance–time graph and hence determine the velocities of the stone at (a) 2 s, (b) 3 s, (c) 4 s.

distance in m	0	4.9	19.6	44.1	78.4	122.5
time in s	0	1	2	3	4	5

4 A simple pendulum oscillates back and forth with the pendulum bob having a displacement x from the vertical position which varies with time t and is described by the equation $x = \sin 20t$. Plot the distance–time graph for the pendulum bob for times between $t = 0$ and $t = 0.5$ s. Hence determine the velocities of the bob at times t of (a) 0.05 s, (b) 0.1 s, (c) 0.2 s, (d) 0.3 s.

5 A tool on a machine has a velocity of 5 mm/s for the first 3 s and then a velocity in the opposite direction of 2 mm/s for 5 s. Sketch the distance–time graph.

15.3 Velocity–time graphs

If the velocity of an object is measured at different times then a velocity–time graph can be drawn. Acceleration is the rate at which the velocity changes. Thus, for the graph shown in figure 15.6, the velocity changes from v_1 to v_2 when the time changes from t_1 to t_2. Thus the acceleration over that time interval is

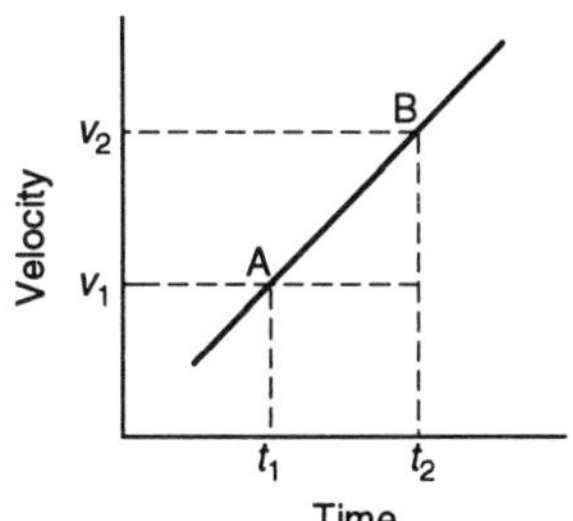

Figure 15.6 *Velocity–time graph (straight line)*

$$\text{acceleration} = \frac{v_2 - v_1}{t_2 - t_1}$$

But this is the gradient of the graph. Thus

acceleration = gradient of the velocity–time graph

In figure 15.6 we have a straight line between the points A and B. Thus, since the gradient is the same for all points between A and B we have a uniform acceleration between those points. Consider the situation when the graph is not straight line, as in figure 15.7. The velocity does not now change by equal amounts in equal intervals of time and the acceleration is no longer uniform. $(v_2 - v_1)$ is the change in velocity in a time of $(t_2 - t_1)$ and thus $(v_2 - v_1)/(t_2 - t_1)$ represents the average acceleration over that time. The smaller we make the times between A and B then the more the average is taken over a smaller time interval and more closely approximates to the instantaneous acceleration. An infinitesimal small time interval means we have the tangent to the curve. Thus if we want the acceleration at an instant of time then we have to determine the gradient of the tangent to the graph at that time, i.e.

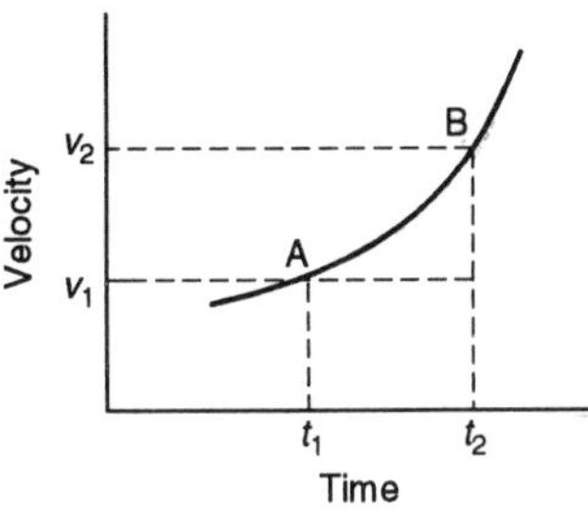

Figure 15.7 *Velocity–time graph (curve)*

instantaneous acceleration = gradient of tangent to the velocity–time graph at that instant

The distance travelled by an object in a particular time interval is given by

distance = average velocity over the time × the time interval concerned

Thus if the velocity changes from v_1 at time t_1 to v_2 at time t_2, as in figure 15.8, then the distance travelled between t_1 and t_2 is represented by the area marked on the graph. But this area is equal to the area under the graph line between t_1 and t_2. Thus

distance travelled between t_1 and t_2 = area under the graph between these times

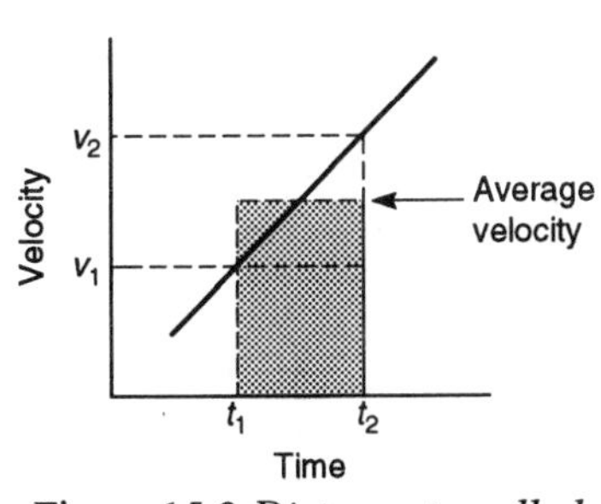

Figure 15.8 *Distance travelled*

When the velocity is zero or there is uniform acceleration and so a straight line velocity–time graph, the areas can be easily calculated. However, if the acceleration is not uniform then the techniques outlined in section 12.5 can be used, i.e. counting squares or the mid-ordinate rule.

Example

For the velocity–time graph shown in figure 15.9, determine the acceleration at points A, B and C and estimate the distance covered in the 8 s.

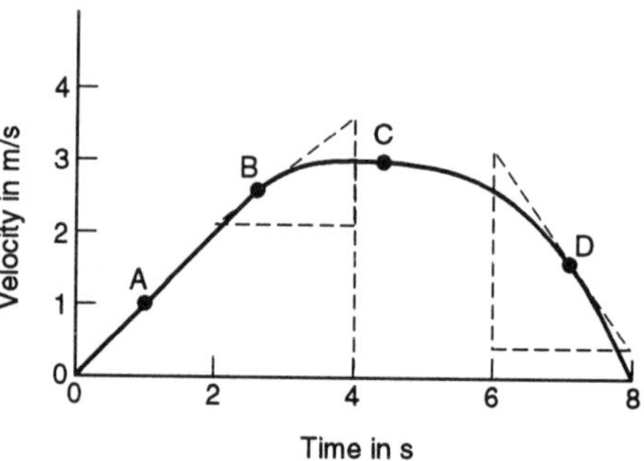

Figure 15.9

Initially the graph is a straight line of constant gradient. The gradient, and hence the acceleration, at A is 2/2 = 1 m/s^2. At B the graph is no longer a straight line. The tangent to the curve at B has a gradient, and hence an acceleration, of about (3.7 – 2.2)/(4 – 2) = 0.75 m/s^2. At C the velocity is not changing with time. Thus there is a uniform velocity with zero acceleration. At D the velocity is decreasing in a non-uniform manner with time. The tangent to the curve at that point has a gradient of about (0.4 – 3.4)/(8 – 6) = –1.5 m/s^2.

The distance covered in the 8 s is the area under the graph. Using the mid-ordinate rule with the time interval divided into four strips gives the distance as $1 \times 2 + 2.8 \times 2 + 3.0 \times 2 + 1.8 \times 2 = 17.2$ m

Revision

6 A train, starting from rest, accelerates at a uniform rate for 20 s to reach a velocity 18 m/s. It then travels with this velocity for 160 s before braking results in a uniform retardation to bring the train to rest in 10 s. Sketch the velocity–time graph and hence determine (a) the acceleration during the initial 0 to 20 s time interval, (b) the retardation during the final 10 s, (c) the total distance travelled in the motion.

7 The velocity of a car varies with time and measurements give the following data. Determine (a) the acceleration at 5 s and the (b) the distance travelled in the 10 s.

velocity in m/s	0	3.0	5.5	7.5	9.0	10
time in s	0	2	4	6	8	10

8 The velocity v of an object varies with time t and is described by the equation $v = 2t^2 + 3$, with v being in m/s and t in seconds. Sketch the velocity–time graph from $t = 0$ to $t = 5$ s and hence determine the acceleration at $t = 3$ s and the total distance covered in the 5 s.

9 A stone is dropped down a well and takes 4 s to reach the water. Sketch the velocity–time graph for the object, taking the acceleration due to gravity to be 10 m/s^2.

Problems

1 An object moving in a straight line gives the following distance–time data. Plot the distance–time graph and hence determine the velocities at times of (a) 2 s, (b) 3 s, (c) 4 s.

distance in mm	0	10	36	78	136	210
time in s	0	1	2	3	4	5

2 An object moving in a straight line gives the following distance–time data. Plot the distance–time graph and hence determine the velocities at times of (a) 5 s, (b) 8 s, (c) 11 s.

distance in m	0	1	15	31	49	75	110
time in s	0	1	4	6	8	10	12

3 The distance–time data for an object moving along a straight line is described by the equation $s = 4t^2 + 2$, the distance s being in millimetres and the time t in seconds. Plot the distance–time graph and hence determine the velocities when (a) $t = 2$ s, (b) $t = 3$ s, (c) $t = 4$ s.

4 The distance–time data for an object moving along a straight line is described by the equation $s = t^2$, the distance s being in millimetres and the time t in seconds. Plot the distance–time graph and hence determine the velocities when (a) $t = 2$ s, (b) $t = 3$ s, (c) $t = 4$ s.

5 The distance–time data for an object moving along a straight line is described by the equation $s = 4t^2 + 2t$, the distance s being in millimetres and the time t in seconds. Plot the distance–time graph and hence determine the velocities when (a) $t = 2$ s, (b) $t = 3$ s, (c) $t = 4$ s.

6 An object starts from rest and maintains a uniform acceleration of 0.5 m/s^2 for 4 s. It then moves with a uniform velocity for 3 s before a uniform retardation which brings it to rest in 2 s. Plot the velocity–time graph and hence determine the retardation during the last 2 s and the total distance covered.

7 A stone is thrown vertically upwards. The following are the velocities at different times. Plot a velocity–time graph and hence determine (a) the acceleration at the time $t = 1$ s, (b) the acceleration at $t = 2.5$ s, (c) the distance travelled to the maximum height where the velocity is zero.

velocity in m/s	15	10	5	0	−5	−10	−15
time in s	0	0.5	1.0	1.5	2.0	2.5	3.0

8 A car starts from rest and travels along a straight road. Its velocity changes at a uniform rate from 0 to 50 km/h in 20 s. It then continues for 50 s with uniform velocity before reducing speed at a constant rate to zero in 10 s. Sketch the velocity–time graph and hence determine (a) the initial acceleration, (b) the final retardation and (c) the total distance travelled.

9 The velocity of an object varies with time. The following are values of the velocities at a number of times. Plot a velocity–time graph and hence estimate the acceleration at a time of 2 s and the total distance travelled in the 5 s.

velocity in m/s	0	1.2	2.4	3.6	3.6	3.6
time in s	0	1	2	3	4	5

10 An object has a maximum acceleration of 4 m/s^2 and a maximum retardation of 8 m/s^2. Determine the shortest time the object will take to move through a distance of 4000 m. Hint: the velocity–time graph will be triangular of area 4000 m.

11 The velocity of an object changes with time. The following are values of the velocities at a number of times. Plot the velocity–time graph and hence determine the acceleration at a time of 4 s and the total distance travelled during the 20 s.

velocity in m/s	0	8	10	16	20	20
time in s	0	4	8	12	16	20

12 The velocity of an object changes with time. The following are values of the velocities at a number of times. Plot the velocity–time graph and hence determine the acceleration at a time of 2 s and the total distance travelled between 1 s and 4 s.

velocity in m/s	25	24	21	16	9	0
time in s	0	1	2	3	4	5

13 State whether the distance–time graphs and the velocity–time graphs will be straight line with a non-zero gradient or zero gradient or curved in the following cases:

(a) The velocity is constant and not zero.
(b) The acceleration is constant and not zero.
(c) The distance covered in each second of the motion is the same.
(d) The distance covered in each successive second doubles.

16 Alternating waveforms

16.1 Introduction

The term *alternating* voltage or current is used when the polarity or direction of flow of the voltage or current alternates, continually changing with time. Alternating voltages and currents can have many different waveforms. Figure 16.1 shows some examples. (a) is a rectangular waveform, (b) a sinusoidal waveform, (c) a triangular waveform. The rectangular waveform in the figure starts with a positive current which then abruptly switches to a negative current, which then abruptly switches to a positive current, which then ... and so on. The sinusoidal waveform in the figure has a current which oscillates from positive values to negative values to positive values, to ... and so on. The triangular waveform shows a similar form of behaviour. This chapter is a consideration of the mean value and the root-mean-square values of such alternating waveforms.

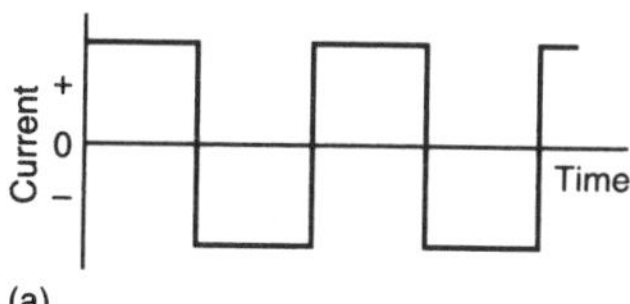

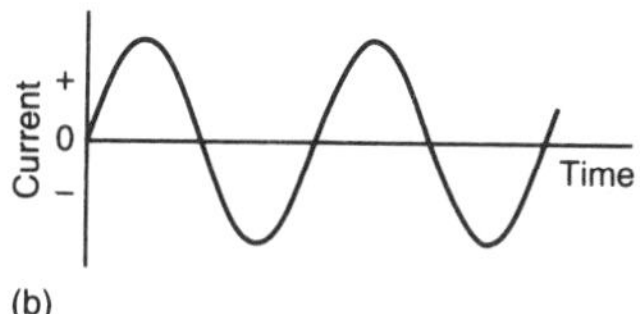

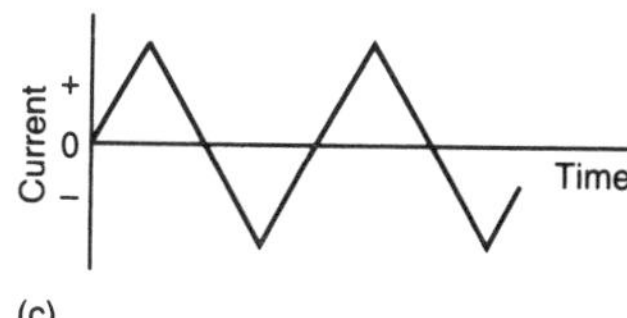

Figure 16.1 *Alternating waveforms*

16.2 Average value

The average value for an alternating current, or voltage, waveform is the value of the horizontal graph line, i.e. the constant current or voltage line, which can be drawn and for which the area under it is the same as the area under the waveform. Figure 16.2 illustrates this. The area under the average line is that of a rectangle and is the average value × the range of time values over which the average is considered. Thus

Figure 16.2 *Average value*

average value × the range of time values over which the average is considered = area under the waveform between these time values

and so

$$\text{average value} = \frac{\text{area under waveform}}{\text{range considered}}$$

We can use the *mid-ordinate rule* to determine the area under a waveform. Thus the waveform is divided into a number of equally spaced time interval strips and then a vertical line, an ordinate, is drawn at the middle of each strip and the value of the current at that instant taken. Then, since the area of a strip is its mid-ordinate value × width of the strip and the widths are the same for each strip,

area under waveform = sum of mid-ordinate values × width of strips

But the range considered is just the number of strips × width of the strips. Hence we can write

$$\text{average value} = \frac{\text{sum of mid-ordinate values} \times \text{width of strips}}{\text{number of mid-ordinates} \times \text{width of strips}}$$

and so

$$\text{average value} = \frac{\text{sum of mid-ordinate values}}{\text{number of mid-ordinates}}$$

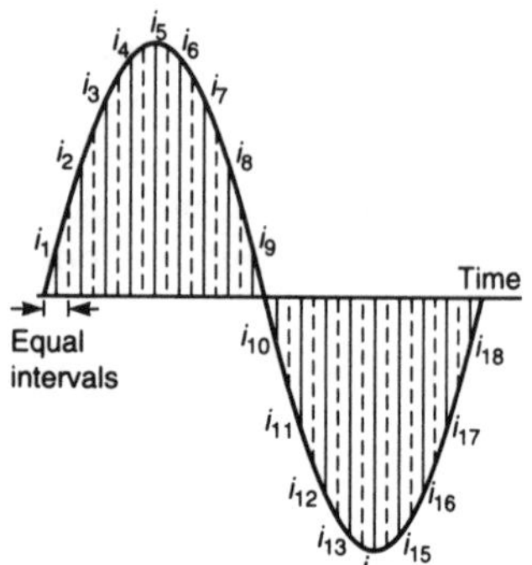

Figure 16.3 *Obtaining the average*

Figure 16.3 shows a sinusoidal waveform with the instantaneous values of the current being shown for the mid-ordinates of a number of equally spaced times. The average value over one cycle will be the sum of all these mid-ordinate values divided by the number of values taken. But, because the waveform has a negative half cycle which is just the mirror image of the positive half cycle, then the average over a full cycle must be zero. For every positive value there will be a corresponding negative value.

Over one half cycle the average value of the sinusoidal waveform is

$$I_{av} = \frac{i_1 + i_2 + i_3 + \ldots + i_9}{9}$$

When the maximum value of the current is 1, readings taken from the graph for the currents are:

i_1	i_2	i_3	i_4	i_5	i_6	i_7	i_8	i_9
0.17	0.50	0.77	0.94	1.00	0.94	0.77	0.50	0.17

The average of these values is

$$\text{average} = \frac{0.17 + 0.50 + 0.77 + 0.94 + 1.00 + 0.94 + 0.77 + 0.50 + 0.17}{9}$$

The average is thus 0.64. The accuracy of the average value is improved by taking more values for the average. The average value for a sinusoidal waveform of maximum value I_m over half a cycle is then found to be 0.6371. This is the value of $2/\pi$ and so for a current of maximum value I_m we have

$$I_{av} = \frac{2I_m}{\pi} = 0.637I_m$$

In a similar way we could have derived the average value for a sinusoidal voltage as

$$V_{av} = \frac{2V_m}{\pi} = 0.637V_m$$

Example

Determine the average value of the half cycle of the current waveform shown in figure 16.4.

Dividing the half cycle into segments of width 1 s (indicated in the figure by the broken lines), then the mid-ordinates are indicated by the solid lines in the figure. The values of these are

Mid-ordinate in ms	0.5	1.5	2.5	3.5	4.5
Current in A	1.0	3.0	4.0	4.0	2.0

Figure 16.4

The average value is thus

$$\text{average} = \frac{1.0 + 3.0 + 4.0 + 4.0 + 2.0}{5} = 2.8 \text{ A}$$

Revision

1 Determine the average values of the segments of waveforms shown in figure 16.5.

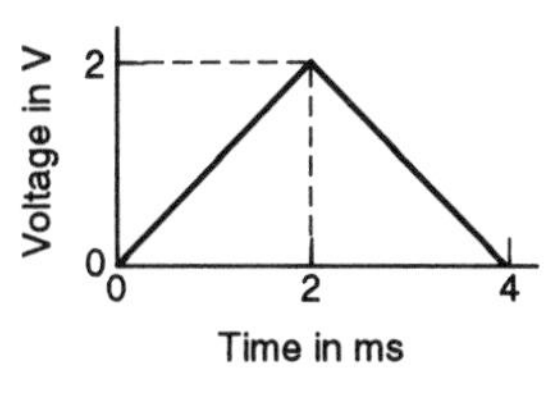

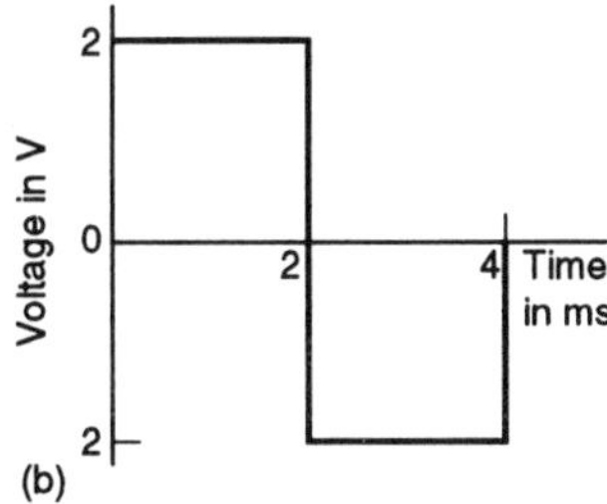

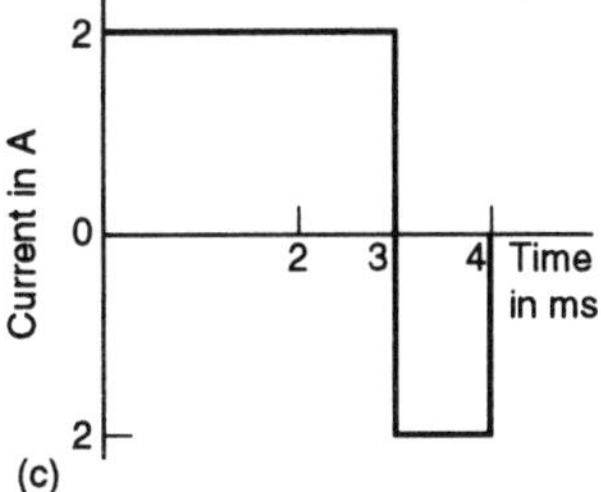

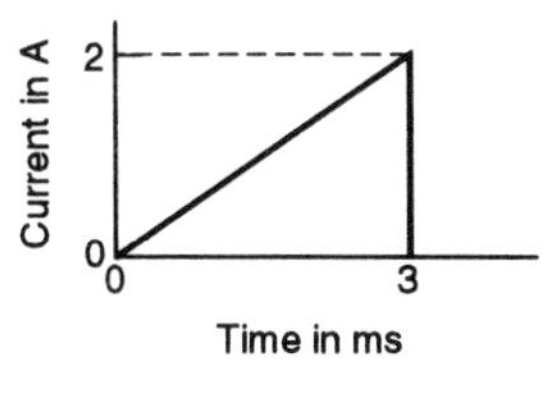

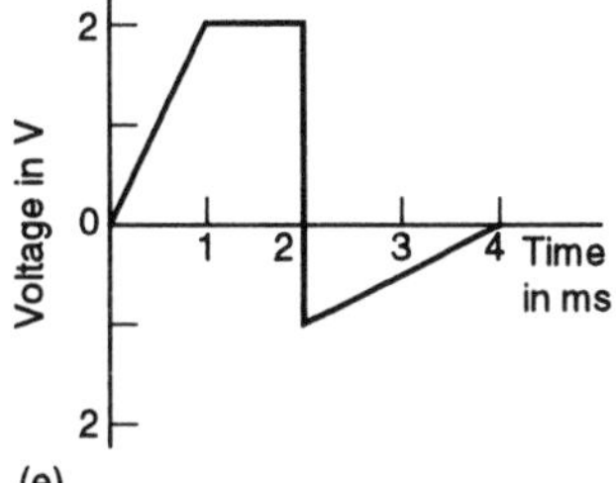

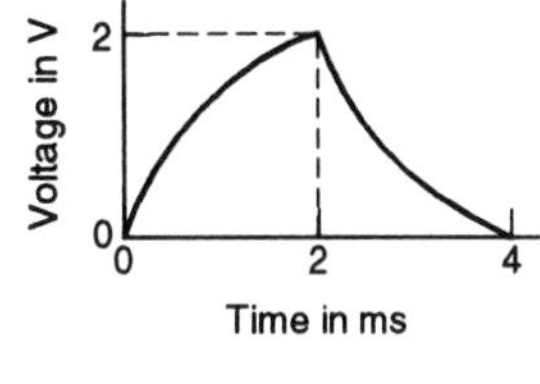

(d) (e) (f)

Figure 16.5

2 Show that for a triangular waveform, of the form shown in figure 16.5(a), with a maximum value of V_m that over half a cycle the average value is $0.5V_m$.

16.3 Root-mean-square value

Since we are frequently concerned with the power developed by a current passing through a circuit component, a useful measure of an alternating current is in terms of the direct current that would give the same power dissipation in a resistor. For an alternating current, the power at an instant of time is i^2R, where i is the current at that instant and R is the resistance. Thus to obtain the power developed by an alternating current over a cycle, we need to find the average power developed over that time. In terms of mid-ordinates we add together all the values of power given at each mid-ordinate of time in the cycle and divide by the number of mid-ordinates considered. Because we are squaring the current values, negative currents give positive values of power. Thus the powers developed in each half cycle add together. For the sinusoidal waveform shown in figure 16.3, the average power P_{av} developed in one complete cycle is the sum of the mid-ordinate powers divided by the number of mid-ordinates used and so

$$P_{av} = \frac{i_1^2R + i_2^2R + i_3^2R + \ldots + i_{18}^2R}{18} = \frac{i_1^2 + i_2^2 + i_3^2 + \ldots + i_{18}^2}{18}R$$

For a direct current I to give the same power as the alternating current, we must have $I^2R = P_{av}$. Thus the square of the equivalent direct current I^2 is the average (mean) value of the squares of the instantaneous currents during the cycle. This equivalent current is known as the *root-mean-square current* I_{rms}.

I_{rms} = √(mean of the sum of the values of the squares of the alternating current)

In the above we considered the power due to a current through a resistor. We could have considered the power in terms of the voltage developed across the resistor. The power at an instant when the voltage is v is v^2/R. The equivalent direct voltage V to give the same power is thus when V^2/R is equal to the average power developed during the cycle. Thus, as before, we obtain

V_{rms} = √(mean of the sum of the values of the squares of the alternating voltage)

Consider the use of the of the mid-ordinate rule to obtain values of the current, or voltage, for the sinusoidal waveform given in figure 16.3. When the maximum value is 1, we have for each of the mid-ordinates:

i_1	i_2	i_3	i_4	i_5	i_6	i_7	i_8	i_9
0.17	0.50	0.77	0.94	1.00	0.94	0.77	0.50	0.17

i_{10}	i_{11}	i_{12}	i_{13}	i_{14}	i_{15}	i_{16}	i_{17}	i_{18}
−0.17	−0.50	−0.77	−0.94	−1.00	−0.94	−0.77	−0.50	−0.17

The square of these current values are:

i_1^2	i_2^2	i_3^2	i_4^2	i_5^2	i_6^2	i_7^2	i_8^2	i_9^2
0.03	0.25	0.59	0.88	1.00	0.88	0.59	0.25	0.03

i_{10}^2	i_{11}^2	i_{12}^2	i_{13}^2	i_{14}^2	i_{15}^2	i_{16}^2	i_{17}^2	i_{18}^2
0.03	0.25	0.59	0.88	1.00	0.88	0.59	0.25	0.03

The sum of these values is 9.00 and thus the mean value is 9.00/18 = 0.50. The root-mean-square current is therefore $\sqrt{0.50} = 1/\sqrt{2}$. Thus, with a maximum current of I_m,

$$I_{rms} = \frac{I_m}{\sqrt{2}}$$

Similarly

$$V_{rms} = \frac{V_m}{\sqrt{2}}$$

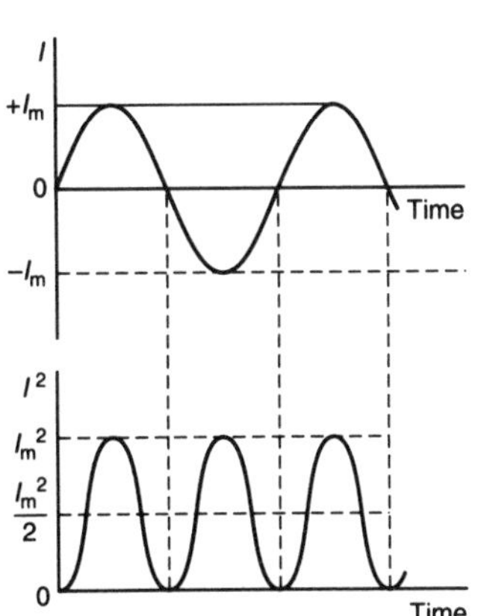

Figure 16.6 *The root-mean-square current*

where V_m is the maximum voltage.

We could have arrived at the above results for the sinusoidal waveform by considering the form of the graph produced by plotting the square of the current values, or the voltage values. Figure 16.6 shows the graph. The squares of the positive and negative currents are all positive quantities and so the resulting graph oscillates between a maximum value of I_m^2 and 0. The mean value of the i^2 graph is $I_m^2/2$, the i^2 graph being symmetrical about this value. The mean power is thus $RI_m^2/2$ and so the root-mean-square current is $I_m/\sqrt{2}$. Thus, for a sinusoidal waveform, the root-mean-square current is the maximum current divided by $\sqrt{2}$.

For alternating currents, or voltages, with other waveforms the relationships between the root-mean-square values and the maximum values are different. Figure 16.7 shows some examples of waveforms with their average and root-mean-square values in terms of their maximum values. The average values given are for over half a cycle. The ratio of the root-mean-square value to the average value over half a cycle is called the *form factor* for a waveform and is an indication of the shape of the waveform.

$$\text{form factor} = \frac{\text{rms value}}{\text{average value over half a cycle}}$$

Waveshape	Average values	RMS value	Form factor
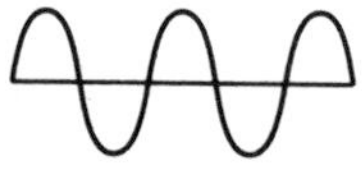 (sine wave)	$0.637\ V_m$	$0.707\ V_m$	1.11
 (half-wave rectified sine)	$0.319\ V_m$	$0.500\ V_m$	1.57
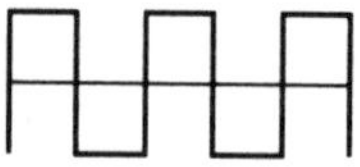 (square wave)	V_m	V_m	1.00
 (triangular wave)	$0.500\ V_m$	$0.577\ V_m$	1.15

Figure 16.7 *Form factors*

Example

Using the mid-ordinate rule the following current values were obtained for the half cycle of a waveform, the other half cycle being just a mirror image. Estimate (a) the average value over half a cycle, (b) the root-mean-square value, (c) the form factor.

Current values in A 2 2 2 2 2 4 4 4

(a) The average value is

$$I_{av} = \frac{2+2+2+2+2+4+4+4}{8} = 2.75\ \text{A}$$

(b) The root-mean-square value is

$$I_{rms} = \sqrt{\frac{4+4+4+4+4+16+16+16}{8}} = 2.92\ \text{A}$$

(c) The form factor is

$$\text{form factor} = \frac{\text{rms value}}{\text{average value}} = \frac{2.92}{2.75} = 1.06$$

Revision

3 Determine the root-mean-square current for a triangular waveform which gave the following mid-ordinate values over a full cycle:

current in mA 5 15 25 15 5 –5 –15 –25 –15 –5

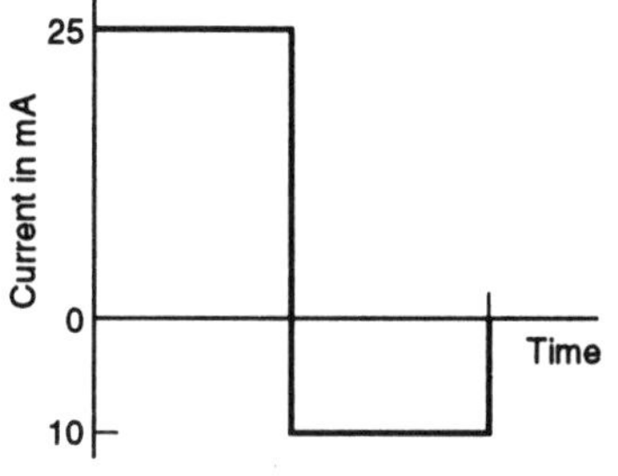

Figure 16.8

4 Determine the root-mean-square voltage for an irregular waveform which gave the following mid-ordinate values over a full cycle:

voltage in V 5 10 12 8 2 –5 –10 –12 –8 –2

5 Show that the root-mean-square value of a triangular waveform of the form shown in figure 16.5(a) and having a maximum value of V_m is given by $V_m/\sqrt{3}$.

6 Determine the average value, the root-mean-square value and the form factor of the waveform giving the current shown in figure 16.8.

Problems

1 A sinusoidal alternating current has a maximum value of 2 A. What is the average value over (a) half a cycle, (b) a full cycle?

2 A rectangular shaped alternating voltage has a value of 5 V for half a cycle and –5 V for the other half. What is the average value over (a) half a cycle, (b) a full cycle?

3 Show that the average value for a square waveform, as in figure 16.5(b), over half a cycle is equal to the maximum value.

4 Show that the average value for a sawtooth waveform over half a cycle, as in figure 16.5(d), is half the maximum value.

5 A sinusoidal alternating current has a maximum value of 4 V. What is the root-mean-square value?

6 A rectangular shaped alternating current has a maximum value of 4 A for half a cycle and –4 A for the other half. What is the root-mean-square value?

7 A half cycle of a waveform gave the following voltage values for the mid-ordinates, the other half cycle being a mirror image. What are (a) the half cycle average value, (b) the root-mean-square value, and (c) the form factor?

Voltages in V 1 3 5 7 9 9 9 9

8 A half cycle of a waveform gave the following current values for the mid-ordinates, the other half cycle being a mirror image. What are (a)

the half cycle average value, (b) the root-mean-square value, and (c) the form factor?

Current in mA 5 15 25 25 15 5

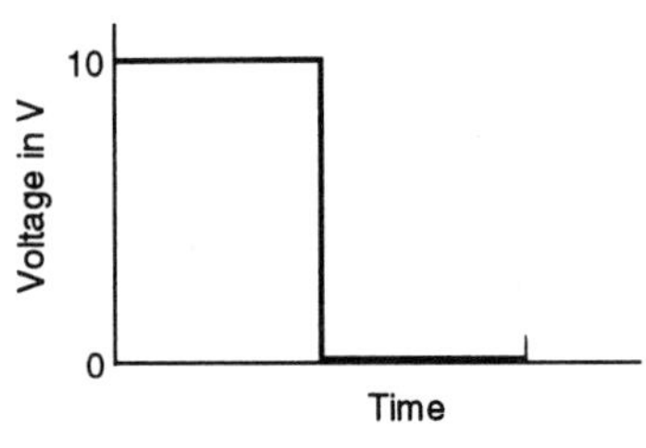

Figure 16.9

9 Show that the root-mean-square value for a square waveform, as in figure 16.5(b), over half a cycle is equal to the maximum value.

10 Show that the root-mean-square value for a sawtooth waveform over half a cycle, as in figure 16.5(d), is the maximum value divided by $\sqrt{3}$.

11 Determine the average value, the root-mean-square value and the form factor for thè voltage with the waveform shown in figure 16.9.

12 Determine the root-mean-square value of a square wave voltage waveform with a maximum value of 1 V.

13 Determine the root-mean-square value of a sinusoidal voltage waveform with a maximum value of 2 V.

14 Determine the root-mean-square value of a sinusoidal current waveform with a maximum value of 10 mA.

15 Determine the maximum value of a sinusoidal current waveform with a root-mean-square value of 100 mA.

16 Determine the maximum value of a square wave voltage waveform with a root-mean-square value of 10 V.

Section D Calculus

The aims of this section are to enable the reader to:

- Identify differentiation in terms of the gradient of a graph and rate of change.
- Differentiate basic equations from first principles and use standard forms.
- Differentiate algebraic, trigonometric, exponential and logarithmic equations.
- Determine maxima and minima.
- Identify integration as the inverse of differentiation and also as the area under a graph.
- Integrate algebraic, trigonometric and exponential equations.
- Use integration to determine the area under graphs.
- Solve engineering problems involving the use of differentiation.
- Solve angineering problems involving the use of integration

This section is an introduction to calculus. Chapter 17 is an introduction to differentiation, with chapter 18 showing its application to the determination of maxima and minima. Chapter 19 is an introduction to integration with integration considered both as the inverse of the differentiation process and as the area under a graph. Chapters 20, 21 and 22 illustrate the application of calculus in engineering.

Chapter 17 assumes a dexterity with algebraic equations and an ability to handle algebraic, trigonometric, exponential and logarithmic functions. It also assumes a knowledge of graphs. Thus chapters 1, 2, 3, 4, 5 and 11 are assumed. Chapter 18 follows on from chapter 17. Chapter 19 assumes that chapter 17 has been completed. Chapter 20 assumes that chapters 17, 18, and 19 have been covered and links closely with chapter 15. Chapter 21 assumes that chapters 17, 18 and 19 have been covered. Chapter 22 assumes that chapter 19 has been covered.

17 Differentiation

17.1 Introduction

Differentiation is a mathematical technique which can be used to determine the rate at which functions change at some particular point and hence the gradients of the tangents to graphs at a particular point. For example, we might have an equation describing how the distance covered in a straight line by a moving object varies with time. We could plot a graph of distance against time and determine the velocity at some instant as the gradient of the tangent to the curve at that instant. By taking a number of such gradient measurements we could then determine how the velocity varied with time. However, differentiation is a technique which enables the velocity to be obtained from the equation without drawing the graph and tangents.

This chapter is an introduction to differentiation. Chapters 18, 20 and 21 illustrate ways in which differentiation is used in engineering.

17.2 Gradients

Consider the problem of determining the gradient of a tangent at a point on a graph. It might, for example, be a distance–time graph for a moving object or the current–time graph for the current in an electrical circuit.

Suppose we want to determine the gradient at point A on the curve shown in figure 17.1. We can select another point B on the curve and join them together and then find the gradient of the line AB. The value of the gradient determined in this way will obviously depend on where we locate the point B. If we let B slide along the curve towards A then the closer B is to A the more the line approximates to the tangent at point A. Thus the line AB_1 is a better approximation to the tangent than the line AB. The method we can use to determine the gradient of the tangent at a point A on a curve is thus:

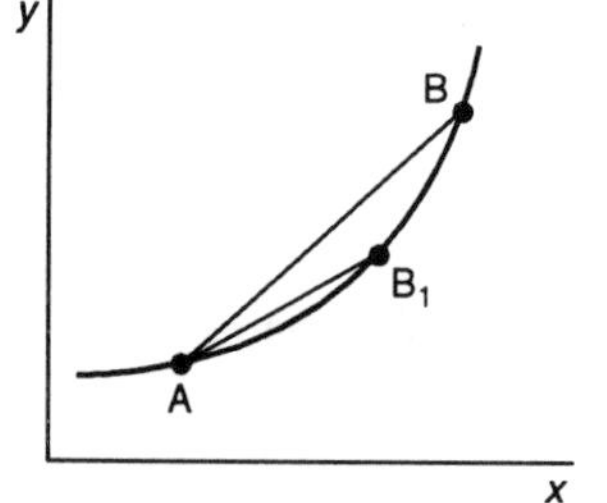

Figure 17.1 *Gradient at A*

1 Take another point B on the same curve and determine the gradient of the line joining A and B.
2 Then move B closer and closer to A. In the limit as the distance between A and B becomes infinitesimally small then the gradient of the line becomes the gradient of the tangent at A.

Consider A having the coordinates (x_1, y_1). If we select a point B with the coordinates (x_2, y_2) then the gradient of the line AB is

$$\text{gradient} = \frac{y_2 - y_1}{x_2 - x_1}$$

We thus have the difference in the value of y between points A and B divided by the difference in the value of x between the points. We can write this difference in the value of x as δx and this difference in the value of y as δy. The δ symbol in front of a quantity means 'a small bit of it' or 'an interval of'. Thus the equation can be written as

$$\text{gradient} = \frac{\delta y}{\delta x}$$

An alternative symbol which is often used is Δx, with the Δ symbol being used to indicate that we are referring to a small bit of the quantity x. These forms of notation do *not* mean that we have δ or Δ multiplying x. The δx or Δx should be considered as a single symbol representing a single quantity.

As we move B closer to A then the interval δx is made smaller. The gradient of the line AB then becomes closer to the tangent to the curve at point A. Eventually when the difference in x between A and B, i.e. δx, tends to zero then we have the gradient of the tangent at point A. We can write this as

$$\lim_{\delta x \to 0} \frac{\delta y}{\delta x} = \frac{dy}{dx}$$

This reads as: the limiting value of $\delta y/\delta x$ as δx tends to a zero value equals dy/dx. A *limit* is a value to which we get closer and closer as we carry out some operation. Thus dy/dx is the value of the gradient of the tangent to the curve at A. Since the tangent is the instantaneous rate of change of y with x at that point then dy/dx is the instantaneous rate of change of y with respect to x. dy/dx is called the *derivative* of y with respect to x. The process of determining the derivative for a function is called *differentiation*. The notation dy/dx should not be considered as d multiplied by y divided by d multiplied by x, but as a single symbol representing the gradient of the tangent and so the rate of change of y with x.

1.3.1 Determining derivatives

Suppose we have the two variables x and y which are related by the equation $y = x^2$. As x varies then y varies in such a way that the y is always equal to the square of the value of x. Consider a point A on the curve (figure 17.2) with the coordinates (x, y). Now consider a point B on the curve. We will take its coordinates to be $(x + \delta x, y + \delta y)$. The gradient of the line AB is then

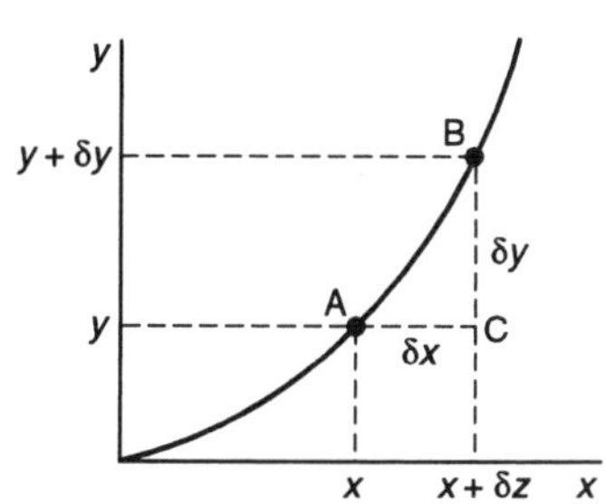

Figure 17.2 *Gradient at A for* $y = x^2$

$$\text{gradient} = \frac{BC}{AC} = \frac{\delta y}{\delta x}$$

Points A and B both lie on the same curve. Thus at point A, we must have

$$y = x^2$$

At B, we must have

$$y + \delta y = (x + \delta x)^2 = x^2 + 2x\,\delta x + (\delta x)^2$$

Subtracting these two equations gives

$$\begin{array}{lrl} & y + \delta y & = x^2 + 2x\,\delta x + (\delta x)^2 \\ \text{minus} & y & = x^2 \\ \hline & \delta y & = \quad 2x\,\delta x + (\delta x)^2 \end{array}$$

Therefore

$$\frac{\delta y}{\delta x} = 2x + \delta x$$

If we want the gradient at $x = 4$ and take a value for δx of 2, i.e. point B is at $x = 4 + 2 = 6$, then the gradient of the line AB is $2 \times 4 + 2 = 10$. If we now move point B closer to A so that we now have $\delta x = 1$ then the gradient of the line AB is $2 \times 4 + 1 = 9$. If we move point B yet closer to A so that $\delta x = 0.5$ then the gradient is the line AB is $2 \times 4 + 0.5 = 8.5$. With $\delta x = 0.1$ we have the gradient of the line AB as $2 \times 4 + 0.1 = 8.1$. As δx tends to the value 0 then the gradient of the line at $x = 4$ tends to the value 8, i.e. $2x$.

In the limit as we make δx smaller and smaller then δx tends to the value 0. Thus, in the above equation, $\delta y/\delta x$ tends to the value $2x$ as δx tends to zero. The gradient of the tangent at point A, coordinates (x, y), is thus given as

$$\text{gradient} = \lim_{\delta x \to 0} \frac{\delta y}{\delta x} = \frac{\mathrm{d}y}{\mathrm{d}x} = 2x$$

Thus we have a gradient at $x = 4$ ($y = 16$, since $y = x^2$) of $2 \times 4 = 8$. The gradient tells us the rate of change of y compared with x and so at $x = 4$ we have y changing 8 times as fast as x. The gradient at $x = 6$ ($y = 36$, since $y = x^2$) is $2 \times 6 = 12$. Thus at $x = 6$ we have y changing 12 times as fast as x.

The procedure adopted above to determine the derivative of y with respect to x when we have $y = x^2$ is

$$\frac{\mathrm{d}y}{\mathrm{d}x} = \lim_{\delta x \to 0} \frac{\delta y}{\delta x} = \lim_{\delta x \to 0} \frac{\text{change in } y \text{ when } x \text{ changes by } \delta x}{\delta x}$$

$$= \lim_{\delta x \to 0} \frac{(x + \delta x)^2 - x^2}{\delta x}$$

$$= \lim_{\delta x \to 0} \frac{x^2 + 2x\,\delta x + (\delta x)^2 - x^2}{\delta x}$$

$$= \lim_{\delta x \to 0} (2x + \delta x)$$

$$= 2x$$

We can use this technique with other functions. The following example and the next section give some examples of using this technique.

Example

The variable y varies with another variable x according to the equation $y = 2x + x^2$. Determine the derivative of y with respect to x and hence the value of the gradient of the graph of y against x at $x = 1$.

Using the procedure outlined above

$$\frac{\mathrm{d}y}{\mathrm{d}x} = \lim_{\delta x \to 0} \frac{\text{change in } y \text{ when } x \text{ changes by } \delta x}{\delta x}$$

$$= \lim_{\delta x \to 0} \frac{\{2(x+\delta x) + (x+\delta x)^2\} - \{2x + x^2\}}{\delta x}$$

$$= \lim_{\delta x \to 0} (2 + 2x + \delta x)$$

$$= 2 + 2x$$

At $x = 1$, the gradient of the graph y plotted against x will have the value $2 + 2 \times 1 = 4$.

Revision

1 Differentiate from first principles, i.e. determine $\mathrm{d}y/\mathrm{d}x$, for the following equations and hence determine the gradients of their graphs at $x = 0$ and $x = 1$.

(a) $y = 3x^2$, (b) $y = x^2 + x$, (c) $y = 3x + 1$, (d) $y = \frac{1}{2x}$, (e) $y = 4$,

(f) $y = 3x$, (g) $y = \frac{1}{4}x$

17.3 Derivative of $y = ax^n$

Consider the equation $y = ax$, where a is some constant. We can find the derivative $\mathrm{d}y/\mathrm{d}x$ using the procedures outlined above. Thus

$$\frac{\mathrm{d}y}{\mathrm{d}x} = \lim_{\delta x \to 0} \frac{\text{change in } y \text{ when } x \text{ changes by } \delta x}{\delta x}$$

$$= \lim_{\delta x \to 0} \frac{a(x+\delta x) - ax}{\delta x}$$

$$= \lim_{\delta x \to 0} a$$

$$= a$$

This what we would expect since $y = ax$ describes a straight-line graph with a gradient of a which is the same for all values of x.

Now consider the equation $y = ax^2$, where a is some constant.

$$\frac{dy}{dx} = \lim_{\delta x \to 0} \frac{\text{change in } y \text{ when } x \text{ changes by } \delta x}{\delta x}$$

$$= \lim_{\delta x \to 0} \frac{a(x + \delta x)^2 - ax^2}{\delta x}$$

$$= \lim_{\delta x \to 0} (2ax + a\,\delta x)$$

$$= 2ax$$

Now consider the equation $y = ax^3$.

$$\frac{dy}{dx} = \lim_{\delta x \to 0} \frac{\text{change in } y \text{ when } x \text{ changes by } \delta x}{\delta x}$$

$$= \lim_{\delta x \to 0} \frac{a(x + \delta x)^3 - ax^3}{\delta x}$$

$$= \lim_{\delta x \to 0} \{3ax^2 + 3ax\,\delta x + a(\delta x)^2\}$$

$$= 3ax^2$$

If we summarise the above, and add further derivatives calculated in the same way, we can establish a pattern.

$y = ax$ $\qquad \dfrac{dy}{dx} = a$

$y = ax^2$ $\qquad \dfrac{dy}{dx} = 2ax$

$y = ax^3$ $\qquad \dfrac{dy}{dx} = 3ax^2$

$y = ax^4$ $\qquad \dfrac{dy}{dx} = 4ax^3$

$y = ax^5$ $\qquad \dfrac{dy}{dx} = 5ax^4$

The pattern in the above data is that if we have $y = ax^n$, where a and n are constants, then

$$\frac{dy}{dx} = nax^{n-1}$$

This relationship applies for positive, negative and fractional values of n. Note that if we have just $y = a$, i.e. $n = 0$, then the above equation gives $dy/dx = 0$. The equation $y = a$ represents a horizontal straight line and the gradient of it is 0. Also note that when a function is multiplied by a constant, as for example the a in the above equations, then the derivative is multiplied by the same constant.

Example

Determine the derivative of $y = x^5$ and hence the gradient of its graph at $x = 1$.

Using the equation derived above,

$$\frac{dy}{dx} = nax^{n-1} = 5x^{5-1} = 5x^4$$

At $x = 1$ then $dy/dx = 5$. Hence the gradient at $x = 1$ is 5.

Example

Determine the derivative of $y = 4x^2$ and hence the gradient of its graph at $x = 2$.

Using the equation derived above,

$$\frac{dy}{dx} = nax^{n-1} = 2 \times 4x^{2-1} = 8x$$

At $x = 2$ then $dy/dx = 16$. Hence the gradient at $x = 2$ is 16.

Example

Determine the derivative of $y = 3/x$ and hence the gradient of its graph at $x = 2$.

This equation can be written as $y = 3x^{-1}$. Thus, using equation derived above,

$$\frac{dy}{dx} = nax^{n-1} = -1 \times 3x^{-1-1} = -\frac{3}{x^2}$$

At $x = 2$ then $dy/dx = -3/4$ and so the gradient at $x = 2$ is $-3/4$.

Example

Determine the derivative of $y = 2x^{3/4}$ and hence determine the gradient of its graph at $x = 4$.

Using the equation derived above,

$$\frac{dy}{dx} = nax^{n-1} = \tfrac{3}{4} \times 2x^{3/4-1} = \tfrac{3}{2}x^{-1/4}$$

At $x = 4$ then $dy/dx = 1.5 \times 0.71 = 1.1$. Hence the gradient is 1.1.

Example

The volume V of a cube is given by $V = L^3$, where L is the length of a side. Determine the rate of change of volume of a cube with respect to its side length.

Using the equation derived above,

$$\frac{dy}{dx} = nax^{n-1} = 3L^{3-1} = 3L^2$$

The rate of change of the volume with respect to the side length is $3L^2$.

Revision

2 Show from first principles that dy/dx for $y = x^{-1}$ is $-x^{-2}$.

3 Determine, using the rule derived above, the derivatives of:

(a) $y = 2x^3$, (b) $y = x^6$, (c) $y = 2x^{-2}$, (d) $y = x^{3/2}$, (e) $y = 2x^{1/4}$,

(f) $y = 3x^{10}$, (g) $y = 3x^{-1/5}$, (h) $y = 0.1x^4$

4 Determine the gradient of the graph of y against x for $y = 5x^2$ at $x = 1$.

5 Determine the gradient of the graph of y against x for $y = 2x^3$ at $x = 2$.

6 Determine the gradient of the graph of y against x for $y = 2/x$ at $x = 2$.

7 Determine the rate of change of the area of a circle with respect to its radius when the radius is 10 mm.

17.3.1 Differentiation of a sum of functions

Consider the problem of determining the derivative of $y = x + x^2$.

$$\frac{dy}{dx} = \lim_{\delta x \to 0} \frac{\text{change in } y \text{ when } x \text{ changes by } \delta x}{\delta x}$$

$$= \lim_{\delta x \to 0} \frac{\{(x + \delta x) + (x + \delta x)^2\} - \{x + x^2\}}{\delta x}$$

$$= \lim_{\delta x \to 0} (1 + 2x + \delta x)$$

$$= 1 + 2x$$

But this is just what we would obtain if we took the sum of the derivatives of the separate terms in the equation $y = x + x^2$. Thus, for $y = x$ we have $dy/dx = 1$ and for $y = x^2$ we have $dy/dx = 2x$. The sum of these two derivatives is $1 + 2x$.

To illustrate it with another example, consider the problem of determining the derivative of $y = 3x^3 - 4x^2$.

$$\frac{dy}{dx} = \lim_{\delta x \to 0} \frac{\text{change in } y \text{ when } x \text{ changes by } \delta x}{\delta x}$$

$$= \lim_{\delta x \to 0} \frac{\{3(x+\delta x)^3 - 4(x+\delta x)^2\} - \{3x^3 - 4x^2\}}{\delta x}$$

$$= \lim_{\delta x \to 0} \{9x^2 + 9x\,\delta x + 3(\delta x)^2 - 8x - 4\,\delta x\}$$

$$= 9x^2 - 8x$$

But this is just what we would obtain if we took the sum of the derivatives of the separate terms in the equation $y = 3x^3 - 4x^2$. Thus, for $y = 3x^3$ we have $dy/dx = 9x^2$ and for $y = -4x^2$ we have $dy/dx = -8x$. The sum of these two derivatives is $9x^2 - 8x$.

Example

Determine the derivative of $y = 2x^2 + x + 4$.

For $y = 2x^2$ we have $dy/dx = 4x$. For $y = x$ we have $dy/dx = 1$. For $y = 4$ we have $dy/dx = 0$. Hence for $y = 2x^2 + x + 4$ we have

$$\frac{dy}{dx} = 4x + 1$$

Example

Determine the derivative of $y = 3x^4 - 2x^3 + 3x$.

For $y = 3x^4$ we have $dy/dx = 12x^3$. For $y = -2x^3$ we have $dy/dx = -6x^2$. For $y = 3x$ we have $dy/dx = 3$. Hence for $y = 3x^4 - 2x^3 + 3x$ we have

$$\frac{dy}{dx} = 12x^3 - 6x^2 + 3$$

Revision

8 Determine, using the rule for the derivative of a sum, the derivatives of the following:

(a) $y = x^2 - 5$, (b) $y = 2x^4 + 3x^2$, (c) $y = 4x^3 - 2x^2 + 3x$,

(d) $y = 4x^3 - 3x^2 + 4x - 2$, (e) $y = 2x^5 + 3x^3 - 2x$, (f) $y = 4x^5 - 4$,

(g) $y = \frac{2}{x} + \frac{5}{x^2}$, (h) $y = 2x^2 - \frac{3}{x}$

17.4 Derivatives of trigonometric functions

Consider the determination of how the gradients of the graph of $y = \sin x$ (figure 17.3(a)) vary with x. Examination of the graph shows that as x increases from 0 to $\pi/2$ then the gradient, which is positive, gradually decreases to become zero at $x = \pi/2$. As x increases from $\pi/2$ to π then the gradient, which is now negative, becomes steeper and steeper to reach a maximum value at $x = \pi$. As x increases from π to $3\pi/2$ then the gradient, which is negative, decreases to become zero at $x = 3\pi/2$. As x increases from $3\pi/2$ to 2π the gradient, which is now positive again, increases to become a maximum at $x = 2\pi$. Figure 17.3(b) shows the result that is obtained by plotting the gradients against x. The figure shows a cosine curve. Thus, the derivative of $y = \sin x$ is

$$\frac{dy}{dx} = \cos x$$

We can show that this is the case if we consider, for $y = \sin x$, a small increase in x of δx. The corresponding increase in the value of y is δy where

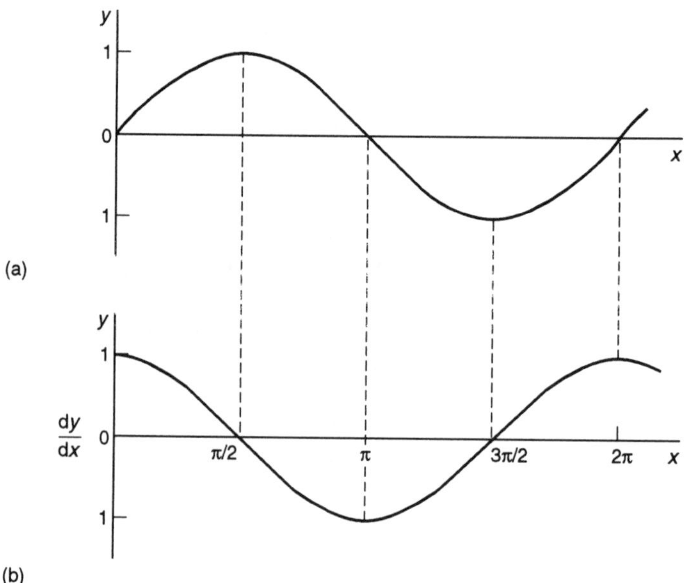

Figure 17.3 *Gradients of $y = \sin x$*

$$y + \delta y = \sin(x + \delta x)$$

Thus

$$\frac{dy}{dx} = \lim_{\delta x \to 0} \frac{\text{change in } y \text{ when } x \text{ changes by } \delta x}{\delta x}$$

$$= \lim_{\delta x \to 0} \frac{\sin(x + \delta x) - \sin x}{\delta x}$$

Suppose we consider a value for x of 1 rad and let δx be 0.1 rad. We then have

$$\frac{\sin 1.1 - \sin 1}{0.1} = 0.50$$

This is close to the value of cos 1, which is 0.54. If we had let $\delta x = 0.1$ rad then we would have obtained

$$\frac{\sin 1.01 - \sin 1}{0.01} = 0.54$$

In the limit as δx tends to a zero value then the derivative tends to the value of cos x. We could carry out similar calculations for other angles and show that the derivative of $y = \sin x$ is $dy/dx = \cos x$.

If we had considered $y = \sin ax$, where a is a constant, then we would have obtained for the derivative

$$\frac{dy}{dx} = a \cos ax$$

Now consider $y = \cos x$. Figure 17.4(a) shows the graph. Consider the gradients at various points along the graph. Between $x = 0$ and $x = \pi/2$ the gradient, which is negative, becomes steeper and steeper and reaches a maximum value at $x = \pi/2$. Between $x = \pi/2$ and $x = \pi$ the gradient, which is negative decreases until it becomes zero at $x = \pi$. Between $x = \pi$ and $x = 3\pi/2$ the gradient, which is positive, increases until it becomes a maximum at $x = 3\pi/2$. Between $x = 3\pi/2$ and $x = 2\pi$ the gradient, which is positive, decreases to become zero at $x = 2\pi$. Figure 17.4(b) shows how the gradient varies with x. The result is an inverted sine graph. Thus, for $y = \cos x$

$$\frac{dy}{dx} = -\sin x$$

Similar calculations to those carried out with sin x when x changes by dx can be used to show that the above relationship holds.

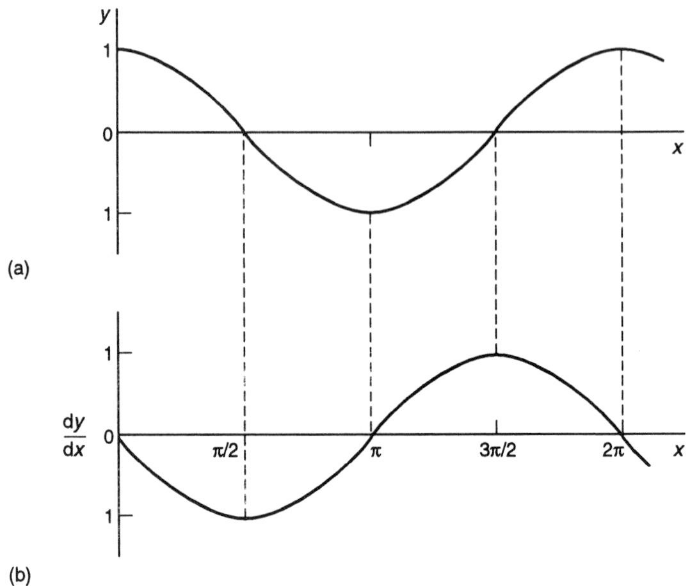

Figure 17.4 *Gradients of y = cos x*

If we had considered $y = \cos ax$, where a is a constant, then we would have obtained for the derivative

$$\frac{dy}{dx} = -a \sin ax$$

Example

Determine the derivative of the function $y = \sin 3x$.

Using the equation derived above for the derivative of $y = \sin ax$,

$$\frac{dy}{dx} = a \cos ax = 3 \cos 3x$$

Example

The current i in an electrical circuit varies with time t and is given by $i = \cos 100t$. Derive an equation for the rate of change of the current with time.

Using the equation derived above for the derivative of $y = \cos ax$, we obtain

$$\frac{di}{dt} = -100 \sin 100t$$

Revision

9 Determine the derivatives of the following:

(a) $y = \sin x$, (b) $y = \sin 5x$, (c) $y = \cos 4x$, (d) $y = \cos 3x$,

(e) $y = 2 \sin 3x$, (f) $y = 10 \cos 4x$

10 The displacement x of an oscillating object varies with time t according to the equation $y = \cos 50t$. Derive an equation indicating how the velocity, i.e. dy/dt, varies with time.

11 The current i, in amperes, in an electrical circuit varies with the time t, in seconds, acording to the equation $i = 2 \sin 100t$. Determine the rate of change of current when $t = 0.01$ s.

17.5 Derivatives of exponential functions

Consider the exponential equation $y = e^x$. Consider a small increase in x of δx. The corresponding increase in the value of y is δy where

$$y + \delta y = e^{x+\delta x} = e^x e^{\delta x}$$

Thus

$$\frac{dy}{dx} = \lim_{\delta x \to 0} \frac{\text{change in } y \text{ when } x \text{ changes by } \delta x}{\delta x}$$

$$= \lim_{\delta x \to 0} \frac{e^x e^{\delta x} - e^x}{\delta x} = \lim_{\delta x \to 0} \frac{e^x (e^{\delta x} - 1)}{\delta x}$$

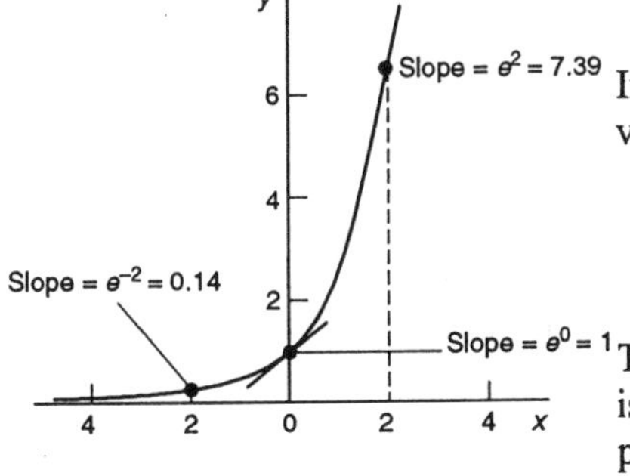

Figure 17.5 $y = e^x$

If we let $\delta x = 0.01$ then $(e^{0.01} - 1)/0.01 = 1.005$. If we take yet smaller values of dx then in the limit we obtain the value 1 for this fraction. Thus

$$\frac{dy}{dx} = e^x$$

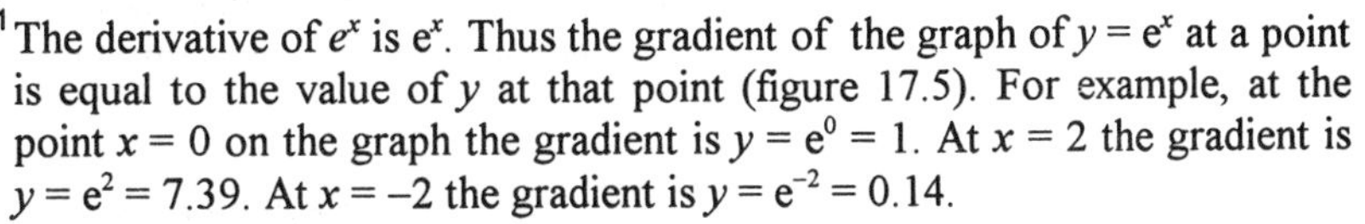

The derivative of e^x is e^x. Thus the gradient of the graph of $y = e^x$ at a point is equal to the value of y at that point (figure 17.5). For example, at the point $x = 0$ on the graph the gradient is $y = e^0 = 1$. At $x = 2$ the gradient is $y = e^2 = 7.39$. At $x = -2$ the gradient is $y = e^{-2} = 0.14$.

If we had $y = e^{ax}$ then

$$\frac{dy}{dx} = a e^{ax}$$

Example

Determine the derivative of $y = e^{3x}$.

Using the equation given above,

$$\frac{dy}{dx} = a\,e^{ax} = 3\,e^{3x}$$

Example

Determine the derivative of $y = e^{-x}$.

Using the equation given above,

$$\frac{dy}{dx} = a\,e^{ax} = -1\,e^{-x}$$

Example

Determine the derivative of $y = 10 - e^{-3x}$.

This is the sum of two terms, hence the derivative is the sum of the derivatives of the two terms. For $y = 10$ we have $dy/dx = 0$ and for $y = -e^{-3x}$ we have $dy/dx = +3\,e^{-3x}$. Thus

$$\frac{dy}{dx} = 3\,e^{-3x}$$

Revision

12 Determine the derivatives of the following:

(a) $y = e^{2x}$, (b) $y = e^{-3x}$, (c) $y = e^{x/2}$, (d) $y = 2\,e^{3x}$, (e) $y = 4 + e^{2x}$,

(f) $y = x + 3\,e^{2x}$, (g) $y = 2\,e^{x} + 3\,e^{2x}$

13 The number of radioactive atoms N in a radioactive element changes with time t according to the equation $N = N_0\,e^{-\lambda t}$, where N_0 is a constant. Determine the rate of change of N with time (this is called the activity). Note that multiplying an exponential by a constant multiplies the derivative by the same constant.

17.6 Derivatives of logarithmic functions

If $y = \ln x$, then we can write this as

$$x = e^{y}$$

Differentiating with respect to y gives

$$\frac{dx}{dy} = e^{y} = x$$

Hence, if we invert this we have

$$\frac{dy}{dx} = \frac{1}{x}$$

Thus the gradient of a graph of $y = \ln x$ is $1/x$. For example, at $x = 4$ the gradient is 1/4. At $x = 1$ the gradient is $1/1 = 1$. Figure 17.6(a) shows the graph of $y = \ln x$ and figure 17.6(b) how the derivative dy/dx varies with x.

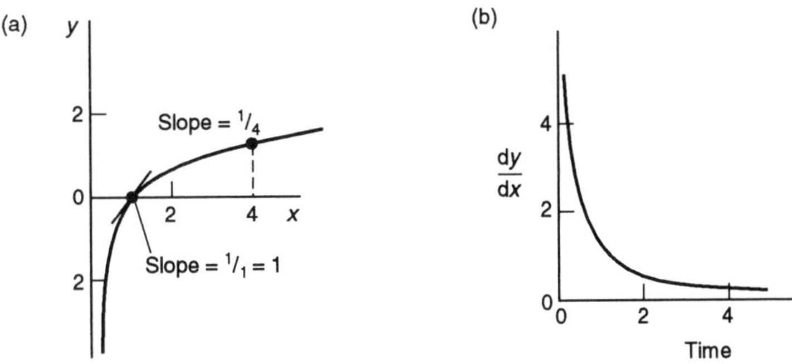

Figure 17.6 $y = \ln x$

Example

Determine the derivative of $y = \ln 2x$.

We can write this equation as $y = \ln 2 + \ln x$. We can then obtain the derivative of each term and add them together to give the required derivative. For $y = \ln 2$ we have $dy/dx = 0$ and for $y = \ln x$ we have $dy/dx = 1/x$. Thus, the derivative for $y = \ln 2x$ is

$$\frac{dy}{dx} = \frac{1}{x}$$

Revision

14 Determine the derivative of $y = \ln 3x$.

17.7 The second derivative

Consider a moving object for which we have a relationship between the displacement s of the object and time t of the form

$$s = ut + \tfrac{1}{2}at^2$$

where u and a are constants. We can plot this equation to give a distance–time graph. If we differentiate this equation we obtain

$$\frac{ds}{dt} = u + at$$

ds/dt is the gradient of the distance–time graph. It also happens to be the velocity. The gradient varies with time. We could thus plot a velocity–time graph, i.e. a ds/dt graph against t. Then differentiating for a second time, to obtain the gradients to this graph, we obtain the acceleration a.

$$\frac{\mathrm{d}}{\mathrm{d}t}\left(\frac{\mathrm{d}s}{\mathrm{d}t}\right) = a$$

The derivative of a derivative is called the *second derivative* and can be written as:

$$\frac{\mathrm{d}}{\mathrm{d}x}\left(\frac{\mathrm{d}y}{\mathrm{d}x}\right) \quad \text{or} \quad \frac{\mathrm{d}^2y}{\mathrm{d}x^2}$$

The first derivative gives information about how the gradients of the tangents change. The second derivative gives information about the rate of change of the gradient of the tangents.

Example

Determine the second derivative of $y = x^3$.

The first derivative is

$$\frac{\mathrm{d}y}{\mathrm{d}x} = 3x^2$$

The second derivative is given by differentiating this equation again:

$$\frac{\mathrm{d}^2y}{\mathrm{d}x^2} = 6x$$

Example

Determine the second derivative of $y = x^4 + 3x^2$.

The first derivative is

$$\frac{\mathrm{d}y}{\mathrm{d}x} = 4x^3 + 6x$$

The second derivative is

$$\frac{\mathrm{d}^2y}{\mathrm{d}x^2} = 12x^2 + 6$$

Revision

15 Determine the second derivatives of the following functions:

(a) $y = x^2$, (b) $y = 2x^2 + x$, (c) $y = x^3 + 2x^2$, (d) $y = \sin x$, (e) $y = e^{2x}$

16 The charge q on the plates of a capacitor is related to time t by the equation $q = CV\,e^{-t/RC}$, where C, V and R are constants. Determine how (a) the current, i.e. dq/dt, changes with time and (b) how the rate of change of current with time, i.e. the second derivative d^2q/dt^2, changes with time.

Problems

1 Working from first principles, determine the derivatives of: (a) $y = 4/x$, (b) $y = 1/x^2$.

2 Determine, using the rules developed in this chapter, the derivatives of the following functions:

(a) $y = 2x^6$, (b) $y = 5x^{-2}$, (c) $y = x^{7/3}$, (d) $y = 0.2x^3$, (e) $y = 2\sqrt{x}$,

(f) $y = 2x^{-3/2}$, (g) $y = 4x^2 + x$, (h) $y = 2x^3 + 3x^2 + 1$, (i) $y = 3x^3 - 5x^2 + 3x$,

(j) $y = \frac{5}{x^3}$, (k) $y = 2x^3 + \frac{1}{x^2}$, (l) $y = \frac{5}{x^2} - \frac{2}{x}$, (m) $y = \sin 4x$,

(n) $y = \cos 2x$, (o) $y = e^x$, (p) $y = e^{2x}$, (q) $y = e^{-2x/3}$, (r) $y = \ln 5x$,

(s) $y = 3\sqrt{x}$, (t) $y = \sin x + \cos 2x$

3 Determine the gradient of the graph of y against x for the equation $y = 3x^2$ at $x = 1$.

4 Determine the gradient of the graph of y against x for the equation $y = 2x^2 + 3x$ at $x = 2$.

5 The displacement s of an object starting from rest and moving with a uniform acceleration a is given by $s = \frac{1}{2}at^2$. Derive an equation for the velocity, i.e. the rate of change of distance with time.

6 Determine the gradients of the graph of $y = \cos 3x$ at the points on the graph (a) $x = \pi/2$, (b) $x = \pi/4$, (c) $x = 0$.

7 Determine the gradient of the graph of $y = e^{4x}$ at $x = 1$.

8 Determine the gradient of the graph of $y = 2 + e^{-x/2}$ at $x = 1$.

9 Determine the rate of change of area of a square with respect to the length of a side when the length is 10 mm.

10 The kinetic energy E of a rotating flywheel depends on the angular velocity ω and is given by the equation $E = \frac{1}{2}I\omega^2$. Determine the rate of change of kinetic energy with angular velocity.

11 The energy E stored by a capacitor depends on the potential difference V across it and is given by the equation $E = \frac{1}{2}CV^2$. Determine the rate of change of energy with respect to potential difference.

12 The height h of an object projected vertically upwards with an initial velocity of 12 m/s depends on the time t and is given by the equation $h = 12t - 5t^2$. Determine the rate at which the height changes with time, i.e. the velocity.

13 The atmospheric pressure p depends on the height h above ground level at which the measurement is made and is given by the equation $p = p_0 e^{-ch}$, where p_0 and c are constants. Determine the rate of change of pressure with height.

14 The instantaneous value of an alternating current i, in amperes, depends on the time t, in seconds, and is given by $i = 10 \sin 120t$. Determine the rate of change of the current with time when $t = 0.01$ s.

15 The instantaneous value of an alternating voltage v, in volts, depends on the time t, in seconds, and is given by $v = 12 \cos 100t$. Determine the rate of change of the voltage with time when $t = 0.02$ s.

16 The e.m.f. E, in millivolts, produced by a particular thermocouple is given by $E = a\theta^2$, where θ is the temperature in °C. If the thermocouple produces 200 mV at 100°C, determine the rate of change of the voltage with temperature at 120°C.

17 For a graph of $y = 2x^3 - 6x^2 - 15x + 5$ determine the points at which the gradient is +3.

18 The volume v, in cubic metres, of water in a reservoir varies with time t, in minutes, after a valve is opened according to $v = 10^4 - 10t^2 - 6t^3$. Determine the rate of change of the water volume at the time t of 2 min.

19 Determine the second derivatives of the following:

(a) $y = 3x^2$, (b) $y = e^{2x}$, (c) $y = \cos 2x$, (d) $y = 3x^2 + 4x$

18 Maxima and minima

18.1 Introduction

In a graph of y against x, where y is a variable that depends on the value of x, the gradient of the graph is given by dy/dx. If dy/dx is positive then y is increasing as x increases. If dy/dx is negative then y is decreasing as x increases. However, if $dy/dx = 0$ then y is neither increasing or decreasing. Points where this occurs are called *stationary points* or *critical points* or *turning points*. At such points we must have the tangent to the graph parallel to the x-axis. In figure 18.1 these are the points A, B, C and D.

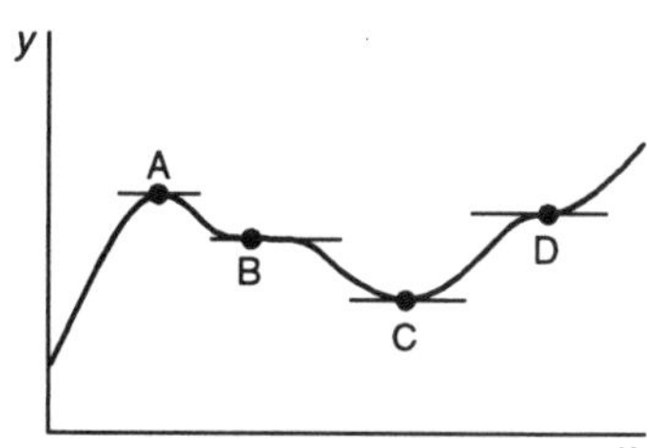

Figure 18.1 *Turning points*

The turning point at A is termed a *maximum*, that at C a *minimum*. At the maximum the value of y is greater than the neighbouring values either side of A. It does not mean that it is necessarily greater than any other value of y for other points on the curve. At the minimum the value of y is smaller than the neighbouring values either side of C. It does not mean that it is necessarily smaller than any other value of y for other points on the curve. Points B and D are termed *points of inflexion*. At such points y is higher on one side of the point than the other.

There are many situations in engineering where we are concerned with establishing maximum or minimum values. Thus with a projectile we might need to determine the maximum height reached. In determining how large a batch of a product to manufacture we might need to establish the number which will have the minimum cost. With an electrical circuit we might need to establish the condition for which maximum power is generated in a load resistance. In these situations we have effectively a graph of one variable against another and have to establish the point on the graph at which there a maximum or a minimum. This chapter is about determining such points.

18.2 Turning points

Consider a graph of y against x when the values of y depend in some way on the values of x. Points on the graph at which $dy/dx = 0$ can be:

1 A local maximum.
2 A local minimum.
3 A point of inflexion.

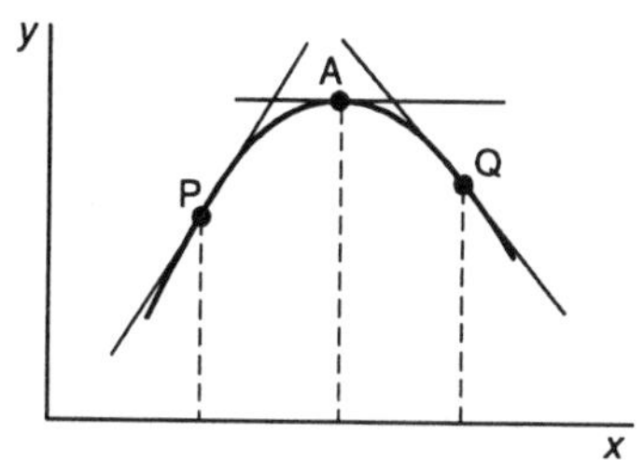

Figure 18.2 *A maximum*

Consider a local maximum, the term local being used because the value of y is only necessarily a maximum for points in the locality and there could be higher values of y elsewhere on the graph. Figure 18.2 shows such a maximum. At the maximum point A we have zero gradient for the tangent, i.e. $dy/dx = 0$. Consider two points P and Q close to A, with P having a value of x less than that at A and Q having a value greater than that at A. The gradient of the tangent at P is positive, the gradient of the tangent at Q is negative. Thus for a maximum we have the gradient changing from being positive prior to the turning point to negative after it.

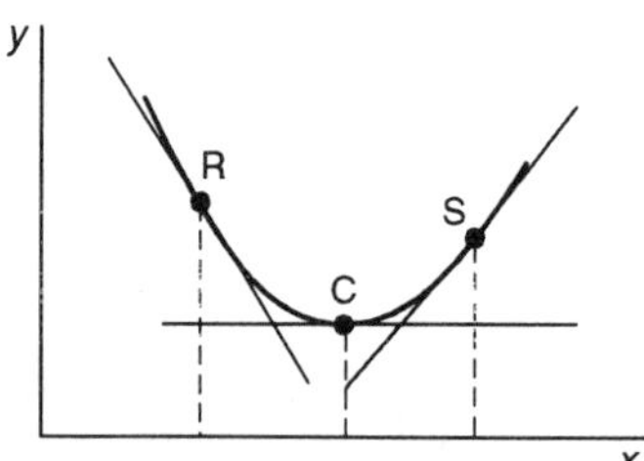

Figure 18.3 *A minimum*

Consider a local minimum, the term local being used because the value of y is only necessarily a minimum for points in the locality and there could be lower values of y elsewhere on the graph. Figure 18.3 shows such a minimum. At the minimum point C we have zero gradient for the tangent, i.e. $dy/dx = 0$. Consider two points R and S close to A, with R having a value of x less than that at C and S having a value greater than that at C. The gradient of the tangent at R is negative, the gradient of the tangent at S is positive. Thus for a minimum we have the gradient changing from being negative prior to the turning point to positive after it.

Consider points of inflexion, as illustrated in figure 18.4. At such points $dy/dx = 0$. However, in neither of the graphs is there a local maximum or minimum. In figure 18.4(a), the gradient at a point T prior to the point is negative and the gradient at a point U after the point is also negative. In figure 18.4(b), the gradient at a point V prior to the point is positive and the gradient at a point W after the point is also positive. For a point of inflexion the sign of the gradient prior to the point is the same as that after the point.

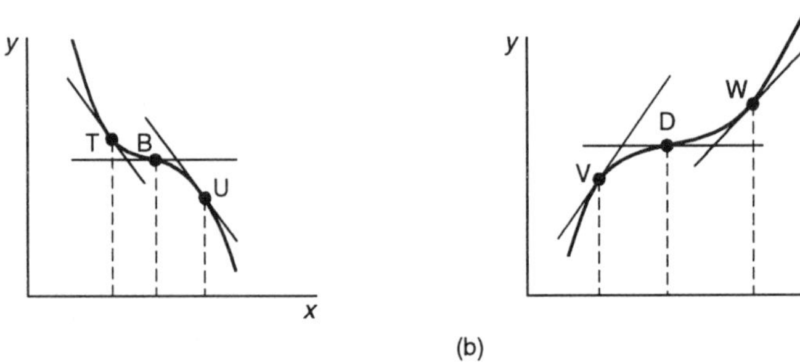

Figure 18.4 *Points of inflexion*

Thus, if we consider the gradients in the vicinity of maxima, minima and points of inflexion, then in moving from points before to after the turning point:

1 At a maximum the gradient changes from being positive to negative.
2 At a minimum the gradient changes from being negative to positive.
3 At a point of inflexion the sign of the gradient does not change.

The gradient at a point on a graph is given by dy/dx. We can thus determine whether a turning point is a maximum, a minimum or a point of inflexion by considering how the value of dy/dx changes for a value of x smaller than the turning point value compared to that for a value of x greater than the turning point value.

There is an alternative method we can use to distinguish between maximum and minimum points. We need to establish how the gradient changes in going from points before to after turning points. Consider, for a maximum, a graph of the gradients plotted against x (figure 18.5(a)). The gradients prior to the maximum are positive and decrease in value to become zero at the maximum. They then become negative and as x

increases become more and more negative. The second derivative d^2y/dx^2 measures the rate of change of dy/dx with x, i.e. the gradient of the dy/dx graph. The gradient of the gradient graph is negative before, at and after the maximum point. Hence at a maximum d^2y/dx^2 is negative.

Consider a minimum (figure 18.5(b)). The gradients prior to the minimum are negative and become less negative until they become zero at the minimum. As x increases beyond the minimum the gradients become positive, increasing in value as x increases. The second derivative d^2y/dx^2 measures the rate of change of dy/dx with x, i.e. the gradient of the dy/dx graph. The gradient of the gradient graph is positive before, at and after the minimum point. Hence at a minimum d^2y/dx^2 is positive. Thus we can distinguish between maximum and minima in that:

1 At a maximum the second derivative is negative.
2 At a minimum the second derivative is positive.

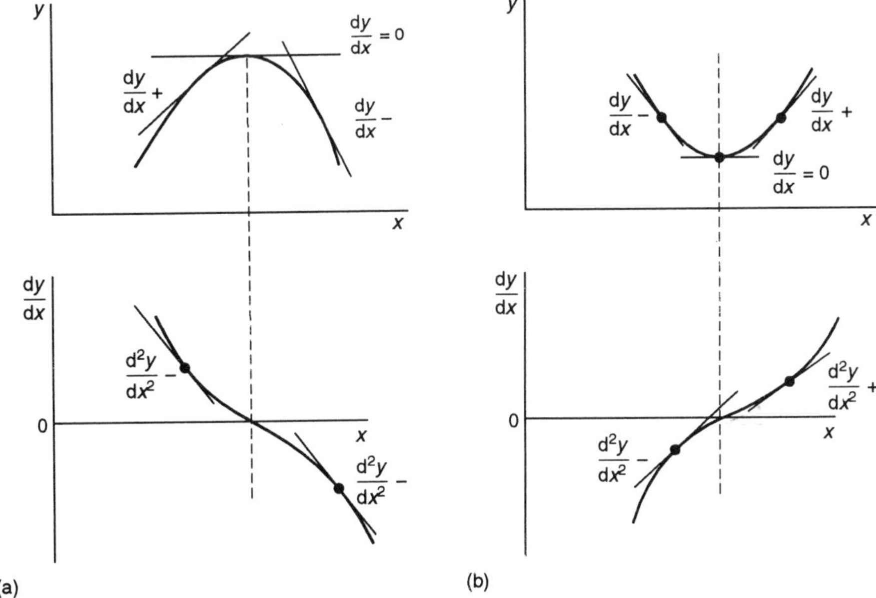

Figure 18.5 *(a) A maximum, (b) a minimum*

Example

Determine, and identify the form of, the turning points on a graph of the equation $y = 2x^3 - 3x^2 - 12x$.

Differentiating the equation gives

$$\frac{dy}{dx} = 6x^2 - 6x - 12$$

Thus the gradient of the graph is zero when $6x^2 - 6x - 12 = 0$. We can rewrite this as

$$6(x^2 - x - 2) = 6(x + 1)(x - 2) = 0$$

Thus the gradient is zero, and hence there are turning points, at $x = -1$ and $x = 2$.

To establish the form of these turning points consider the gradients just prior to and just after them. Prior to the $x = -1$ turning point at $x = -2$, the gradient is $6x^2 - 6x - 12 = 6 \times (-2)^2 - 6 \times (-2) - 12 = 12$. After the point at $x = 0$ we have a gradient of -12. Thus the gradient prior to the $x = -1$ point is positive and after the point it is negative. The point is thus a maximum.

Consider the $x = 2$ turning point. Prior to the turning point at $x = 0$ the gradient is -12. After the turning point at $x = 3$ the gradient is $6 \times 3^2 - 6 \times 3 - 12 = 24$. Thus the gradient prior to the $x = 2$ point is negative and after the point it is positive. The point is thus a minimum.

Alternatively we could determine the form of the turning points by considering the sign of the second derivative at the points. The second derivative is obtained by differentiating the dy/dx equation. Thus

$$\frac{d^2y}{dx^2} = 12x - 6$$

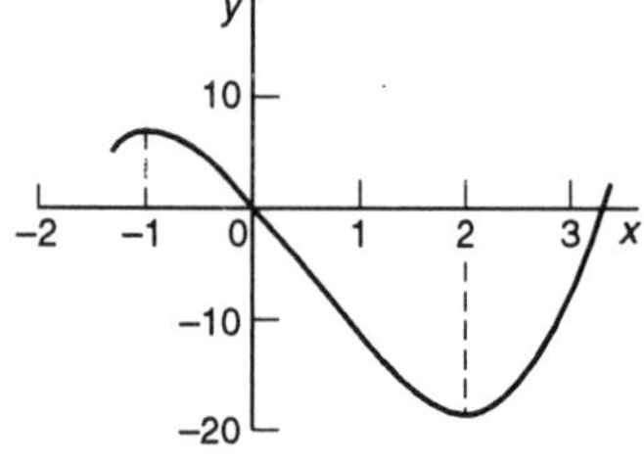

Figure 18.6

At $x = -1$ then the second derivative is $12 \times (-1) - 6 = -18$. The negative value indicates that the point is a maximum. At $x = 2$ the second derivative is $12 \times 2 - 6 = 18$. The positive value indicates that the point is a maximum.

Figure 18.6 shows a graph of the equation $y = 2x^3 - 3x^2 - 12x$, showing the maximum at $x = -1$ and the minimum at $x = 2$.

Example

Determine, and identify the form of, the turning points on a graph of the equation $y = x^3$.

Differentiating the equation gives

$$\frac{dy}{dx} = 3x^2$$

Thus $dy/dx = 0$ when $x = 0$. The turning point is thus at $x = 0$.

Consider the value of the gradient just prior to the turning point. At $x = -1$ then the gradient is $3 \times (-1)^2 = 3$. After the turning point at $x = 1$ then the gradient is $3 \times 1^2 = 3$. The gradient is thus positive prior to the turning point and positive after it. The turning point is thus an inflexion point. Figure 18.7 shows the graph.

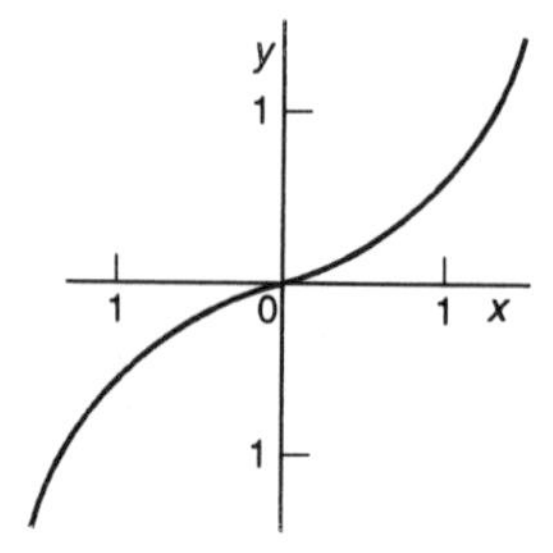

Figure 18.7

Example

The displacement y in metres of an object is related to the time t in seconds by the equation $y = 5 + 4t - t^2$. Determine the maximum displacement.

Differentiating the equation gives

$$\frac{dy}{dt} = 4 - 2t$$

dy/dt is 0 when $t = 2$ s. There is thus a turning point at the displacement $y = 5 + 4 \times 2 - 2^2 = 9$ m.

We need to check that this is a maximum displacement. The gradient prior to the turning point at $t = 1$ has the value $4 - 2 = 2$. After the turning point at $t = 3$ it has the value $4 - 6 = -2$. The gradient changes from a positive value prior to the turning point to a negative value afterwards. It is thus a maximum.

Alternatively we could have established this by determining the second derivative. Differentiating 4 – 2t gives $d^2y/dx^2 = -2$. Thus, the turning point is a maximum.

Example

If the sum of two numbers is 40, determine the values which will give the minimum value for the sum of their squares.

Let the two numbers be x and y. Then we must have $x + y = 40$. We have to find the minimum value of S when we have $S = x^2 + y^2$. We need an equation which expresses the sum in terms of just one variable. Thus substituting from the previous equation gives

$$S = x^2 + (40 - x)^2 = x^2 + 1600 - 80x + x^2$$

$$= 2x^2 - 80x + 1600$$

Differentiating this equation, then

$$\frac{dS}{dx} = 4x - 80$$

The value of x to give a zero value for dS/dx is when $4x - 80 = 0$ and so when $x = 20$.

We can check that this is the value giving a minimum by considering the values of dS/dx prior to and after the point. Thus prior to the point at $x = 19$ we have $dS/dx = 4 \times 19 - 80 = -4$. After the point at $x = 21$ we have $dS/dx = 4 \times 21 - 80 = 4$. Thus dS/dx changes from being negative to positive. The turning point is thus a minimum. Alternatively we could check that this is a minimum by obtaining the second derivative. Differentiating $4x - 80$ gives $d^2S/dx^2 = 4$. Since this is positive then we have a minimum. Thus the two numbers which will give the required minimum are 20 and 20.

Example

Determine the maximum area of a rectangle with a perimeter of 32 cm.

If the width of the rectangle is w and its length L then the area $A = wL$. But the perimeter has a length of 32 cm. Thus $2w + 2L = 32$. We need an equation expressing the area in terms of just one variable. Thus, if we eliminate w from the area equation by substitution,

$$A = L(16 - L) = 16L - L^2$$

Hence

$$\frac{dA}{dL} = 16 - 2L$$

dA/dL is zero when $16 - 2L = 0$ and so when $L = 8$.

We can check that this gives a maximum area by considering values of the gradient at values of L below and above 8. At $L = 7$ the gradient is 2 and at $L = 9$ it is -2. It is thus a maximum. Alternatively we could have considered the second derivative. Since $d^2A/dL^2 = -2$ and so is negative, we have a maximum.

For a maximum area we must therefore have $L = 8$ cm, and after substituting this value in $2w + 2L = 32$, $w = 8$ cm.

Revision

1 Determine, and identify the form of, the turning points on graphs of the following equations:

(a) $y = x^2 - 6x + 4$, (b) $y = 2x^3 - 3x^2 - 12x + 8$, (c) $y = 3x^2 - x^3$,

(d) $y = 2x^3 + 3x^2 - 36x + 7$, (e) $y = x^2 + 3x$, (f) $y = 5x - 10x^2$,

(g) $y = \sin x$

2 The rate r at which a chemical reaction proceeds is related to the quantity x of chemical present by the equation $r = k(a - x)(b + x)$. Determine the maximum rate.

3 A water tank is to be made of sheet metal. It is to have a square base and be open at the top. What is the least area of metal that can be used if the tank is to have a volume of 32 m^3?

4 The bending moment M of a uniform beam of length L at a distance x from one end is given by the equation $M = \frac{1}{2}wLx - \frac{1}{2}wx^2$, where w is the

weight per unit length of beam. Determine the point at which the bending moment is a maximum.

5 Determine the two numbers whose sum is 40 and for which the product is a maximum.

6 A stone is thrown vertically upwards. Its height h, in metres, depends on the time t, in seconds, and is given by the equation $h = 40t - 5t^2$. After what time is the height a maximum?

7 A cylindrical container with closed ends is to have a volume of 0.25 m^3. What radius and height will it have if the area of sheet metal used in its construction is to be a minimum?

8 A wire of length 60 cm is to be bent to form a rectangle. What should be the length of its sides if the area is to be a maximum?

9 Determine the dimensions an open box can have for maximum volume if constructed with a square base from a sheet of metal of area 100 cm^2.

10 The power P supplied to an electrical circuit is given by $P = IE - I^2r$, where I is the current taken by the circuit, E the e.m.f. of the supply and r its internal resistance. Determine the condition for maximum power transfer.

11 An electrical voltage v, in volts, depends on the time t, in seconds, and is given by $v = 30 \cos 500t + 15 \sin 500t$. Determine the maximum value of the voltage.

12 The e.m.f. E produced by a thermocouple is related to the temperature θ by $E = a\theta + b\theta^2$. Determine the temperature at which the e.m.f. is a maximum.

Problems

1 Determine, and identify the form of, the turning points on graphs of the following equations:

(a) $y = 4x^2 - 8x + 3$, (b) $y = 2x^2 + 6x + 10$, (c) $y = x^2$, (d) $y = x^3 - x$,

(e) $y = x^3 + x^2$, (f) $y = 2x^3 - 3x^2 - 36x - 15$, (g) $y = x^2 - \frac{16}{x}$,

(h) $y = 2x + \frac{8}{x} + 3$, (i) $y = 3 - x^3$, (j) $y = x^2 - 2x$, (k) $y = 3x^2 - 6x + 3$,

(l) $y = x^3 - 3x + 7$, (m) $y = 4x^3 + 9x^2 - 12x + 7$,

(n) $y = x^3 - 3x^2 - 24x + 5$

2 The profit P gained by a company from the sales of product X depends on the number n sold, and is given by $P = 40n - 200 - \frac{1}{2}n^2$. Determine the number that have to be sold to maximise the profit.

3 When one number is added to twice a second number then the result is 24. What should the numbers be if their product is to be a maximum?

4 The production costs C per day for steel tube depends on the length L, in metres, produced per day and is given by the equation

$$C = 30L + \frac{300\,000}{L}$$

Determine the length that has to be produced per day to keep the costs to a minimum.

5 The speed v, in metres per second, of a car when in first gear is related to the time t, in seconds, by $v = 5 + 3t - t^2$. Determine the maximum speed of the car.

6 An open rectangular tray is constructed from a rectangular sheet measuring 200 mm by 300 mm by cutting equal size squares from each corner and then folding up the sides to form the tray. What should be the side lengths of the squares cut from each corner if the volume of the tray is to be a maximum?

7 A open cylindrical tank is to have a volume of 2 m^3. What should be the dimensions of the cylinder if the area of the sheet used to form it is to be a minimum?

8 A rocket moves vertically upwards with its distance h, in metres, from the ground related to the time t by $h = 500t - 5t^2$. Determine the greatest height reached.

9 An open tank is to have a square base and vertical sides. It is to be lined with plastic sheet. Show that the minimum area of plastic sheet will be needed when the depth of the tank is twice the side length of its base.

10 The cost C of producing n units of a product is given by a setting-up cost of £500 which applies regardless of how many items are produced and a cost which depends on the number of units produced and is given, in £, by $0.1n + 0.001n^2$. Determine the production level which will minimise the average cost per item, i.e. the value of C/n, produced.

19 Integration

19.1 Introduction

Integration can be considered to be the mathematical process which is the reverse of the process of differentiation. It also turns out to be a process for finding an area under a graph. The term *integral* is used for the outcome of integration. This chapter can be considered to be a basic introduction to integration.

As an illustration of the application of integration in engineering, consider the situation where the velocity v of an object varies with time t, say $v = 2t$. Since velocity is the rate of change of distance x with time we can write this as

$$\frac{\mathrm{d}x}{\mathrm{d}t} = 2t$$

Thus we know how the gradient of the distance–time graph varies with time. Integration is the method we can use to determine from this how the distance varies with time. We thus start out with the gradients and find the graph responsible for them, the reverse of the process used with differentiation where we start with a graph and find the gradients.

But we could have considered the same problem in a different way. Suppose the velocity variation with time was represented as a velocity–time graph. The distance travelled in a particular time can be obtained from the graph by determining the area under the graph between those times (see section 15.3). Integration can be used to determine such areas.

19.2 Integration as the reverse of differentiation

Suppose we have an equation $y = x^2$. When this equation is differentiated we obtain the derivative of $\mathrm{d}y/\mathrm{d}x = 2x$. Differentiation is the determination of the relationship for the gradient of a graph. We can define *integration* as the mathematical process which reverses differentiation, i.e. given the gradient relationship, finding the equation which was responsible for it. Thus, in this case, when given $2x$ we need to find the equation which on being differentiated gives $2x$. Thus integrating $2x$ should give us x^2. However, the derivative of $x^2 + 1$ is also $2x$, likewise the derivative of $x^2 + 2$, the derivative of $x^2 + 3$, and so on. Figure 19.1 shows part of the family of graphs which all have the gradients given by $2x$. Thus, for each of the graphs, at a particular value of x, such as $x = 1$, they all give the same gradient of $2x = 2$. Thus in the integration of $2x$ we are not sure whether there is a constant term or not, or what value it might have. Hence a constant C has to be added to the result. Thus the outcome of the integration of $2x$ has to be written as being $x^2 + C$. The integral in such circumstances is, because of there being a constant, referred to as the *indefinite integral*.

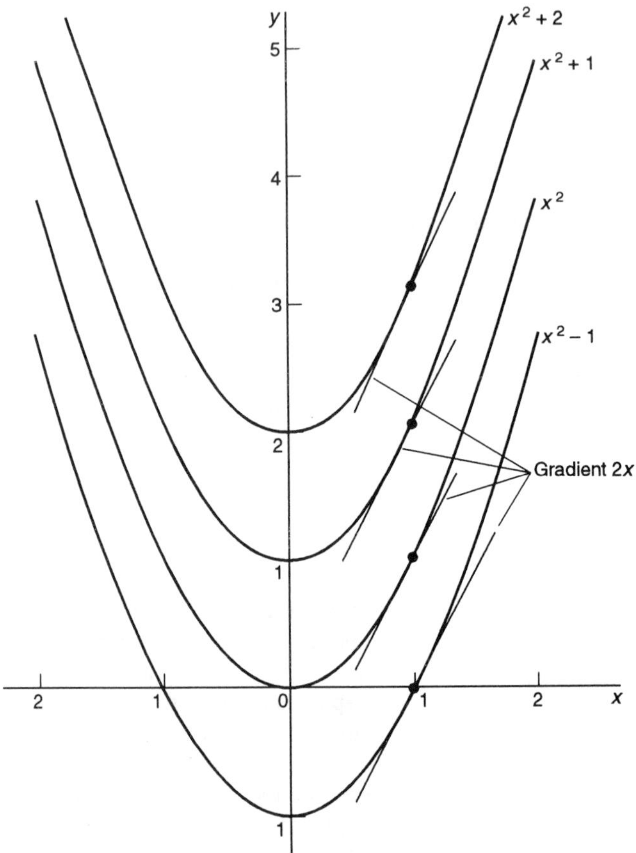

Figure 19.1 $dy/dx = 2x$

To indicate the process of integration a special symbol $\int$ with dx is used. The $\int$ sign indicates that integration is to be carried out and the dx that x is the variable we are integrating with respect to. Thus the integration referred to above can be written as

$$\int 2x \, dx = x^2 + C$$

The above thus reads as 'the integral of $2x$ with respect to x is x^2 plus a constant'.

Example

By considering integration as the reverse of differentiation, what is the integral of $3x^2$?

We have to find what function when differentiated gives $3x^2$. If we differentiate x^3 we obtain $3x^2$. If we differentiate $x^3 + C$, where C is some constant, we also obtain $3x^2$. Thus the integral of $3x^2$ is $x^3 + C$.

$$\int 3x^2\,dx = x^3 + C$$

Thus we could have $y = x^3$, or $y = x^3 + 1$, or $y = x^3 + 2$, or $y = x^3 - 1$, etc. All these graphs have gradient equal to $3x^2$. Because we do not know which one, we include the constant C.

Revision

1 By considering integration as the reverse of differentiation, determine the integrals of the following with respect to x:

(a) $4x$, (b) $6x^2$, (c) $x + 2$, (d) $\sin x$, (e) e^x

19.2.1 Integrals of common functions

The integrals of functions can be determined by considering what equation will give the function when differentiated. For example, consider

$$\int x^n\,dx$$

Considering integration as the inverse of differentiation, the question becomes as to what function gives x^n when differentiated. The derivative of x^{n+1} is $(n + 1)x^n$. Thus the derivative of $x^{n+1}/(n + 1)$ is x^n. Hence

$$\int x^n\,dx = \frac{x^{n+1}}{n+1} + C$$

This is true for positive, negative and fractional values of n other than $n = -1$, i.e. the integral of x^{-1}. For the integral of x^{-1}, i.e.

$$\int \frac{1}{x}\,dx$$

then since the derivative of $\ln x$ is $1/x$,

$$\int \frac{1}{x}\,dx = \ln x + C$$

This only applies if x is positive, i.e. $x > 0$. If x is negative, i.e. $x < 0$, then the integral of $1/x$ in such a situation is *not* $\ln x$. This is because we cannot have the logarithm of a negative number as a real quantity.

Consider the integral of the exponential function e^x, i.e.

$$\int e^x\,dx$$

The derivative of e^x is e^x. Thus

$$\int e^x\,dx = e^x + C$$

Consider the integral of trigonometric functions. For sin x, i.e.

$$\int \sin x \, dx$$

we have, since the derivative of cos x is −sin x,

$$\int \sin x \, dx = -\cos x + C$$

For the integral of cos x, i.e.

$$\int \cos x \, dx$$

since the derivative of sin x is cos x, we have

$$\int \cos x \, dx = \sin x + C$$

The above, with the inclusion of a constant a are listed below in table 19.1.

Table 19.1 Common integrals

Function being integrated	Outcome of the integration
$\int ax^n \, dx$	$\frac{ax^n}{n+1} + C$, n not −1
$\int \frac{1}{x} \, dx$	$\ln x + C$
$\int e^{ax} \, dx$	$\frac{1}{a} e^{ax} + C$
$\int \sin ax \, dx$	$-\frac{1}{a} \cos ax + C$
$\int \cos ax \, dx$	$\frac{1}{a} \sin ax + C$

Example

Evaluate the integrals:

(a) $\int x^4 \, dx$, (b) $\int x^{1/2} \, dx$, (c) $\int x^{-4} \, dx$, (d) $\int x^{-1} \, dx$,

(e) $\int \cos 4x \, dx$, (f) $\int e^{2x} \, dx$, (g) $\int 5 \, dx$

(a) Using $\int x^n \, dx = \frac{x^{n+1}}{n+1} + C$, then

$$\int x^4 \, dx = \frac{x^{4+1}}{4+1} + C = \frac{x^5}{5} + C$$

(b) Using $\int x^n \, dx = \frac{x^{n+1}}{n+1} + C$, then

$$\int x^{1/2} \, dx = \frac{x^{1/2+1}}{1/2+1} + C = \frac{x^{3/2}}{3/2} + C = \tfrac{2}{3} x^{3/2} + C$$

(c) Using $\int x^n \, dx = \frac{x^{n+1}}{n+1} + C$, then

$$\int x^{-4} \, dx = \frac{x^{-4+1}}{-4+1} + C = \frac{x^{-3}}{-3} + C = -\tfrac{1}{3} x^{-3} + C$$

(d) Using the relationship giving in the table

$$\int x^{-1} \, dx = \ln x + C$$

(e) Using $\int \cos ax \, dx = \frac{1}{a} \sin x + C$, then

$$\int \cos 4x \, dx = \tfrac{1}{4} \sin 4x + C$$

(f) Using $\int e^{ax} \, dx = \frac{1}{a} e^x + C$, then

$$\int e^{2x} \, dx = \tfrac{1}{2} e^{2x} + C$$

(g) Using $\int x^n \, dx = \frac{x^{n+1}}{n+1} + C$, with $n = 0$ since $x^0 = 1$, then

$$\int 5 \, dx = \int 5x^0 \, dx = 5x + C$$

Revision

2 Determine the integrals:

(a) $\int x^3 \, dx$, (b) $\int 2x^3 \, dx$, (c) $\int x^{-7} \, dx$, (d) $\int x^{-1/2} \, dx$, (e) $\int x^{1/3} \, dx$,

(f) $\int \sin 3x \, dx$, (g) $\int \cos 2x \, dx$, (h) $\int e^{4x} \, dx$, (i) $\int e^{-5x} \, dx$

19.2.2 Integral of a sum

The derivative of, for example, $x^2 + x$ is the derivative of x^2 plus the derivative of x, i.e. it is $2x + 1$. The integral of $2x + 1$ is thus $x^2 + x + C$. Thus the integral of the sum of a number of functions is the sum of their separate integrals.

Example

Determine the integral $\int (x^3 + 2x + 4)\,dx$.

We can write this as

$$\int (x^3 + 2x + 4)\,dx = \int x^3\,dx + \int 2x\,dx + \int 4\,dx$$

Hence the integral is

$$\frac{x^4}{4} + P + x^2 + Q + 4x + R$$

where P, Q and R are constants. We can combine these constants into a single constant C. Hence the integral is

$$\frac{x^4}{4} + x^2 + 4x + C$$

Revision

3 Determine the following integrals:

(a) $\int (5x + 3)\,dx$, (b) $\int (4x^3 + 3x^2 + 2)\,dx$, (c) $\int (\sin x + \cos 2x)\,dx$,

(d) $\int (2 + e^{-3x})\,dx$, (e) $\int \left(\frac{1}{x} + 3x\right)\,dx$, (f) $\int (3x + \sin 4x)\,dx$

19.2.3 Finding the constant

The solution given by the above integrations is a general solution and includes a constant. As was indicated in figure 19.1 the integration of $2x$ gives $y = x^2 + C$. This solution indicates a family of possible equations which could give $dy/dx = 2x$. We can, however, find a particular solution if we are supplied with information giving specific coordinate values which have to fall on the graph curve. Thus, in this case, we might be given the condition that the solution must pass through the point (1, 1). This means that the condition of $y = 1$ when $x = 1$ must fit the equation $y = x^2 + C$. This can only be the case when $C = 0$. Hence the solution is $y = x^2$.

Example

Determine the equation of a graph if it has to pass through the point (0, 2) and has a gradient given by

$$\frac{dy}{dx} = 3x + 2$$

To obtain the general solution, i.e. the family of curves which fit the above gradient equation, we integrate. Thus

$$y = \int (3x+2)\,dx = \tfrac{3}{2}x^2 + 2x + C$$

The particular curve we require must pass through the point (0, 2). Thus we must have $y = 2$ when $x = 0$. Putting this data into the equation gives $2 = 0 + 0 + C$. Hence $C = 2$ and so the particular solution is

$$y = \tfrac{3}{2}x^2 + 2x + 2$$

Revision

4 Determine the equations of the following curves, given the equation for the gradient of a curve and a point through which the curve must pass:

(a) $\dfrac{dy}{dx} = 2x + 3$, curve passes through (2, 3),

(b) $\dfrac{dy}{dx} = 9x^2 + 2x - 2$, curve passes through (1, 3),

(c) $\dfrac{dy}{dx} = 4x - 9$, curve passes through (2, −5),

(d) $\dfrac{dy}{dx} = 2 - \dfrac{1}{x^2}$, curve passes through (1, 3)

5 The velocity v, in m/s, of an object varies with time t, in seconds, and is given by $v = 10 + 2t$. As $v = dx/dt$, determine how the distance x varies with time if at $t = 0$ the distance is 2 m.

6 The velocity v, in m/s, of an object varies with time t, in seconds, and is given by $v = 2t + 5$. As $v = dx/dt$, determine how the distance x varies with time if at $t = 0$ the distance is 0.

19.3 Integration as the area under a graph

Consider a moving object and its graph of velocity v against time t. The distance travelled between times of t_1 and t_2 is the area under the velocity–time graph between those times (see section 15.3). If we divide the area into a number of equal width strips then we can represent this area under the velocity–time graph as being the sum of the areas of these equal width strip areas, as illustrated in figure 19.2. If t is the value of the time at the centre of a strip of width δt and v the velocity at this time, then a strip has an area of $v\,\delta t$. Thus the area under the graph between the times t_1 and t_2 is equal to the sum of the areas of all such strips between the times t_1 and t_2,

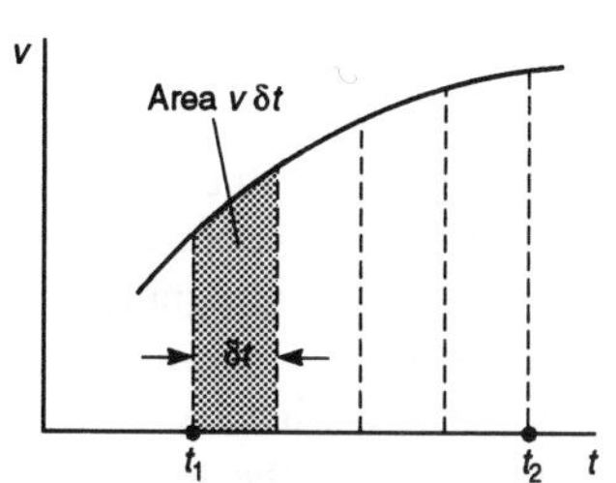

Figure 19.2 *Velocity–time graph*

distance travelled = area under graph between t_1 and t_2

= sum of the areas of all the strips between t_1 and t_2

We can write this summation as

$$x = \sum_{t=t_1}^{t=t_2} v\,\delta t$$

The Σ sign is used to indicate that we are carrying out a summation of a number of terms. The limits between which this summation is to be carried out are indicated by the information given below and above the sign. If we make δt very small, i.e. let δt tend to 0, then we denote it by dt. The sum is then the sum of a series of very narrow strips and is written as

$$x = \lim_{\delta t \to 0} \sum_{t=t_1}^{t=t_2} v\,\delta t = \int_{t_1}^{t_2} v\,\mathrm{d}t$$

The sign $\int$ is an "S" for summation and the t_1 and t_2 are said to be the limits of the range of the variable t. Here x is the *integral* of the v with time t between the limits t_1 and t_2. The process of obtaining x in this way is termed *integration*. Because the integration is between specific limits it is referred to as a *definite integral*.

The definitions of integration in terms of the reverse of differentiation and as the area under a graph describe the same concept. Suppose we increase the area under a graph of y plotted against x by one strip. Then the increase in the area δA is the area of this strip. Thus

$$\text{increase in area } \delta A = y\,\delta x$$

So we can write

$$\frac{\delta A}{\delta x} = y$$

In the limit as δx tends to 0 then we can write dA/dx and so

$$\frac{\mathrm{d}A}{\mathrm{d}x} = y$$

With integration defined as the inverse of differentiation then the integration of the above equation gives the area A, i.e.

$$A = \int y\,\mathrm{d}x$$

This is an indefinite integral, which is the same as that given by the definition for integration as the area under a graph when limits are imposed. An *indefinite integral* has no limits and the result has a constant of integration. Integration between specific limits gives a *definite integral*.

Consider the integration of y with respect to x when we have $y = 2x$. This has no specified limits and so is an indefinite integral, with the solution (as the function which differentiated would give $2x$)

$$\int 2x \, dx = x^2 + C$$

Now consider the area under the graph of $y = 2x$ between the limits of $x = 1$ and $x = 3$. We can write this as the definite integral

$$\int_1^3 2x \, dx$$

We can consider this as representing the area between $x = 3$ and $x = 0$ minus the area between $x = 1$ and $x = 0$, i.e.

$$\int_0^3 2x \, dx - \int_0^1 2x \, dx$$

The integral of $2x$ is $x^2 + C$. Thus we have the area as being that given by $x^2 + C$ when we have $x = 3$ minus that given by $x^2 + C$ when $x = 1$, i.e.

$$(9 + C) - (1 + C) = 8 \text{ square units}$$

Such a calculation is more usually written as

$$\int_1^3 2x \, dx = [x^2 + C]_1^3$$

The square brackets round the $x^2 + C$ are used to indicate that we have to impose the limits of 3 and 1 on it. Thus the integral is the value of $x^2 + C$ when $x = 3$ minus the value of $x^2 + C$ when $x = 1$.

$$\int_1^3 2x \, dx = (9 + C) - (1 + C) = 8 \text{ square units}$$

The constant term C vanishes when we have a definite integral.

Note that in the above example an area below the x-axis is negative. If the total shaded area is required then the parts below and above the x-axis must be found separately and then added, disregarding the sign of the area.

Example

Determine the area between a graph of $y = x + 1$ and the x-axis between $x = -2$ and $x = 4$.

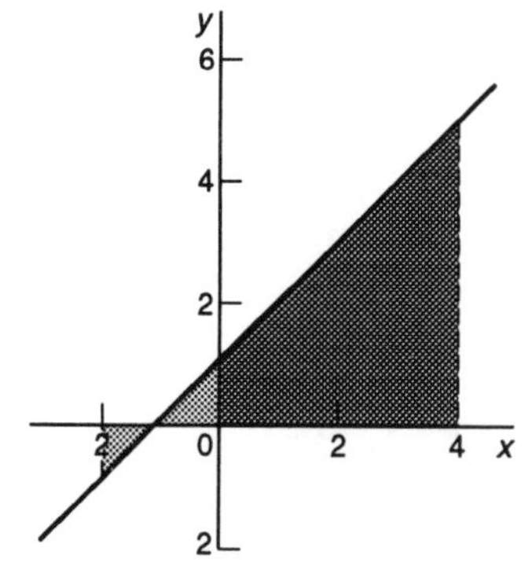

Figure 19.3

Figure 19.3 shows the graph. The area required is that between the values of x of -2 and 4. We can break this area down into a number of elements. The area under the graph between $x = 0$ and $x = 4$ is that of a rectangle 4×1 plus a triangle $\frac{1}{2}(4 \times 4)$ and so is +12 square units. The area between $x = -1$ and 0 is that of a triangle $\frac{1}{2}(1 \times 1) = 0.5$ square units and the area between $x = -1$ and $x = -2$ is a triangular area below the axis and so is negative and given by $\frac{1}{2}(1 \times 1) = -0.5$ square units. Hence the total area under the graph is $+12 + 0.5 - 0.5 = 12$ square units.

Alternatively we can consider this area as the integral

$$\text{area} = \int_{-2}^{4} (x+1)\,dx = \left[\frac{x^2}{2} + x + C\right]_{-2}^{4}$$

and so

$$\text{area} = (8 + 4 + C) - (2 - 2 + C) = 12 \text{ square units}$$

Example

Determine the value of the integral $\int_{-2}^{4} e^{2x}\,dx$.

We can consider that this integral represents the area under the graph between $x = -2$ and $x = 4$ of e^{2x} plotted against x.

$$\int_{-2}^{4} e^{2x}\,dx = \left[\tfrac{1}{2}e^{2x} + C\right]_{-2}^{4} = \tfrac{1}{2}e^{8} - \tfrac{1}{2}e^{-4} = 1490.479 - 0.009$$

The answer is thus 1490.470.

Example

Determine the value of the integral $\int_{0}^{\pi/3} \cos 2x\,dx$.

We can consider that this integral represents the area under the graph between $x = 0$ and $x = \pi/3$ of $\cos 2x$ plotted against x.

$$\int_{0}^{\pi/3} \cos 2x\,dx = \left[\tfrac{1}{2}\sin 2x + C\right]_{0}^{\pi/3} = \tfrac{1}{2}\sin 2\pi/3 - \tfrac{1}{2}\sin 0 = 0.433$$

Revision

7 Determine, by integration, the area under the following graphs between the specified limits:

(a) $y = x + 5$ between $x = -2$ and $x = 2$,

(b) $y = 4x + 1$ between $x = 1$ and $x = 2$,

(c) $y = 3x^2 + 2x$ between $x = 0$ and $x = 2$,

(d) $y = 6x^2 + 4x + 2$ between $x = -1$ and $x = 1$,

(e) $y = \sin x$ between $x = 0$ and $x = \pi/2$,

(f) $y = \cos x$ between $x = 0$ and $x = \pi/4$,

(g) $y = e^{-x}$ between $x = 0$ and $x = 2$,

(h) $y = 3 \sin 2x$ between $x = 0$ and $x = \pi/3$,

(i) $y = 8 - 2x^2$ between $x = -2$ and $x = 2$,

(j) $y = x^2 - x + 2$ between $x = -1$ and $x = 2$

8 The work done w by a force is the area under a graph of force F against displacement x in the direction of the force. Determine the work done when $F = 2x$ and x changes from 0 to 2, i.e. the value of the integral

$$\int_0^2 2x \, dx$$

9 The work done w by a force is the area under a graph of force F against displacement x in the direction of the force. Determine the work done when $F = 2x^2 + 3x$ and x changes from 2 to 5, i.e. the value of

$$\int_2^5 (2x^2 + 3x) \, dx$$

Problems

1 Determine the following integrals:

(a) $\int x^5 \, dx$, (b) $\int x^{-3} \, dx$, (c) $\int \sqrt{x} \, dx$, (d) $\int \frac{1}{\sqrt{x}} \, dx$, (e) $\int 2x \, dx$,

(f) $\int e^{-3x} \, dx$, (g) $\int e^{x/2} \, dx$, (h) $\int \cos x \, dx$, (i) $\int \sin 4x \, dx$,

(j) $\int \sin 2x \, dx$, (k) $\int \cos 2x \, dx$, (l) $\int (2x + x^2) \, dx$, (m) $\int (x^3 + 2x + 4) \, dx$,

(n) $\int (3 + e^{-2x}) \, dx$, (o) $\int (\sin 2x + \cos 3x) \, dx$

2 Determine the equations of the following curves, given the equation for the gradient of a curve and a point through which the curve must pass:

(a) $\frac{dy}{dx} = 3x^2$, the curve passes through (1, 3),

(b) $\frac{dy}{dx} = 4x - 2$, the curve passes through (2, 3),

(c) $\frac{dy}{dx} = 6x^2 + 10x - 5$, the curve passes through (−3, 5),

(d) $\frac{dy}{dx} = 2$, the curve passes through (2, 3)

3 The current i, in milliamps, through a capacitor varies with time t, in seconds, according to the equation $i = 3 \sin 100t$. If $i = dq/dt$, determine how the charge q on the capacitor will vary with time if $q = 0$ at $t = 0$.

4 The velocity v of an object varies with time t according to the equation $v = 10t + 1$. Determine how the displacement x varies with time if $x = 0$ when $t = 0$.

5 A moving iron ammeter has a current sensitivity given by

$$\frac{d\theta}{dI} = cI$$

Determine how the angular deflection θ of the pointer depends on the current I if $\theta = 0$ at $I = 0$.

6 Determine the values of the following integrals:

(a) $\int_1^4 x^2\,dx$, (b) $\int_0^{\pi/2} \sin 2x\,dx$, (c) $\int_0^4 x\,dx$, (d) $\int_{\pi/6}^{\pi/3} \cos x\,dx$, (e) $\int_0^3 \frac{1}{x}\,dx$,

(f) $\int_{-1}^{2} (x^3 - 3x^2 - 2x + 2)\,dx$, (g) $\int_1^3 (x^2 + 4)\,dx$

7 Determine the areas under the following graphs with the x-axis between the limits given:

(a) $y = x^2 - 3x$ between $x = 3$ and $x = 5$,

(b) $y = x^3$ between $x = 0$ and $x = 1$,

(c) $y = 4$ between $x = 1$ and $x = 3$,

(d) $y = 3x^2 + 1$ between $x = 1$ and $x = 4$,

(e) $y = x^3 - 3x + 2$ between $x = 1$ and $x = 3$,

(f) $y = 4x^3$ between $x = -2$ and $x = 2$,

(g) $y = \cos x$ between $x = 0$ and $x = \pi/4$,

(h) $y = 3x^2 + 2x + 4$ between $x = -1$ and $x = 3$,

(i) $y = 4 \cos x$ between $x = 0$ and $x = \pi/2$,

(j) $y = 4\,e^{3x}$ between $x = 0$ and $x = 1$.

20 Straight-line motion

20.1 Introduction

There are many situations in engineering where we are concerned with the rate at which some quantity changes or the area under a graph. An obvious example is in the case of straight-line motion where we are concerned with distances, velocities and accelerations. Velocity is the rate at which distance is covered and acceleration the rate at which velocity changes (see chapter 6 for the definitions and chapter 15 for the representation of motion using graphs). We may want, for example, to determine a velocity given a distance–time equation, i.e. obtain a rate of change, or perhaps the reverse of this and obtain a distance–time equation given a velocity–time equation. We might want to determine the distance travelled between two times by calculating the area under a velocity–time graph between those times. This chapter aims to show how calculus can be used to solve engineering problems involving straight-line motion.

20.2 Obtaining rates of change

The *velocity* of a particle is the rate at which its distance in a straight line from a fixed point changes with time. Thus if we have the distance–time graph shown in figure 20.1, then when the displacement changes by δx in a time δt the average velocity over that time interval is $\delta x/\delta t$. As δt tends to 0 then $\delta x/\delta t$ tends to a limiting value dx/dt which is the gradient of the tangent to the distance–time graph at a time t, point A on the graph, and so the instantaneous velocity v at the time t. Thus, since we have

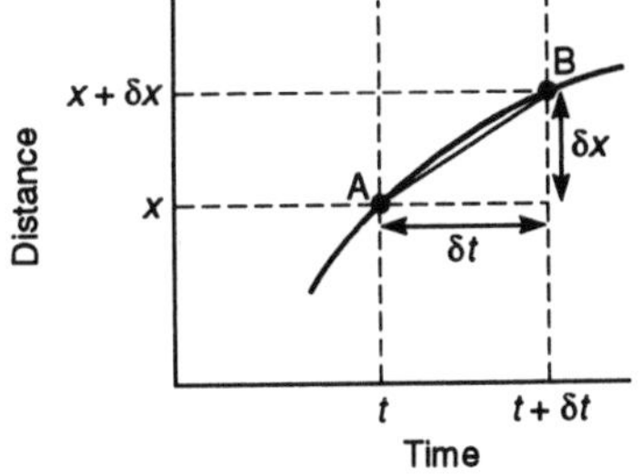

Figure 20.1 *Distance–time graph*

$$\lim_{\delta t \to 0} \frac{\delta x}{\delta t} = \frac{dx}{dt}$$

then the velocity at an instant of time is

$$v = \frac{dx}{dt}$$

dx/dt will be positive if x increases as t increases and negative if x decreases as t increases.

The *acceleration* of a particle is the rate at which the velocity changes. Thus if we have the velocity–time graph shown in figure 20.2, then when the velocity changes by δv in a time δt the average acceleration over that time interval is $\delta v/\delta t$. As δt tends to 0 then $\delta v/\delta t$ tends to a limiting value dv/dt which is the gradient of the tangent to the velocity–time graph at time t, point A on the graph, and so the acceleration a at that time t. Thus, since we have

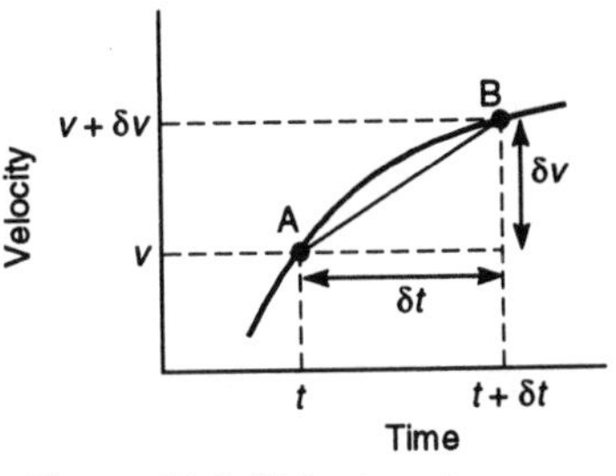

Figure 20.2 *Velocity–time graph*

$$\lim_{\delta t \to 0} \frac{\delta v}{\delta t} = \frac{dv}{dt}$$

then the acceleration at an instant of time is

$$a = \frac{dv}{dt}$$

dv/dt will be positive if v increases as t increases and negative if v decreases as t increases. The term *retardation* is used for a negative acceleration.

We can rewrite this equation for acceleration as a second derivative. Since the velocity $v = dx/dt$ and the acceleration a is the rate of change of velocity with time, then we can write

$$a = \frac{d}{dt}\left(\frac{dx}{dt}\right) = \frac{d^2x}{dt^2}$$

Thus velocity is the first derivative of displacement with time and acceleration the first derivative of velocity with time or the second derivative of displacement with time.

Example

The distance x covered by an object moving in a straight line varies with the time t according to the equation $x = 2t^2$. Determine (a) how the velocity varies with time, (b) how the acceleration varies with time, and (c) the velocity and acceleration after a time of 2 s if, in the above equation, the distance is in metres and the time in seconds.

(a) Using $v = dx/dt$ then we can obtain the velocity from the distance–time equation by differentiation. Thus

$$v = \frac{dx}{dt} = 4t$$

(b) Using $a = dv/dt$ then we can obtain the acceleration from the velocity–time equation by differentiation. Thus

$$a = \frac{dv}{dt} = 4$$

(c) The velocity $v = 4t$ and so $v = 4 \times 2 = 8$ m/s. The acceleration $a = 4$. Thus it does not vary with time and is constant at 4 m/s^2.

Example

The displacement x of a particle from some fixed point when oscillating with simple harmonic motion is given by the equation $x = A \cos \omega t$. Determine how (a) the velocity v and (b) the acceleration a vary with time t.

(a) Using $v = dx/dt$ then we can obtain the velocity from the distance–time equation by differentiation. Thus

$$v = \frac{dx}{dt} = -A\omega \sin \omega t$$

(b) Using $a = \mathrm{d}v/\mathrm{d}t$ then we can obtain the acceleration from the velocity–time equation by differentiation. Thus

$$a = \frac{\mathrm{d}v}{\mathrm{d}t} = -A\omega^2 \cos \omega t$$

Since $x = A \cos \omega t$ then we can rewrite this equation as $a = -\omega^2 x$.

Example

The distance x covered by an object moving in a straight line varies with the time t according to the equation $x = 2t - t^3$. Determine (a) how the velocity varies with time, (b) how the acceleration varies with time, and (c) the velocity and acceleration after a time of 2 s if, in the above equation, the distance is in metres and the time in seconds.

(a) Using $v = \mathrm{d}x/\mathrm{d}t$ then we can obtain the velocity from the distance–time equation by differentiation. Thus

$$v = \frac{\mathrm{d}x}{\mathrm{d}t} = 2 - 3t^2$$

(b) Using $a = \mathrm{d}v/\mathrm{d}t$ then we can obtain the acceleration from the velocity–time equation by differentiation. Thus

$$a = \frac{\mathrm{d}v}{\mathrm{d}t} = -6t$$

(c) The velocity is $v = 2 - 3t^2 = 2 - 12 = -10$ m/s. The negative sign means that it is moving in the opposite direction to that in which we measure distance outwards from the fixed reference point, i.e. it is moving back towards the point and not away from it. The acceleration $a = -6t = -6 \times 2 = -12$ m/s^2. The negative sign means that there is a retardation and the object is slowing down.

Revision

1 The distance x covered by an object moving in a straight line varies with the time t according to the equation $x = 10t - t^2$. Determine (a) how the velocity varies with time, (b) how the acceleration varies with time, and (c) the velocity and acceleration after a time of 2 s if, in the above equation, the distance is in metres and the time in seconds.

2 The distance x covered by an object moving in a straight line varies with the time t according to the equation $x = \sqrt{t}$. Determine (a) how the velocity and (b) the acceleration vary with time.

3 When an object is thrown vertically upwards, its height h varies with time t and is given by the equation $h = 4t - 5t^2$. Determine the velocity after 0.2 s.

4 Determine the velocity and the acceleration after 2 s for an object for which the displacement x, in metres, is related to the time t, in seconds, by $x = 10 + 20t - 2t^2$.

5 Determine the equations for the velocity and acceleration of a body after a time t if its displacement x varies with time t and is given by the equation $x = \sin 2t + \cos 3t$.

6 A ball rolls from rest down an inclined plane. If the distance travelled down the plane x, in metres, is related to time t, in seconds, by the equation $x = 3t^2$ determine (a) how the velocity varies with time, (b) how the acceleration varies with time and (c) the velocity and the acceleration after the ball has rolled 3 m.

20.3 Reversing differentiation

As illustrated in the previous section of this chapter, we can obtain the velocity–time relationship for an object in motion by differentiating the distance–time relationship. Integration can be considered the reverse of differentiation. Thus, given the equation relating the velocity and time for an object in motion, e.g.

$$v = \frac{dx}{dt} = 4t$$

then we can find the distance–time relationship by integration. In this case,

$$x = \int v\,dt = \int 4t\,dt$$

If we want to find the distance travelled between two times, t_1 and t_2, then the integral is written as

$$x = \int_{t_1}^{t_2} v\,dt = \int_{t_1}^{t_2} 4t\,dt$$

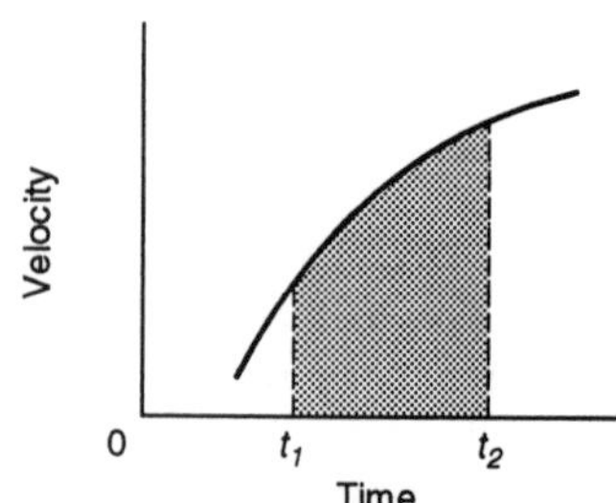

Figure 20.3 *Velocity-time graph*

This represents the area under a velocity–time graph in that time interval (figure 20.3).

We can obtain the acceleration–time relationship for an object in motion by differentiating the velocity–time relationship. Since integration is the reverse of differentiation, given an equation relating the acceleration and time for an object in motion, e.g.

$$a = \frac{dv}{dt} = 4$$

then we can find the velocity–time relationship by integration. In this case,

$$v = \int a\,dt = \int 4\,dt$$

If we want the velocity between two times, t_1 and t_2, then the integral is written as

$$v = \int_{t_1}^{t_2} a\,dt = \int_{t_1}^{t_2} 4\,dt$$

Example

The acceleration a, in m/s^2, of an object depends on the time t, in seconds, and is given by the equation $a = t + 2$. Determine the equation describing how the velocity varies with time if the velocity is 2 m/s at the time $t = 0$.

We can write the acceleration as dv/dt and so we have

$$\frac{dv}{dt} = t + 2$$

Integrating this equation gives

$$v = \int (t+2)\,dt = \tfrac{1}{2}t^2 + 2t + C$$

We are given the information that $v = 2$ m/s at $t = 0$. Thus $C = 2$. Hence the equation is

$$v = \tfrac{1}{2}t^2 + 2t + 2$$

Example

The velocity v, in m/s, of an object is related to the time t, in seconds, by the equation $v = t^2 + 2t + 1$. Determine the distance travelled by the object between $t = 2$ s and $t = 3$ s.

The velocity v can be written as dx/dt and so we have

$$v = \frac{dx}{dt} = t^2 + 2t + 1$$

The distance–time relationship can be obtained by integrating this equation. Since we require the distance travelled between two times then the integral becomes

$$x = \int_{t_1}^{t_2} v\,dt = \int_2^3 (t^2 + 2t + 1)\,dt = \left[\tfrac{1}{3}t^3 + t^2 + t + C\right]_2^3$$

Thus $x = (9 + 9 + 3 + C) - (\tfrac{8}{3} + 4 + 2 + C) = 12.3$ m

Example

The acceleration a, in m/s^2, of a moving object varies with time t, in seconds, and is given by the equation $a = 4t + 1$. Determine (a) the velocity–time equation and (b) the distance–time equation, given that $v = 2$ m/s at $t = 0$ and $x = 5$ m at $t = 0$.

(a) The acceleration a can be written as $a = \mathrm{d}v/\mathrm{d}t$ and so

$$a = \frac{\mathrm{d}v}{\mathrm{d}t} = 4t + 1$$

Thus integrating this gives

$$v = \int (4t + 1)\,\mathrm{d}t = 2t^2 + t + C$$

We have $v = 2$ m/s at $t = 0$ and so $2 = 0 + 0 + C$. Hence the equation is

$$v = 2t^2 + t + 2$$

(b) The velocity v can be written as $\mathrm{d}x/\mathrm{d}t$ and so we have

$$v = \frac{\mathrm{d}x}{\mathrm{d}t} = 2t^2 + t + 2$$

Thus integrating this gives

$$x = \int (2t^2 + t + 2)\,\mathrm{d}t = \tfrac{2}{3}t^3 + \tfrac{1}{2}t^2 + 2t + C$$

We have $x = 5$ m at $t = 0$ and so $5 = 0 + 0 + 0 + C$. Thus the equation is

$$x = \tfrac{2}{3}t^3 + \tfrac{1}{2}t^2 + 2t + 5$$

Revision

7 The velocity v, in m/s, of a moving object varies with time t, in seconds, and is given by the equation $v = 8t - t^2$. Determine the distance travelled in the time from $t = 0$ to $t = 3$ s.

8 The velocity v, in m/s, of a moving object varies with time t, in seconds, and is given by the equation $v = 2t + 5$. Determine the distance travelled in the time from $t = 0$ to $t = 4$ s.

9 The velocity v, in m/s, of a moving object varies with time t, in seconds, and is given by the equation $v = 5t^2 + 3t^2$. Determine the distance travelled in the time from $t = 1$ to $t = 2$ s.

10 The velocity v, in m/s, of a moving object varies with time t, in seconds, and is given by the equation $v = 4t - 2t^2$. If at time $t = 0$ the displacement is 1 m, what will be the displacement at 4 s?

11 The acceleration a, in m^2/s, of a moving object varies with time t, in seconds, and is given by the equation $a = 2t + 1$. Determine the velocity–time relationship given that the velocity is 10 m/s at $t = 2$ s.

12 The acceleration a, in m/s^2, of a moving object varies with time t, in seconds, and is given by the equation $a = 6t$. Determine (a) the velocity–time equation and (b) the distance–time equation, given that $v = 4$ m/s at $t = 0$ and $x = 10$ m at $t = 0$.

Problems

1 Determine the velocity and acceleration after a time t for an object which has a displacement x from a fixed point which changes with time according to the equation $x = 2t^2 - 3t + 4$.

2 Determine the velocity and acceleration after a time t for an object which has a displacement x from a fixed point which changes with time according to $x = 10 \cos 2t$.

3 Determine the velocity and acceleration after a time t for an object which has a displacement x from a fixed point which changes with time according to the equation $x = 1 - e^{-2t}$.

4 Determine the velocity and acceleration after a time t for an object which has a displacement x from a fixed point which changes with time according to the equation $x = 3t^3 - t$.

5 Determine the velocity and acceleration after a time t for an object which has a displacement x from a fixed point which varies with time according to the equation $x = 1 + 6t - 2t^2$.

6 Determine the velocity and the acceleration after 2 s for an object for which the displacement x, in metres, is related to the time t, in seconds, by $x = 20 + 10t - 2t^2$.

7 Determine the velocity and the acceleration after 2 s for an object for which the displacement x, in metres, is related to the time t, in seconds, by $x = 20t - 4t^2$.

8 A particle has a displacement x, in metres, which varies with time t, in seconds, according to $x = 10 + 20t - t^2$. At what time will the velocity be zero?

9 The velocity v, in m/s, of a moving object depends on the time t, in seconds, and is given by the equation $v = 2t + 1$. Determine the acceleration.

10 A particle has a displacement x, in metres, which varies with time t, in seconds, according to $x = 20t + 12t^2 - 2t^3$. At what time will the acceleration be zero?

11 The velocity v, in m/s, of a moving object varies with time t, in seconds, and is given by the equation $v = 36 - 4t$. If at time $t = 1$ s the displace-

ment is 20 m derive an equation for how the displacement varies with time.

12 The velocity v, in m/s, of a moving object varies with time t, in seconds, and is given by the equation $v = 4t + 6t^2$. If at time $t = 0$ the displacement is 20 m what will be the displacement at 4 s?

13 The velocity v, in m/s, of a moving object varies with time t, in seconds, and is given by the equation $v = 4t + 1$. If at time $t = 0$ the displacement is 0 what will be the displacement at 1 s?

14 The velocity v, in m/s, of a moving object varies with time t, in seconds, and is given by the equation $v = 2\ e^{-t}$. If at time $t = 0$ the displacement is 0 what will be the displacement at 4 s?

15 The velocity v, in m/s, of a moving object varies with time t, in seconds, and is given by the equation $v = 48 - 3t$. If at time $t = 0$ the displacement is 0 what will be the time when the object is back at its starting point?

16 The acceleration a, in m/s^2, of a moving object varies with time t, in seconds, and is given by the equation $a = 4t^2$. Determine the velocity–time relationship given that the velocity is 5 m/s at $t = 1$ s.

17 The acceleration a, in m/s^2, of a moving object varies with time t, in seconds, and is given by the equation $a = 6t + 2$. Determine the velocity–time relationship given that the velocity is 10 m/s at $t = 1$ s.

18 The acceleration a, in m/s^2, of a moving object varies with time t, in seconds, and is given by the equation $a = 2\sqrt{t}$. Determine the velocity–time relationship given that the velocity is 10 m/s at $t = 4$ s.

21 Capacitors & inductors

21.1 Introduction

This chapter is an introduction to the behaviour of capacitors and inductors in electrical circuits. This involves considering the rate at which charge is moved through circuit elements and the rates at which voltages and currents change.

21.1.1 Current

Electrical *current*, i, is the rate of movement of charge q with time t. Thus if δq moves in a time δt then the average current over that time is $\delta q/\delta t$. In the limit as δt tends to zero, the current i at an instant is given by

$$i = \frac{\mathrm{d}q}{\mathrm{d}t}$$

The unit of charge is defined so that a current of 1 ampere (A) is a movement of charge at the rate of 1 coulomb (C) per second.

If we require to know the charge moved through a circuit when a current flows then we require the reverse of the above. Since integration is the reverse of differentiation then

$$q = \int i\,\mathrm{d}t$$

Example

Derive an equation for the current through a circuit element if the charge q, in coulombs, entering the element is related to the time t, in seconds, by the equation $q = 4t$.

Current is the rate of movement of charge. Thus

$$i = \frac{\mathrm{d}q}{\mathrm{d}t} = 4\text{ A}$$

Example

Determine the charge that has flowed through a circuit between times of $t = 1$ s and $t = 2$ s when the current i, in amperes, is given by $2\,\mathrm{e}^{-0.1t}$.

Using the equation given above, then

$$q = \int i\,\mathrm{d}t = \int_1^2 2\,\mathrm{e}^{-0.1t}\,\mathrm{d}t = \left[2 \times \left(-\frac{1}{0.1}\right)\mathrm{e}^{-0.1t} + C\right]_1^2$$

$$= -20\,e^{-0.2} + 20\,e^{-0.1} = -16.37 + 18.10 = 1.73\text{ C}$$

Revision

1 Derive an equation for the current through a circuit element if the charge q, in coulombs, entering the element is related to the time t, in seconds, by $q = 10t^2$.

2 Derive an equation for the current through a circuit element if the charge q, in coulombs, entering the element is related to the time t, in seconds, by $q = 4\,e^{-3t}$.

3 Derive an equation for the current through a circuit element if the charge q, in coulombs, entering the element is related to the time t, in seconds, by $q = 2t + e^{-5t}$.

4 Derive an equation for the current through a circuit element if the charge q, in coulombs, entering the element is related to the time t, in seconds, by $q = 2 \cos 5t$.

5 Determine the charge that has flowed through a circuit between times of $t = 0$ s and $t = 1$ s when the current i, in amperes, is given by $3 + e^{-0.1t}$.

6 Determine the charge that has flowed through a circuit between times of $t = 1$ s and $t = 2$ s when the current i, in amperes, is given by $2t$.

21.2 Capacitors

The circuit element called a capacitor is essentially just a pair of parallel conducting plates separated by an insulator. When a voltage is applied across a capacitor then one of the plates becomes negatively charged and the other positively charged as a result of charge flowing on to one of the plates and leaving the other plate (figure 21.1). The amount of charge gained by one plate is the same as that lost by the other plate. It is found that the amount of charge q on a capacitor plate is directly proportional to the potential difference v between the plates. Hence we can write $q \propto v$ and so

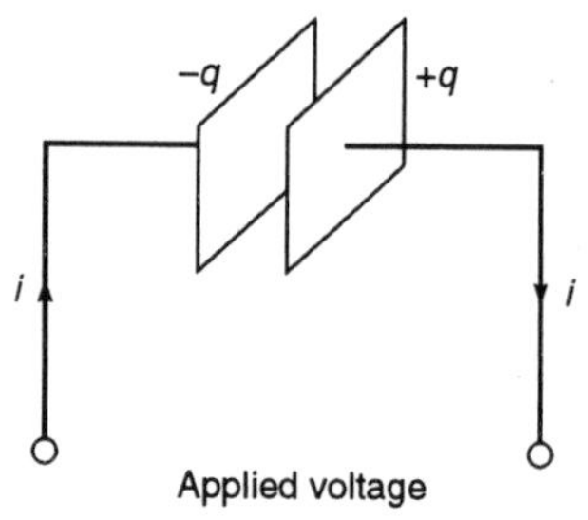

Figure 21.1 *Capacitor*

$$q = Cv$$

where C, the constant of proportionality, is termed the *capacitance*. Capacitance has the unit of a farad (F) when the voltage is in volts (V) and the charge in coulombs (C).

Consider a small element of charge δq moving onto one plate of the capacitor and off the other. Then the resulting change in potential difference δV is given by

$$\delta q = C\,\delta v$$

If we divide both sides of this equation by the time interval δt during which the change is taking place, then

$$\frac{\delta q}{\delta t} = C\frac{\delta v}{\delta t}$$

If we consider the changes when δt tends to a zero value, then we can write

$$\frac{\mathrm{d}q}{\mathrm{d}t} = C\frac{\mathrm{d}v}{\mathrm{d}t}$$

Since current i is the rate of movement of charge, we can write the above equation as

$$i = \frac{\mathrm{d}q}{\mathrm{d}t} = C\frac{\mathrm{d}v}{\mathrm{d}t}$$

Thus the current in the circuit at some instant of time is the product of the capacitance and the rate at which the potential difference across the capacitor is changing with time.

Example

The voltage v, in volts, across a capacitor is continuously adjusted so that it varies with time t, in seconds, according to the equation $v = 2t$. Derive an equation indicating how the current varies with time for a capacitance of 1 mF.

With $v = 2t$ then we have

$$i = C\frac{\mathrm{d}v}{\mathrm{d}t} = 0.001 \times 2 = 0.002 \text{ A}$$

The current is thus maintained constant at 2 mA.

Example

The voltage v, in volts, across a capacitor varies with time t, in seconds, according to the equation $v = 10 \sin 3t$. How will the current vary with time if the capacitance is 0.5 mF?

With $v = 10 \sin 3t$ then

$$i = C\frac{\mathrm{d}v}{\mathrm{d}t} = 0.005 \times 30 \cos 3t = 0.15 \cos 3t \text{ A}$$

The current is thus given by a cosine curve.

Revision

7 The voltage v, in volts, across a capacitor varies with time t, in seconds, according to the equation $v = 10t$. How will the current vary with time if the capacitance is 0.5 mF?

8 The voltage v, in volts, across a capacitor varies with time t, in seconds, according to the equation $v = 4 \sin 10t$. How will the current vary with time if the capacitance is 1 mF?

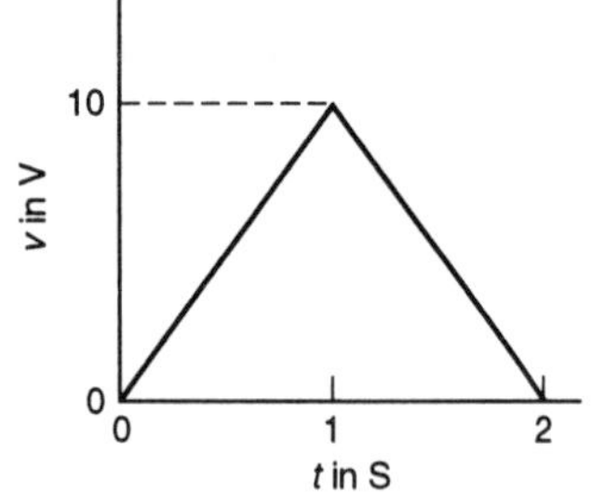

Figure 21.2

9 The voltage v, in volts, across a capacitor varies with time t, in seconds, according to the equation $v = 4 \cos 10t$. How will the current vary with time if the capacitance is 1 mF?

10 The voltage v, in volts, across a capacitor varies with time t, in seconds, as shown in figure 21.2. How will the current vary with time if the capacitance is 1 mF?

21.3 Inductors

An inductor is essentially just a coil of wire. When there is a current passing through a wire then a magnetic field is produced. We talk of there being magnetic flux generated by the current. The magnetic flux is proportional to the current. When the current through the coil changes then the magnetic flux generated by the current changes. The magnetic flux passing through the coil thus changes. Hence, as a result of electromagnetic induction, an e.m.f. is induced in the coil. This induced e.m.f. is in such a direction as to oppose the change producing it. Thus if it was produced by an increasing current the direction of the e.m.f. will be such as to slow down the increase. If it was produced by a decreasing current then the direction of the e.m.f. is such as to slow the decrease in current. The induced e.m.f. is proportional to the rate of change of magnetic flux. Hence, since the amount of magnetic flux is proportional to the current we have the induced e.m.f. proportional to the rate of change of current. We can thus write

$$\text{e.m.f.} \propto -\frac{\mathrm{d}i}{\mathrm{d}t}$$

The minus sign is because the e.m.f. opposes the change in current responsible for it. This equation can then be written as

$$\text{e.m.f.} = -L\frac{\mathrm{d}i}{\mathrm{d}t}$$

L, the constant of proportionality, is called the *inductance*. The unit of inductance is the henry (H) when the e.m.f. is in volts (V), the current in amperes (A) and the time in seconds (s).

When we have a current through a resistance in a circuit then the potential difference across the resistor is given by $V = RI$. However, if we are concerned with a pure inductance, i.e. one which has only inductance and no resistance, then since $R = 0$ there can be no potential difference across the inductor due to its resistance. To maintain a current through the inductor then the source must supply a potential difference v across the inductor which is just sufficient to cancel out the induced e.m.f. Thus

$$v = L\frac{\mathrm{d}i}{\mathrm{d}t}$$

Then we have (v + induced e.m.f.) = 0. The potential difference applied across the inductor is thus proportional to the rate of change of current through it.

Example

The current i, in amperes, through an inductor of inductance 2 H varies with time t in seconds, being related to the time by the equation $i = 2t$. Derive a relationship for the voltage across the inductor.

Using the equation derived above,

$$v = L\frac{\mathrm{d}i}{\mathrm{d}t} = 2 \times 2 = 4 \text{ V}$$

The voltage is thus constant at 4 V.

Example

The current i, in amperes, through an inductor of inductance 0.2 H varies with time t in seconds, being related to the time by the equation $i = 2 \sin 10t$. Derive a relationship for the voltage across the inductor.

Using the equation $v = L\,\mathrm{d}i/\mathrm{d}t$,

$$v = L\frac{\mathrm{d}i}{\mathrm{d}t} = 0.2 \times 2 \times 10 \cos 10t = 4 \cos 10t \text{ V}$$

The voltage is given by a cosine graph, the current by a sine graph. The voltage and current are thus out of phase.

Example

The voltage v, in volts, applied to an inductor of inductance 0.1 H varies with time t, in seconds, being related to the time by the equation $v = 20\ \mathrm{e}^{-10t}$. Derive a relationhip for how the current varies with time if the current is zero when $t = 0$.

We need to reverse the equation $v = L\,\mathrm{d}i/\mathrm{d}t$ for this problem. Hence

$$i = \frac{1}{L}\int v\,\mathrm{d}t = \frac{1}{0.1}\int 20\,\mathrm{e}^{-10t}\,\mathrm{d}t = -20\ \mathrm{e}^{-10t} + \mathrm{C}$$

When $t = 0$ then $i = 0$. Thus $0 = -20 + C$. Hence $C = 20$. Thus the equation is

$$i = = -20\ e^{-10t} + 20 = 20(1 - e^{-10t})\ A$$

Revision

11 The current i, in amperes, through an inductor of inductance 2 H varies with time t in seconds, being related to the time by the equation $i = 5t$ Derive a relationship for the voltage across the inductor.

12 The current i, in amperes, through an inductor of inductance 0.1 H varies with time t, in seconds, according to $i = 2(1 - e^{-100t})$. Determine how the potential difference across the inductor varies with time.

13 The current i, in amperes, through an inductor of inductance 0.01 H varies with time t, in seconds, and is described by $i = 10\,e^{-2t}$. Determine how the potential difference across the inductor varies with time.

14 The voltage v, in volts, across an inductor of inductance 1 H varies with time t, in seconds, and is described by the equation $v = 9t^2$. Determine how the current through the inductor varies with time if the current is zero at $t = 0$.

Problems

1 Derive an equation for the current through a circuit element if the charge q, in coulombs, entering the element is related to the time t, in seconds, by $q = 2t^2$.

2 Derive an equation for the current through a circuit element if the charge q, in coulombs, entering the element is related to the time t, in seconds, by $q = 5\ e^{-2t}$.

3 Derive an equation for the current through a circuit element if the charge q, in coulombs, entering the element is related to the time t, in seconds, by $q = 2t + 5\ e^{-2t}$.

4 Derive an equation for the current through a circuit element if the charge q, in coulombs, entering the element is related to the time t, in seconds, by $q = 5 \sin 2t$.

5 Determine the charge that has flowed through a circuit between times of $t = 1$ s and $t = 2$ s when the current i, in amperes, varies with time and is given by the equation $0.2\ e^{-0.5t}$.

6 Determine the charge that has flowed through a circuit between times of $t = 0$ and $t = 2$ s when the current i, in amperes, is given by the equation $i = 0.2t$.

7 The voltage v across a capacitor varies with time t. Derive equations for how the current varies with time if (a) $v = a$, with a being a constant, (b) $v = at$, with a being a constant, (c) $v = V \cos \omega t$.

8 The voltage v, in volts, across a capacitor varies with time t, in seconds, according to the equation $v = 2 \sin 100t$. How will the current vary with time if the capacitance is 1 mF?

9 The voltage v, in volts, across a capacitor varies with time t, in seconds, according to the equation $v = 4t^2$. How will the current vary with time if the capacitance is 2 mF?

10 The current i, in amperes, through an inductor of inductance 0.1 H varies with time t in seconds, being related to the time by the equation $i = 0.2t$. Derive a relationship for the voltage across the inductor.

11 The current i, in amperes, through an inductance of 0.01 H varies with time t, in seconds, according to the equation $i = 10\,e^{-3t}$. Determine how the potential difference across the inductor varies with time.

12 The current i, in amperes, through an inductance of 0.2 H varies with time t, in seconds, according to the equation $i = 4\,e^{-20t} - 2\,e^{-40t}$. Determine how the potential difference across the inductor varies with time.

13 The current applied to an inductor increases at a constant rate for 1 s and then becomes constant. Determine the waveform of the voltage across the inductor.

14 The voltage v, in volts, across an inductor of inductance 1 H varies with time t, in seconds, and is described by the equation $v = 4t$. Determine how the current through the inductor varies with time if the current is zero at $t = 0$.

22 Work

22.1 Introduction

Work is the term given to the process by which energy is transferred by the point of application of a force moving through a distance. With a constant force F and a displacement x in the direction of the force, then the work done W on a body, i.e. the energy transferred to it, is given by

$$W = Fx$$

The unit of work is the joule (J) when the force is in newtons and the displacement in metres.

This chapter is a brief consideration of the work done when the force applied to an object varies with displacement.

22.2 Work done as an integral

Consider the graph shown in figure 22.1 for the force applied to an object and how it varies with the displacement of that object. For a small displacement δx we can consider the force to be effectively constant at F. Thus the work done for that displacement is $F\,\delta x$. This is the area of the strip under the force–distance graph. If we want the work done in changing the displacement from x_1 to x_2 then we need to determine the sum of all such strips between these displacements, i.e. the total area under the graph between the ordinates for x_1 and x_2. Thus

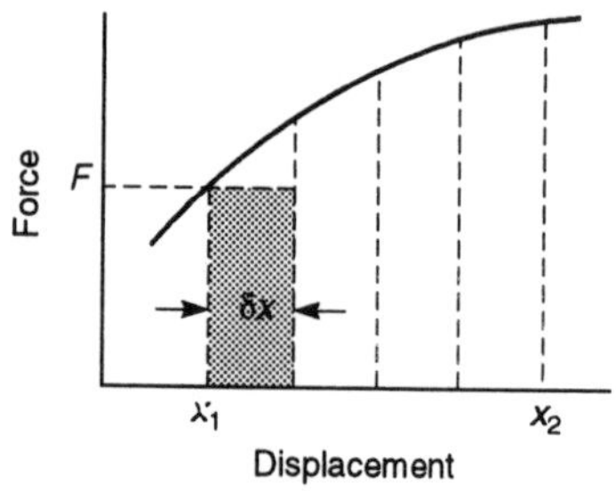

$$\text{work done} = \sum_{x=x_1}^{x=x_2} F\,\delta x$$

Figure 22.1 *Force-distance graph*

If we make the strips tend towards zero thickness then the above summation becomes the integral, i.e.

$$\text{work done} = \int_{x_1}^{x_2} F\,\mathrm{d}x$$

Example

Determine the work done in stretching a spring when a force F is applied and causes a displacement change in its point of application, i.e. an extension, from 0 to x if $F = kx$, where k is a constant.

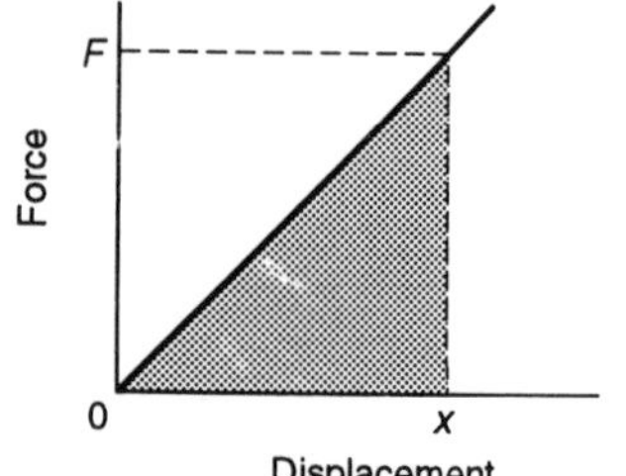

Figure 22.2

Figure 22.2 shows the force–distance graph. The work done is the area under the graph between 0 and x. This is the area of a triangle and so the work done is $\frac{1}{2}Fx$. Since $F = kx$ we can write this as $\frac{1}{2}kx^2$.

We could have solved this problem by integration. Thus

$$\text{work done} = \int_0^x F\,\mathrm{d}x = \int_0^x kx\,\mathrm{d}x = \left[\tfrac{1}{2}kx^2 + C\right]_0^x = \tfrac{1}{2}kx^2$$

Revision

1 Determine the work done in stretching a spring when a force F is applied and causes a displacement change in its point of application, i.e. an extension, from 0 to x if $F = kx^2$, where k is a constant.

2 An object falling under gravity from rest in air experiences a resistive force of $0.6s$, where s is the distance it has fallen. Determine the work done by the falling object against this resistive force when it falls through a distance of 10 m.

3 A horizontal force F, in newtons, acts on an object and causes the object to move from its starting position at $s = 0$ through a distance of 3 m in the direction of the force. If $F = 2s + 3$, determine the work done.

22.2.1 Liquids and gases

Consider work being done as a result of a piston being moved, as in figure 22.3, to reduce the volume of a gas. The work done in moving the piston through a small distance δx when the force is F is $F\,\delta x$. But the gas is subject to a pressure as a result of the force. Since pressure is force per unit area, then if the force acts over an area A the pressure $p = F/A$. Thus we can write for the work done

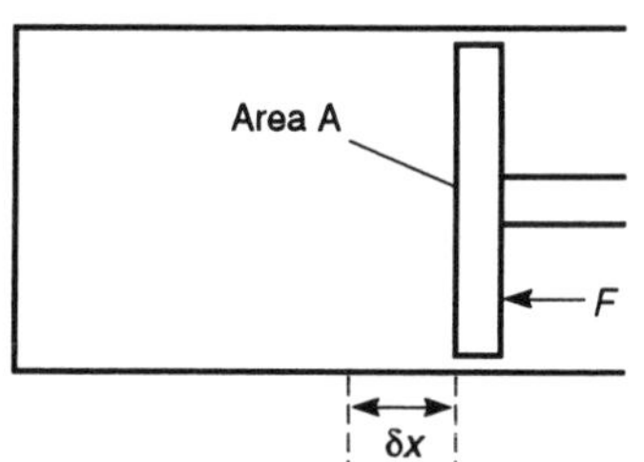

Figure 22.3 *Compressing a gas*

$$\text{work done} = F\,\delta x = pA\,\delta x$$

But $A\,\delta x$ is the change in volume δV of the gas. Hence

$$\text{work done} = p\,\delta V$$

The total work done in changing the volume of a gas from V_1 to V_2 is thus

$$\text{work done} = \sum_{V_1}^{V_2} p\,\delta V$$

If we consider δV tending to zero then we can write

$$\text{work done} = \int_{V_1}^{V_2} p\,\mathrm{d}V$$

Example

Determine the equation for the work done in compressing a gas from a volume V_1 to V_2 if the gas obeys Boyle's law, i.e. pV = a constant.

For the gas we have $pV = k$ and hence $p = k/V$. The work done is thus

$$\text{work done} = \int_{V_1}^{V_2} p\,\mathrm{d}V = \int_{V_1}^{V_2} \frac{k}{V}\mathrm{d}V = [\ln V + C]_{V_1}^{V_2}$$

Hence the work done is $\ln V_2 - \ln V_1$.

Revision

4 Determine the equation for the work done in compressing a gas from a volume V_1 to V_2 if the gas obeys the law $pV^{5/3}$ = a constant.

Problems

1 Determine the work done in stretching a spring when a force F is applied and causes a displacement change in its point of application, i.e. an extension, from 0 to x if $F = k\sqrt{x}$, where k is a constant.

2 A horizontal force F, in newtons, acts on an object and causes the object to move from its starting position at $s = 0$ through a distance of 2 m in the direction of the force. If $F = 3s^2 + 1$, determine the work done.

3 An object is projected upwards against gravity in air. It experiences a resistive force of $2s$, where s is the distance vertically upwards from the point of projection. Determine the work done against the resistance in the object moving vertically upwards from its start position to a height h.

4 A horizontal force F, in newtons, acts on an object and causes the object to move from its starting position at $s = 0$ through a distance of 4 m in the direction of the force. If $F = 4s + 3$, determine the work done.

5 The force F between two charged particles is given by $F = k/x^2$, where x is the distance between them and k a constant. Determine the work done in fetching a charged particle from infinity, where $F = 0$, up to a distance r from another charged particle.

6 Determine the equation for the work done in compressing a gas from a volume V_1 to V_2 if the gas obeys the law $pV^{1.4}$ = a constant.

Answers

Chapter 1

Revision

1 (a) $\frac{5}{6}$, (b) $\frac{8}{15}$, (c) $\frac{3}{10}$, (d) $\frac{7}{6}$, (e) $\frac{13}{20}$

2 (a) $\dfrac{2b+3a}{ab}$, (b) $\dfrac{R_2+R_1}{R_1R_2}$, (c) $\dfrac{3y-5x}{xy}$, (d) $\dfrac{3a+2b}{6}$

3 (a) $\frac{1}{15}$, (b) $\frac{8}{21}$, (c) $\dfrac{x}{15}$, (d) $\dfrac{ab}{12}$, (e) $\frac{2}{15}$, (f) $\dfrac{4a}{3b}$, (g) $\frac{7}{4}$

4 (a) 2^7, (b) a^7, (c) x^9, (d) a^5b^{10}, (e) 2^6x^6

5 (a) a^2, (b) 2^2, (c) a^{-1}, (d) $54x$, (e) 2^{-1}

6 (a) $a^{7/2}$, (b) $x^{3/2}$, (c) $x^{1/2}$, (d) $2^{-1/2}$, (e) $a^{29/6}$, (f) $x^{1.6}$

7 (a) 243, (b) 0.03125, (c) 1.621, (d) 1.442, (e) 2.83, (f) 4.50, (g) 0.885, (h) 0.61

8 (a) $x = 5$, (b) $x = 2$, (c) $x = 1$, (d) $x = 24$, (e) $x = 6$, (f) $x = 2$, (g) $x = 2$, (h) $x = 7$, (i) $x = -\frac{7}{2}$, (j) $x = \frac{5}{6}$, (k) $x = 16$, (l) $x = 4$

9 (a) 5 Ω, (b) 0.5 N/mm, (c) 10 N, (d) 8.7 Ω

10 As question 8

11 (a) $a = \dfrac{v-u}{t}$, (b) $r_1 = \dfrac{F_2r_2}{F_1}$, (c) $x = \dfrac{FL}{AE}$, (d) $L = \dfrac{T^2g}{4\pi^2}$, (e) $X = \sqrt{Z^2-R^2}$, (f) $m = \dfrac{pV}{RT}$, (g) $v = \sqrt{\dfrac{2E}{m}}$, (h) $a = \dfrac{2(s-ut)}{t^2}$, (i) $d = \sqrt{\dfrac{6Z}{b}}$

12 0.000 04 per °C

13 4000 N

14 0.15 m^3

15 N/m^2 (or Pa)

16 N m

17 J kg^{-1} K^{-1}

18 Ω m

19 kg m^2 s^{-2}

Problems

1 (a) $\frac{14}{15}$, (b) $\frac{1}{30}$, (c) $\frac{19}{21}$, (d) $\dfrac{c+b}{ac}$, (e) $\frac{8}{21}$, (f) $\frac{2}{15}$, (g) $\dfrac{x}{3y}$, (h) $\frac{9}{4}$, (i) $\frac{28}{5}$, (j) $\dfrac{4a}{5b}$, (k) $\dfrac{b}{a+b}$

2 (a) 3^8, (b) a^5, (c) a^3, (d) a^2, (e) 2^23^2 or 36, (f) x^{-6}, (g) $a^{11/2}$, (h) $x^{-3/2}$, (i) $a^{5/2}$, (j) $x^{5/2}$, (k) $\frac{1}{16}$, (l) $2^{7/2}$, (m) ab^2, (n) $\dfrac{a+1}{a^2}$, (o) $\dfrac{a}{4b^4}$

3 (a) 128, (b) 0.0625, (c) 2.408, (d) 3.085

4 (a) Yes, (b) No

5 (a) $x = \frac{1}{3}$., (b) $x = 16$, (c) $x = \frac{3}{8}$, (d) $x = \frac{4}{15}$, (e) $x = 11$, (f) $x = 7$, (g) $x = 7$, (h) $x = 8$, (i) $x = \frac{1}{2}$, (j) $x = 2$, (k) $x = 16$, (l) $x = 2$

6 79.8 mm

7 5 s
8 1.02 m
9 J/kg
10 kg/m^3
11 N m
12 (a) $h = \frac{p_1 - p_2}{\rho g}$, (b) $R = \frac{\rho L}{A}$, (c) $r = \frac{V - E}{I}$, (d) $T = I\alpha + mr(a + g)$,
(e) $F = \frac{2m}{t^2}(s - ut)$, (f) $x = \frac{mgL}{EA}$, (g) $R = \frac{EI}{M}$, (h) $m = \frac{2E}{v^2}$,
(i) $r = \sqrt{\frac{3V}{\pi h}}$, (j) $R = \frac{P}{I^2}$, (k) $a_2 = -\frac{m_1 a_1}{m_2}$, (l) $v = \sqrt{\frac{Fr}{m}}$, (m) $d = \frac{\varepsilon_0 \varepsilon_r A}{C}$
13 $T_2 = \frac{V_2 T_1}{V_1}$, 600 K
14 $w = \frac{2R - W}{L}$, 4 kN

Chapter 2

Revision

1 (a) $x = 1$, (b) $x = 2$, (c) $x = 3$, (d) $x = 0.86$, (e) $x = 1$, (f) $x = -\frac{5}{2}$,
(g) $x = -0.83$, (h) $x = 3$, (i) $x = -6$, (j) $x = 4.4$, (k) $x = 1.9$, (l) $x = 6.94$,
(m) $x = 12.2$
2 (a) 0.75 A, (b) 4.35×10^{-3} °C^{-1}, (c) 2.15 Ω
3 (a) $x = 1, y = 1$, (b) $x = 2, y = 1$, (c) $x = 1, y = 3$, (d) $x = 4, y = -1$,
(e) $x = -1, y = 2$, (f) $x = 5, y = -2$, (g) $x = 1, y = 5$, (h) $x = \frac{1}{2}, y = 1$,
(i) $x = 2, y = 1$, (j) $x = 2, y = 4$
4 $R_1 = 3.85$ kN, $R_2 = 1.65$ kN
5 $I_1 = -0.71$ A, $I_2 = 1.14$ A
6 $F_1 = 1$ N, $F_2 = 1$ N
7 $u = 2$ m/s, $a = 3$ m/s^2
8 As given in revision problems 3, 4, 5, 6 and 7

Problems

1 (a) $x = 1$, (b) $x = 3.53$, (c) $x = 2$, (d) $x = -0.70$, (e) $x = 840$, (f) $x = -42$,
(g) $x = -0.25$, (h) $x = 1.26$, (i) $x = -6$
2 $R = 10$ Ω
3 $F = 1$
4 (a) $x = 2, y = -1$, (b) $x = 3, y = 1$, (c) $x = -1, y = 4$, (d) $x = 2, y = 2$,
(e) $x = 3, y = -1$, (f) $x = 4, y = -1$, (g) $x = 1, y = 4$, (h) $x = 3, y = 5$,
(i) $x = 2, y = 6$, (j) $x = 1, y = 3$
5 $a = 2$ N, $b = \frac{1}{2}$
6 $I_1 = 1$ A, $I_2 = 0.5$ A
7 $F_1 = 2$ N, $F_2 = 3$ N
8 $u = 2$ m/s, $a = 0.5$ m/s^2
9 $u = 1$ m/s, $a = 2$ m/s^2
10 $R_1 = R_2 = 40$ N

Chapter 3

Revision

1 Linear (a), (b), (g); quadratic (c), (d), (f), (h).
2 (a) −2, +1, (b) −4, +2, (c) +1, +4, (d) −2, +2, (e) −2, −2, (f) +1, +1,
(g) 0, −7, (h) −1, + 1, (i) −3, +1, (j) +2, +3

3 (a) $x^2 - 5x + 4 = 0$, (b) $x^2 - x - 2 = 0$, (c) $x^2 - 6x = 0$, (d) $x^2 + 5x + 6 = 0$, (e) $x^2 - 2x = 0$, (f) $x^2 + 3x + 2 = 0$, (g) $x^2 - 4x + 3 = 0$
4 (a) –2, –6, (b) –3.65, 1.65, (c) –1.14, 1.46, (d) –1.16, 3.16, (e) –3.13, 0.64, (f) 0.22, –1.55, (g) 1.46, -5.46, (h) 3.41, 0.59
5 As given in revision problem 4.
6 (a) –0.56, 3.56, (b) –2, –6, (c) –4.08, 0.08, (d) –1.16, 3.16, (e) –4.54, 1.54
7 18.8 m or 1.6 m
8 5.41 cm and 8.41 cm
9 431.7 rad/s
10 1.84 m and 18.16 m
11 6 s
12 4.42 cm
13 12 cm
14 (a) Imaginary, (b) two real distinct roots, (c) imaginary, (d) two real coincident roots, (e) two real distinct roots

Problems

1 (a) –2, +1, (b) –1, –3, (c) +1, +3, (d) –1, –1, (e) –3, + 3, (f) –5, + 5, (g) 0, –3
2 (a) $x^2 - 1 = 0$, (b) $x^2 - 4x + 4 = 0$, (c) $x^2 + 4x + 3 = 0$, (d) $x^2 - x = 0$
3 (a) –2, 4, (b) –2.3, 1.3, (c) 0.33, 1, (d) –1.5, 1, (e) –3.45, 1.45, (f) –0.88, 0.68, (g) –2.5, 1.5, (h) 0.31, 1.29, (i) –1.69, 0.44
4 12.8 cm by 7.8 cm
5 10 cm by 3 cm
6 2.73 s and 7.47 s (on the way up and the way down)
7 31.9°C and 326°C
8 18.94 m or 1.06 m
9 2.61 cm
10 0.385 A or –10.39 A
11 16 mm
12 4.6 mm, 17.4 mm
13 8.41 cm, 5.41 cm
14 14 and 19 or –14 and –19
15 4 cm, 7 cm
16 10 cm by 10 cm
17 9 or –10
18 5 cm
19 12 and 13 or –12 and –13
20 14 m, 19 m
21 (a) Two distinct real roots, (b) two distinct real roots, (c) imaginary roots, (d) imaginary roots, (e) two real coincident roots

Chapter 4

Revision

1 (a) The negative sign, (b) N_0, (c) λ small
2 (a) The power is positive, (b) L_0, (c) α high means a high expansion
3 (a) 0, (b) 3 A
4 (a) 2, infinite, (b) 10, 0, (c) 0, 2, (d) 2, 0, (e) –4, 0,(f) 0, 0.5, (g) 0, 4,

(h) 10, 0, (i) 0, 0.2
5 (a) 403.43, (b) 3.48×10^{-3}, (c) 1.49, (d) 0.741, (e) 1.99×10^5, (f) 3.70×10^{-3}
6 5, 13.59, 36.94, 100.43, infinite; 5, 1.84, 0.68, 0.25, 0; 0, 3.16, 4.32, 4.75, 5
7 10, 73.89, 546.00, 4034.29, infinite; 10, 1.35, 0.18, 0.02, 0; 8.65, 9.82, 9.98, 10
8 2, 3.30, 5.44, 8.96, infinite; 2, 1.21, 0.74, 0.45, 0; 0, 0.78, 1.26, 1.55, 2
9 10, 11.05, 12.21, 13.50, infinite; 10, 9.05, 8.19, 7.41, 0; 0, 0.95, 1.81, 2.59, 10
10 (a) 18.10 V, (b) 7.36 V
11 9.96×10^4 Pa
12 (a) 0, (b) $0.86E/R$ A
13 (a) 0, (b) $0.86E$
14 (a) 2^{8x}, (b) 3^{-2x}, (c) e^8, (d) e^{8t}, (e) e^{-2t}, (f) e^{-12t}, (g) $1 + 2\,e^{2t} + e^{4t}$, (h) e^{-3t}, (i) e^{-2t}, (j) e^{-8t}, (k) $5\,e^{3t}$, (l) $0.4\,e^{4t}$

Problems
1 100 V, decreasing
2 (a) Negative sign, (b) C_0, (c) slower
3 (a) 2, infinite, (b) 12, infinite, (c) 3, 0, (d) 10, 0, (e) 2, 0, (f) 0, 2, (g) 0, 12, (h) 0, 2
4 (a) 54.60, (b) 0.18, (c) 1.49, (d) 0.67, (e) 19.48, (f) 2.87, (g) −19.78, (h) 9.50, (i) 0.19, (j) 2.43
5 10, 12.21, 14.92, 18.22, infinite; 10, 8.19, 6.70, 5.49, 0; 0, 1.81, 3.30, 4.51, 10; 10, 11.05, 12.21, 13.50, infinite; 10, 9.05, 8.19, 7.41, 0; 0, 0.95, 1.81, 2.59, 10
6 0.95 C
7 (a) 0, (b) 1.57 A, (c) 2.53 A, (d) 3.11 A, (e) 4 A
8 5.13 V
9 5637 Ω
10 (a) 1.26 A, (b) 1.72 A, (c) 1.90 A
11 0.70 g
12 (a) 8.61×10^4 Pa, (b) 7.41×10^4 Pa
13 (a) 198.0 V, (b) 190.2 V
14 (a) $-E$, (b) $-0.61E$
15 (a) 0, (b) $0.63E$
16 (a) 200°C, (b) 134.1°C, (c) 0°C
17 (a) 100%, (b) 95.1%, (c) 60.7%, (d) 0%
18 + gives growth, − decay.
19 (a) 4^{3x}, (b) 5^{2x}, (c) e^{9x}, (d) e^{-4x}, (e) e^{4t}, (f) $1 + 2\,e^{2t} + e^{4t}$, (g) $4 + 6\,e^{4t} + 9\,e^{8t}$, (h) $4\,e^{-4t}$, (i) $3\,e^{-2t}$, (j) $0.25\,e^{-6t}$

Chapter 5

Revision
1 (a) $x = \lg_{10} 7.8$, (b) $x = \lg_{10} 0.97$, (c) $x = \ln 1.54$, (d) $x = \ln 0.99$, (e) $x = \lg_2 4.5$, (f) $x = \lg_5 1.5$, (g) $x = \lg_{10} 4.5$, (h) $x = \ln 1.8$
2 (a) $7.5 = 10^x$, (b) $0.45 = 10^x$, (c) $9.9 = e^x$, (d) $0.88 = e^x$, (e) $20 = 2^x$, (f) $12 = 5^x$, (g) $3 = e^x$
3 (a) 4, (b) 2, (c) −2, (d) 10 000, (e) 9, (f) 16

4 (a) 0.892, (b) −0.0132, (c) 0.432, (d) −0.0101, (e) 0.875, (f) −0.347, (g) 2.293, (h) −0.128, (i) 1.61, (j) 1.11, (k) 8.067, (l) 3.380, (m) 0.580, (n) 0.273, (o) 11 013
5 (a) 4.800, (b) −0.981, (c) 0.062, (d) −5.203, (e) −11.525, (f) 1.779, (g) 2, (h) −6
6 (a) −0.438, (b) 0.270, (c) 0256, (d) 0.0477
7 (a) 0.926, (b) 1.639, (c) 0.931, (d) −2, (e) −8.64, (f) 2.35, (g) 0.693, (h) 2.386
8 $\frac{\ln 2}{\lambda}$
9 1.31×10^{-4} /°C

Problems
1 (a) $x = \lg_{10} 2.4$, (b) $x = \lg_{10} 2.5$, (c) $x = \lg_{10} 0.68$, (d) $x = \ln 1.3$, (e) $x = \ln 55.6$, (f) $x = \ln 5.8$, (g) $x = \ln 0.35$
2 (a) $34 = 10^x$, (b) $0.54 = 10^x$, (c) $345 = 10^x$, (d) $37 = e^x$, (e) $0.86 = e^x$, (f) $397 = e^x$
3 (a) 0.380, (b) 0.340, (c) −0.167, (d) 0.262, (e) 4.018, (f) 1.758, (g) −1.050, (h) 1.531, (i) −0.268, (j) 2.538, (k) 3.611, (l) −0.151, (m) 5.984
4 (a) −1.478, (b) 2.870, (c) 1.967, (d) −0.972, (e) 16.014, (f) 6.300
5 (a) 0.858, (b) 205 936, (c) 0.242, (d) 0.507
6 (a) 1.532, (b) 1.151, (c) 0.511, (d) 1.783
7 0.527
8 (a) 4.19, (b) 1.76, (c) 0.44, (d) 1.15, (e) −1.56, (f) 0.347, (g) −1.19
9 7.16×10^4
10 9.16×10^{-6} s
11 5.36×10^{-4} s^{-1}
12 36.4

Chapter 6

Revision
1 (a) 64 m, (b) 15 m/s, (c) 8 m, (d) 4 s, (e) 12 m, (f) 2 s, (g) 5 m/s
2 30 m/s
3 400 m
4 20 s
5 (a) $s = \frac{v^2 - u^2}{2a}$, (b) $a = \frac{v^2 - u^2}{2s}$, (c) $u = \sqrt{v^2 - 2as}$
6 (a) 7.5 s, (b) 112.5 m
7 300 m
8 36 s
9 10.1 m, 5.2 m/s
10 0.51 s
11 31.3 m/s

Problems
1 3 m/s
2 0.2 m/s^2
3 1650 m
4 12 m/s
5 500 m

6 20 s
7 (a) $a = \frac{v-u}{t}$, (b) $t = \frac{v-u}{a}$, (c) $u = v - at$
8 (a) $a = \frac{2s-2ut}{t^2}$, (b) $u = \frac{s-\frac{1}{2}at^2}{t}$
9 9.8 m/s
10 1.6 s
11 9.9 m/s
12 4.1 m, 0.92 s
13 2 s, 5 s
14 22.5 m
15 180 m
16 0.50 m

Chapter 7

Revision

1 $V = \sqrt{PR}$
2 15 Ω
3 5 V
4 (a) 2 V, (b) 0.04 W
5 (a) 30 Ω, (b) 2.73 Ω
6 (a) 6.32 Ω, (b) 0.0911 A
7 6 Ω
8 (a) 0.25 A, (b) 1 V, 3 V
9 (a) 1 A, 0.33 A, (b) 5.33 W
10 (a) 22 Ω, 0.55 A, (b) 6 Ω, 2 A, (c) 12.8 Ω, 0.94 A, (d) 50 Ω, 0.24 A
11 (a) 20 Ω, (b) 0.6 A, 0.2 A, 0.4 A, (c) 7.2 W
12 (a) 22 Ω, (b) 5.45 V, 6.55 V, (c) 6.55 W
13 (a) 0.71 A, (b) −0.44 A, (c) −0.70 A

Problems

1 (a) 200 V, (b) 2000 Ω
2 0.05 A
3 (a) 5 V, (b) 0.25 W
4 (a) 20 V, (b) 0.4 W
5 0.2 A, 4 V, 6 V, 10 V
6 (a) 1.43 W, 0.41 W, 1.02 W, (b) 5 W, 2 W, 7 W
7 6.88 mA
8 2.03 V, 9.97 V
9 3.43 Ω, 2 A, 1 A, 0.5 A
10 1.875 A
11 2.22 A
12 (a) 15 Ω, (b) 6.8 Ω
13 (a) 0.2 A, (b) 2.2 A
14 (a) 1.5 A, (b) −1A, (c) 1.2 A, (d) 0.29 A

Chapter 8

Revision

1 (a) 1.13 rad, (b) 2.18 rad, (c) 4.28 rad, (d) 0.35 rad, (e) 2.08 rad, (f) 5.45 rad
2 (a) 17.2°, (b) 108.9°, (c) 200.5°, (d) 30°, (e) 77.1°, (f) 36°

3 (a) 3/5, (b) 4/5, (c) 3/4, (d) 5/3, (e) 5/4, (f) 4/3
4 (a) 0.342, (b) 2.924
5 (a) 0.707, (b) 1.414
6 (a) 0.577, (b) 1.732
7 (a) 0.819, (b) 0.966, (c) 1.732, (d) 1.743, (e) 5.759, (f) 2.145, (g) 0.783, (h) 0.454, (i) 0.309, (j) 1.277, (k) 1.065, (l) 2.070
8 (a) 58.2°, (b) 69.5°, (c) 56.3°, (d) 64.2°, (e) 12.1°, (f) 31.8°, (g) 39.8°, (h) 39.7°
9 (a) 2.88, (b) 5.74, (c) 2.75, (d) 22.64, (e) 9.46, (f) 5.09
10 16.8 m
11 14.5°
12 11.1 m, 4.9 m, 7.0 m
13 6.8°
14 270.8 m, 44.6° W of N
15 66.4 m
16 (a) $\cos\theta$, (b) $\tan\theta$, (c) $\sin\theta$, (d) $\sin^2\theta$
17 As given in the problem
18 As given in the problem
19 As given in the problem
20 As given in the problem
21 (a) $12\sqrt{3}$ m, (b) $\dfrac{8}{\sqrt{3}}$ m, (c) $5\sqrt{3}$ m
22 (a) 10 m, (b) $\dfrac{6}{\sqrt{2}} = 3\sqrt{2}$ m, (c) 5 m
23 As given in the problem
24 (a) 0.819, (b) −0.819, (c) −0.176, (d) −0.342, (e) −0.766, (f) 5.672, (g) −0.940, (h) 0.500, (i) −0.364, (j) −0.259, (k) 0.598, (l) −0.999, (m) 1.778, (n) −0.612
25 $\frac{1}{2}$
26 $-\dfrac{\sqrt{2}}{2}$
27 (a) 59.5°, 239.5°, (b) 53.1°, 126.9°, (c) 134.4°, 225.6°, (d) 210°, 330°, (e) 36.9°, 323.1°
28 (a) 30°, 150°, (b) 120°, 240°, (c) 45°, 225°

Problems

1 (a) 0.391, (b) 0.965, (c) 0.354, (d) 1.788, (e) 1.942, (f) 0.649, (g) 0.479, (h) 0.353, (i) 0.627, (j) 1.332, (k) 2.010, (l) 0.355, (m) 0.643, (n) −0.866, (o) −1.192, (p) −0.174, (q) −0.966, (r) 1.192, (s) −0.985, (t) 0.342, (u) −1.000, (v) −0.643, (w) 0.985, (x) −0.988, (y) 1.898, (z) −3.133
2 (a) 51.3°, (b) 81.4°, (c) 62.2°
3 (a) 2.48, (b) 6.97, (c) 1.41, (d) 19.20, (e) 1.37, (f) 1.23
4 27.5 m
5 2.1 m
6 6.4 m
7 23.0 km
8 4.4 m
9 7.7 m

10 824.3 m
11 5.02 m
12 5.91 m
13 As given in the problem
14 As given in the problem
15 As given in the problem
16 (a) 20.5°, 159.5°, (b) 123.4°, 236.6°, (c) 53.1°, 306.9°, (d) 5.7°, 185.7°, (e) 123.7°, 303.7°, (f) 208.7°, 331.3°, (g) 141.1°, 321.1°, (h) 113.6°, 293.6°, (i) 203.6°, 336.4°, (j) 41.8°, 138.2°
17 (a) 0.5, (b) infinity, (c) 0, (d) 0.5, (e) –1, (f) –0.87, (g) 0.58

Chapter 9

Revision

1 (a) 406.0 m, (b) 177.7 m, (c) 36.2°, (d) 52.4°, (e) 51.3 mm, (f) 51.5 mm, (g) 83.8 mm, (h) 46°
2 5.49 m, 4.71 m
3 28.9 m
4 (a) 63.2° or 116.8°, 91.8° or 38.2°, (b) 56.8° or 123.2°, 81.2° or 14.8°
5 (a) 12.5, (b) 44.4°, (c) 62.9, (d) 16.9, (e) 79.7°, (f) 15.6, (g) 33.1 mm, (h) 63.4 mm
6 100.3°
7 101°, 79°
8 59.0°, 70.5°, 50.5°
9 (a) 353.6 mm^2, (b) 2.30 m^2, (c) 3288.9 mm^2, (d) 5.91 m^2, (e) 738.8 mm^2, (f) 1285.6 mm^2, (g) 344.1 mm^2
10 1184.8 m^2
11 (a) 6 m^2, (b) 24 m^2, (c) 34.98 m^2, (d) 17.3 cm^2, (e) 26.8 cm^2, (f) 14.7 cm^2, (g) 9.8 cm^2
12 4.79 cm^2

Problems

1 (a) 42°, 41.4 mm, 29.9 mm, (b) 45°, 77°, 206.8 mm, (c) 75°, 4.29 cm, 9.81 cm, (d) 110°, 9.24 mm, 46.08 mm, (e) 99.3°, 30.7°, 38.85 mm, (f) 70°, 136.2 mm, 146.2 mm, (g) 56°, 53.3 mm, 53.3 mm, (h) 65°, 138.2 mm, 141.9 mm, (i) 50.9°, 49.1°, 126.6 mm, (j) 53.1°, 96.9°, 99.3 mm or 126.9°, 23.1°, 39.2 mm
2 6.35 m, 5.37 m
3 2.01 m
4 44.9°
5 (a) A = 4.8°, B = 153.9°, C = 21.3°, (b) a = 85.8, B = 50.1°, C = 99.6°, (c) A = 31.5°, B = 48.9°, C = 99.6°, (d) A = 120°, B = 38.2°, C = 21.8°, (e) A = 34°, B = 101.5°, C = 44.5°, (f) a = 9.67, B = 45.5°, C = 54.5°, (g) B = 47.3°, C = 66.7°, a = 49.7 mm, (h) A = 34.1°, B = 44.4°, C = 101.5°, (i) A = 76.9°, B = 29.1°, c = 98.7 mm
6 118.2 km
7 128.4 mm
8 31.2 mm
9 (a) 199.2 mm^2, (b) 919.3 mm^2, (c) 3.94 m^2, (d) 19.7 m^2, (e) 13.42 cm^2, (f) 1.73 m^2, (g) 30 cm^2
10 33.7°, 146.3°

11 34.8 cm^2

Chapter 10

Revision

1 Vectors (b), (e), scalars (a), (c) (d) (f)
2 See figure A.1
3 (a) and (i), (b) and (j), (d) and (h)
4 5 at an angle of 36.9° to **a**
5 8.3 at an angle of 19.9° north of east
6 (a) 6.08 N at 34.7° to the horizontal, (b) 6.71 N at 26.6° west of north, (c) 7.61 N at 28.5° to 5 N force, (d) 10.2 N at 13.0° to 8 N force, (e) 11.6 N at 12° to 7 N force, (f) 17.9 N at 18.9° to 10 N force, (g) 13.5 N at 45.7° to 12 N force
7 3.17 kN at 9.4° to 1.2 kN force
8 21 N, 16 N
9 (a) 11.18 km/h at 26.6° east of north, (b) 4.25 km/h at 48.3° east of north, (c) 6.24 km/h at 76.1° east of north
10 60° to the bank
11 17.4 km/h in southerly direction
12 About 134 kN at 23°
13 About 4.7 N at 49°
14 (a) 4.47 at 26.6° south of west, (b) 4.4 at 37° north of east, (c) 10.75 at 9.0° north of east, (d) 7.61 at 36.5° east of north
15 65 m/s due south
16 15.9 m/s at 57.8° east of south
17 32°
18 15.9 m/s at 32.2° south of east
19 (a) 7.7 N, 6.4 N, (b) 5.1 kN, 14.1 kN, (c) 11.3 N, 4.1 N, (d) 5.2 kN, 29.5 kN
20 28.2 N, 10.3 N
21 (a) 0.78 N at 39.9° south of west, (b) 4.17 N at 4.5° north of east, (c) 6.85 N at 40.7° south of east
22 28.3 m/s, 28.3 m/s
23 375.9 m/s, 136.8 m/s
24 7.07 m, 7.07 m

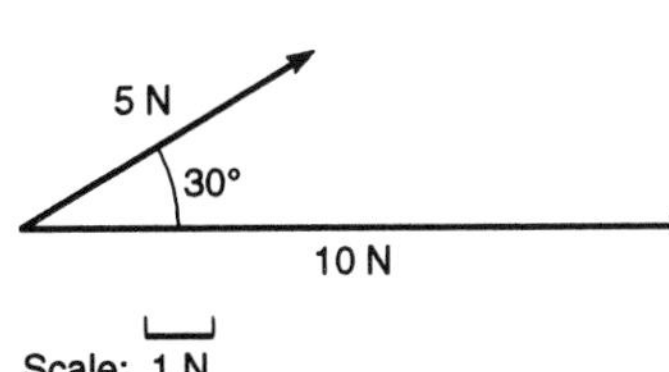

(a)

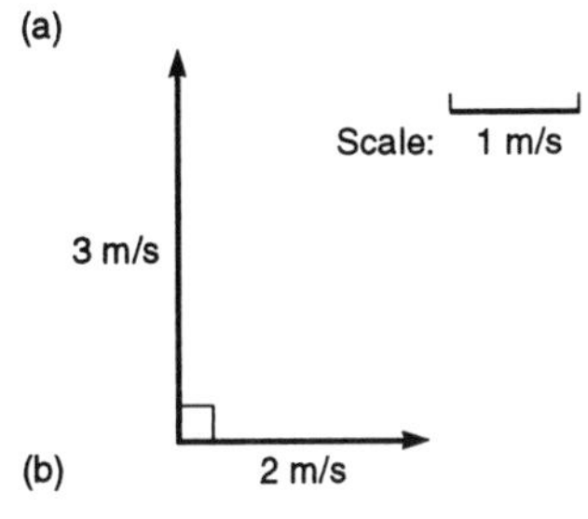

(b)

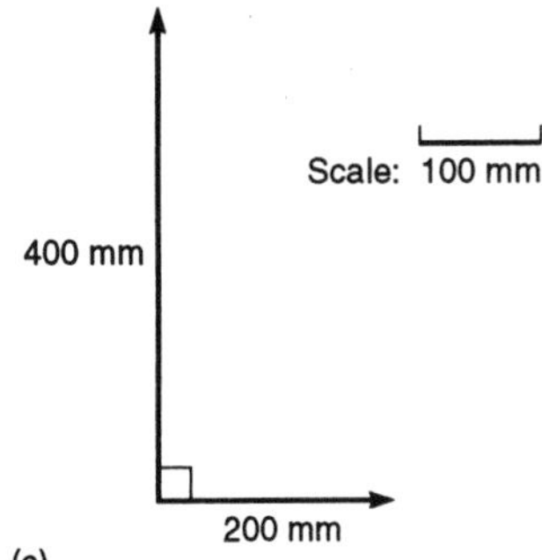

(c)

Figure A.1 *Chapter 10 Revision problem 2*

Problems

1 (a) 3.61 at 56.3° norht of east, (b) 6.45° at 26° west of north, (c) 2.95 at 73° north of east, (d) 3.61 at 56.3° south of west, (e) 2.83 at 3.5° north of east, (f) 5.60 at 30.4° south of east
2 (a) 4.47 kN at 26.5° north of west, (b) 22.3 N at 63.4° from horizontal, (c) 83.2 N at 22.5° from 40 N force, (d) 27.5 N at 16.2° from vertical, (e) 13.5 N at 45.7° from 12 N force, (f) 17.4 kN at 83.4° from 12 kN force
3 14.0 N
4 7.3 N, 16.3 N
5 34.3°
6 44.7 kN at 26.6° to the 40 kN force
7 58° and 48° with respect to the 90 N force
8 22.8 kN at 70.3° to the horizontal

9 (a) 5 km/h at 36.9° east of north, (b) 10.8 m/s at 21.8° to the vertical, (c) 27.5 m/s at 16.2° to 20 m/s velocity, (d) 14.7 km/h at 16.3° west of north, (e) 10 m/s north
10 11.2 m/s at 63.4° to the horizontal
11 239.9 km/h at 17.1° south of east
12 8.6 km at 54.5° north of east
13 9.27 km at 62.8° west of south
14 7.2 N at 9° east of north
15 21.4 km/h at 119.7° west of north
16 5 m/s at 36.9° west of north
17 5.0 N, 8.7 N
18 5.3 N at 33.8° west of north
19 12.3 kN, 8.6 kN
20 34.3 N at 27.8° west of north
21 4.1 N at 6° north of west

Chapter 11

Revision

1 (a) $v = 10 \sin 2\pi 50t$ V, (b) (i) 5.88 V, (ii) 9.51 V, (iii) −5.88 V
2 (a) $i = 50 \sin 2\pi \times 2 \times 10^3 t$ mA, (b) (i) −47.6 mA, (ii) −29.4 mA, (iii) 47.6 mA
3 (a) 0, (b) 10 V, (c) 1.99 V
4 0, 0.20, 0.39, 0.56, 0.72, 0.84, 0.93, 0.99, 1.0
5 5, 2.5, 0, −2.5, −5, −2.5, 0, 2.5, 5
6 (a) 0.87 V, (b) −0.87 V
7 (a) 100 mA, (b) 100 Hz, (c) 0.25 rad or 14.3° lagging
8 (a) 12 V, (b) 50 Hz, (c) 0.5 rad or 28.6° leading
9 See figure A.2
10 $22.4 \sin(\omega t + 1.11)$ mA
11 $6.81 \sin(\omega t + 0.147)$ V
12 $126.2\sqrt{2} \sin(\omega t - 0.071)$ V
13 $8.72\sqrt{2} \sin(\omega t - 0.639)$ A
14 $2.95 \sin(\omega t - 0.50)$ V
15 $6.4 \sin(\omega t + 0.90)$ A
16 $85.0 \sin(\omega t + 0.728)$ mA

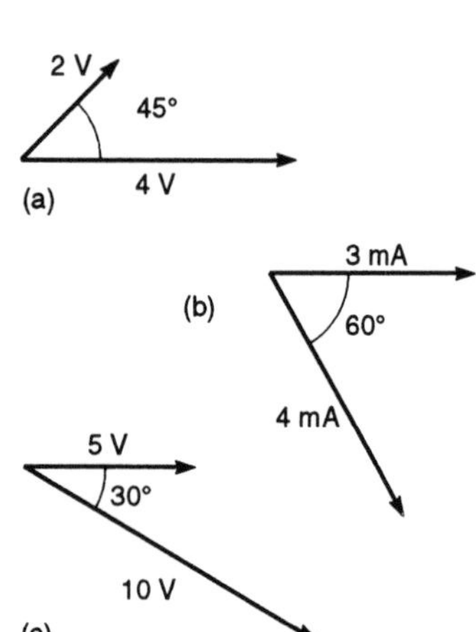

Figure A.2 *Chapter 11 Revision problem 9*

Problems

1 (a) 0, 1.68, 1.82, 0.28, −1.51, −1.92; 0, 1.98, 3.89, 5.65, 7.17, 8.41; 0, 4.55, −17.20, −1.40, 4.95, −2.72
(b) 0, 8.66, 10.00, 8.66, 0; 10.00, 5.00, 0, −5.00, −10.00; 8.66, 8.66, 5.00, 0, −5.00
2 4.79 A
3 1.99 V
4 (a) 500 rad/s, (b) 79.6 Hz, (c) 17.1 V
5 See figure A.3
6 (a) 20 mA, (b) 250 Hz, (c) 0.5 rad or 28.6° lagging
7 (a) 5 V, (b) 500 Hz, (c) 0.2 rad or 11.5° leading
8 $4.47 \sin(\omega t + 0.464)$ V
9 $6.24 \sin(\omega t - 0.281)$ V
10 $6.94 \sin(\omega t + 0.935)$ V

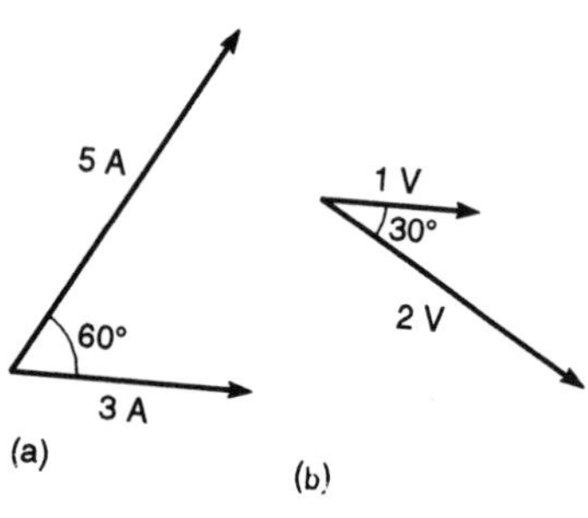

Figure A.3 *Chapter 11 Problem 4*

11 $43.6 \sin(\omega t + 0.648)$ mA
12 $50 \sin(\omega t + 0.927)$ mA
13 $5.66 \sin(\omega t + \pi/4)$ V
14 $25 \sin(\omega t - 0.644)$ V
15 $14.7 \sin(\omega t - 0.5)$ mA
16 $17.8 \sin(\omega t + 0.691)$ mA
17 $4.47 \sin(\omega t + 0.464)$ V
18 $8.77 \sin(\omega t - 0.524)$ V
19 (a) $12.81 \sin(\omega t - 0.151)$ mA, (b) $12.81 \sin(\omega t + 1.198)$ mA
20 $10 \sin(\omega t + \pi/3)$ A
21 $3.1 \sin(\omega t - 0.33)$ A

Chapter 12

Revision

1 See figure A.4
2 See figure A.5
3 See figure A.6
4 See figure A.7
5 See figure A.8
6 See figure A.9
7 (a) $x = 1, y = 2$, (b) $x = 2, y = -1$, (c) $x = -2.2, y = 4.8, x = 0.69, y = 0.48$, (d) $x = -0.4, y = 0.6, x = 0.6, y = 1.4$
8 (a) 5, (b) −2, −1, (c) 3.4, 0.6, (d) −1.8, 0, 2.8, (e) 1, (f) −2, 1, 3
9 (a) 0, (b) +2, −2, (c) −1, −1, (d) −2, −1, (e) no real roots
10 Shifted vertically by 3 − 1 = 2 units
11 (a) 1, (b) 1/5, (c) 1/4, (d) 3/5, (e) −2
12 (a) 5, 3, (b) 1, −3, (c) 2, 1/2, (d) −2, 2
13 (a) $y = 2x + 3$, (b) $y = 4x - 3$, (c) $y = -2x + 1$, (d) $y = 5x$
14 3, 4
15 (a) 4, (b) 5, (c) 13

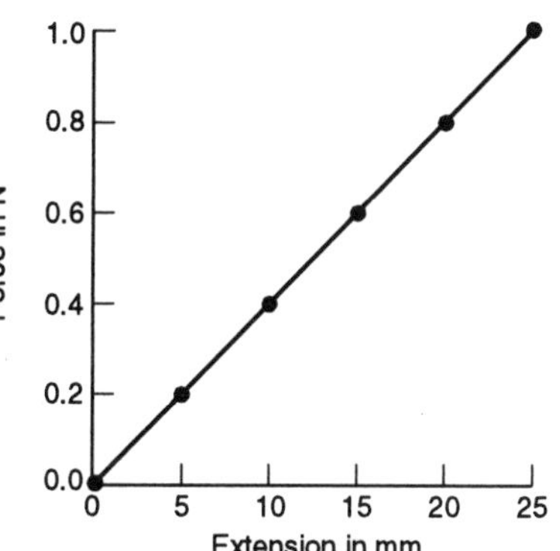
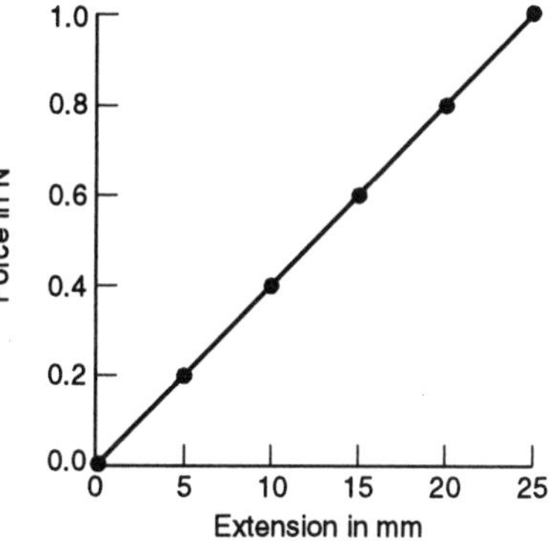

Figure A.4 *Chapter 12 Revision problem 1*

Figure A.5 *Chapter 12 Revision problem 2*

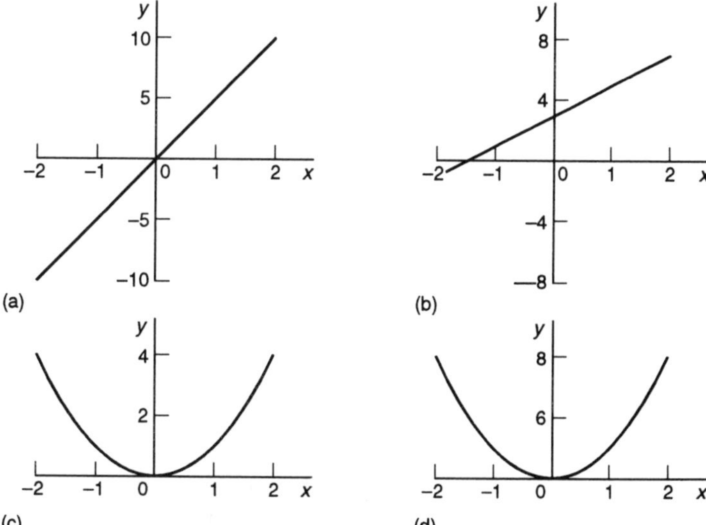

Figure A.6 *Chapter 12 Revision problem 12*

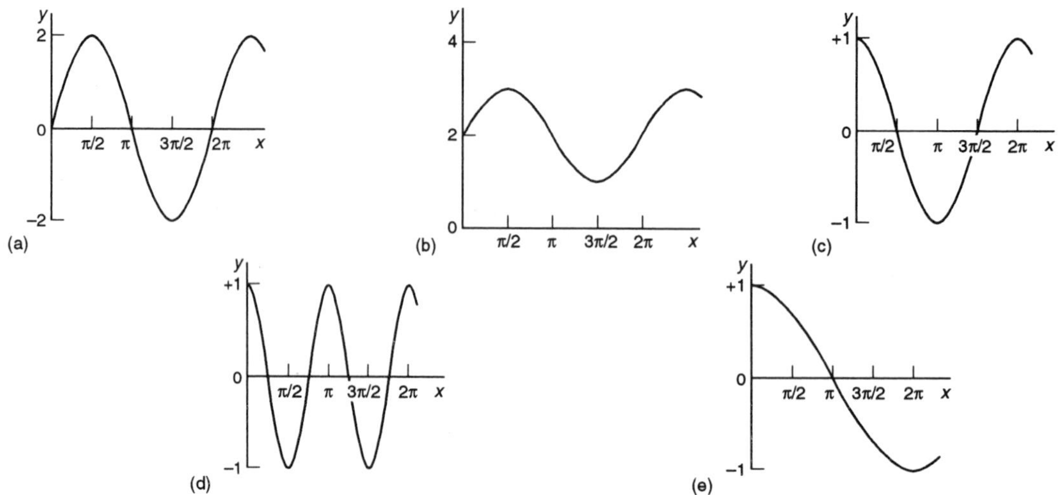

Figure A.7 *Chapter 12 Revision problem 4*

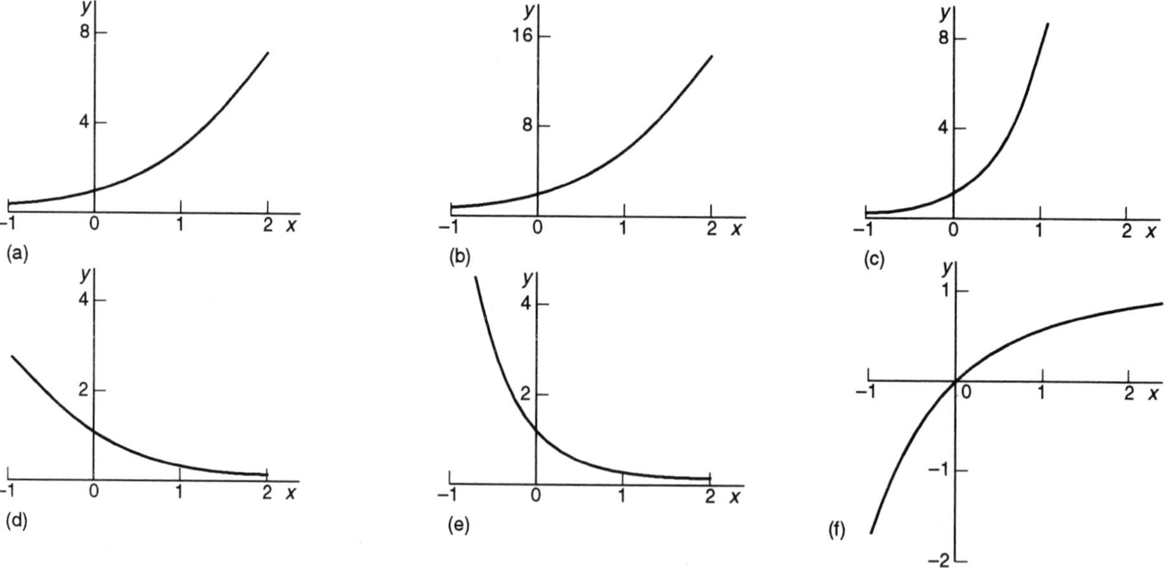

Figure A.8 *Chapter 12 Revision problem 6*

16 (a) 4 square units, (b) 2.5 square units, (c) 1/3 square units, (d) 23/3 square units
17 (a) 8 square units, (b) 8.7 square units, (c) 20 square units
18 51 square units
19 1235 m

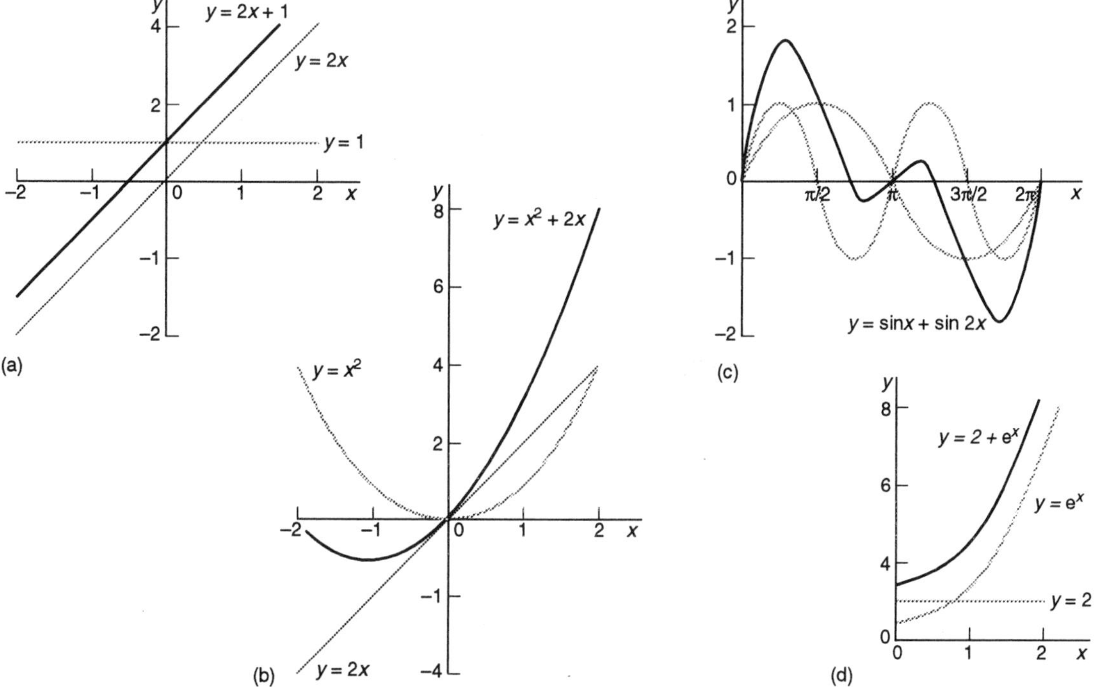

Figure A.9 *Chapter 12 Revision problem 6*

Problems

1 See figure A.10

2 (a) −6, (b) +1, +2, (c) −1, −3, (d) −2.2, +0.7, (e) 0, −3, +1, (f) −1.95, +0.33, +1.52, (g) −1.5, +0.3, +3.7, (h) +0.95, (i) +0.87

3 (a) $x = 1, y = 1$, (b) $x = -1, y = 1, x = 3, y = 9$, (c) $x = 1, y = 2$, (d) $x = 1.63, y = 8, x = -1.63, y = 8$, (e) $x = -1.79, y = 3.21, x = 2.79, y = 7.79$, (f) $x = -1$

4 (a) 0, (b) +3, −3, (c) −4, + 1, (d) no real roots, (e) −2, −2, (f) no real roots, (g) −3, +2

5 See figure A.11

6 (a) $y = x + 2$, (b) $y = 5x - 1$, (c) $y = -3x + 2$, (d) $y = -5x - 2$, (e) $y = 3$, (f) $y = 2x - 5$

7 (a) 7, (b) −2, (c) 3, (d) 8, (e) 6

8 (a) 8.63 square units, (b) 1.01 square units, (c) 2.83 square units, (d) 4 square units, (e) 12 square units, (f) 28 square units

9 21.66 square units

10 40.5 square units with 5 strips, 46.7 square units with many more

11 71.25 square units

12 1.49 square units

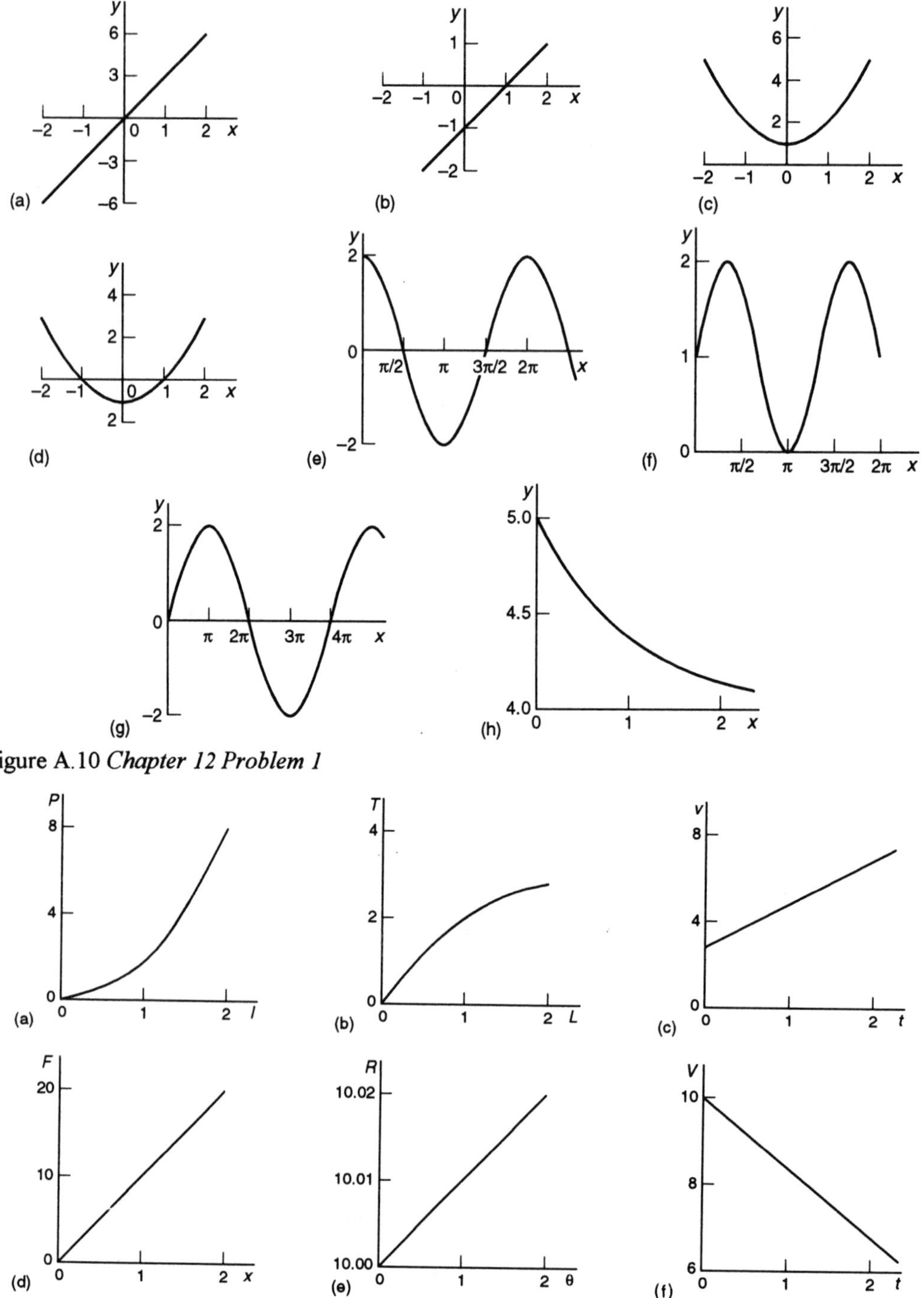

Figure A.10 *Chapter 12 Problem 1*

Figure A.11 *Chapter 12 Problem 5*

Chapter 13

Note that in those problems where the data has been tabulated, the units and multiples used with the data have been assumed to also refer to the variables in the equations.

Revision

1 (a) $V = 0.07\theta + 19$, (b) $n = 0.15V - 3$, (c) $L = -2.91\theta + 2540$, (d) $\sigma = -0.025\theta + 85.2$

2 (a) $y = 3x^2 - 10$, (b) $y = x^2 + 2$, (c) $y = -2x^2 + 5$

3 (a) $y = \frac{1}{x} + 1$, (b) $y = \frac{2}{x} - 1$, (c) $y = -\frac{1}{x} + 2$

4 (a) $y = 3x^2 - x$, (b) $y = x^2 + 2x$, (c) $y = -x^2 + x$

5 (a) $y = x^3$, (b) $y = 2x^{-2}$, (c) $y = -2x^{1/2}$

6 (a) $y = 0.5\ e^{0.3x}$, (b) $y = 3\ e^{-0.2x}$, (c) $y = -2\ e^{0.1x}$

7 As problems 5 and 6

8 (a) log-linear, 1 cycle, (b) log-log, 1 cycle × 1 cycle, (c) log-log, 1 cycle × 2 cycles

Problems

1 (a) $E = 0.45L + 5$, (b) $R = 4.2L$, (c) $v = 0.1t + 4$, (d) $\sigma = -0.098\theta + 160$

2 (a) $y = 2x^2 - 1$, (b) $y = -x^2 + 3$, (c) $y = \frac{3}{x} + 1$, (d) $y = -\frac{2}{x} - 4$, (e) $y = 2x^2 + x$, (f) $y = -x^2 - x$, (g) $y = 0.5x^{1/2}$, (h) $y = -0.6x^3$, (i) $y = 100\ e^{-0.1x}$, (j) $y = 10\ e^{0.3x}$

3 (a) T against $\sqrt{L}$, gradient $2\pi/\sqrt{g}$, intercept 0,
(b) s/t against t, gradient $a/2$, intercept u,
(c) e/θ against θ, gradient b, intercept a,
(d) R against θ, gradient $R_0\alpha$, intercept R_0,
(e) ln i against t, gradient $1/RC$, intercept ln I,
(f) p against $1/V$, gradient k, intercept 0,
(g) y against L^4, gradient $w/8EI$, intercept 0, or lg y against lg L, gradient 4, intercept lg $(w/8EI)$.

4 $R = \frac{1200}{V} + 50$

5 $R = \frac{0.16}{d^2}$

6 $T = 2.0L^{1/2}$

7 $V = 16p^{-1}$

8 $T = 500p^{0.28}$

9 $C = 0.001n^3 + 30$

10 $s = 0.1v^2 + 0.5v$

11 $I = 0.001V^2$

12 $E = 4.1f - 4.3$

13 2 × 2

14 2

15 $T = 0.2R^{3/2}$ (this is known as one of Kepler's laws)

16 $v = 10.2\ e^{-0.1t}$

17 $\theta = 800\ e^{-0.2t}$

18 $Q = 2.6h^{2.5}$

19 $A = 400\ e^{-0.02t}$

20 $T = 50\ e^{0.3\theta}$

Chapter 14

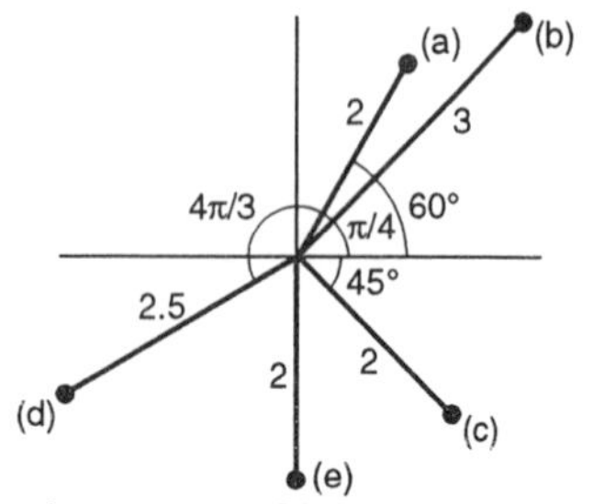

Figure A.12 *Chapter 14 Revision problem 1*

Revision

1 See figure A.12
2 (a) (4.47, 26.6°) or (4.47, 0.46), (b) (2.83, –45°) or (2.83, –π/4), or (2.83, 315°) or (2.83, 7π/4), (c) (4.47, 116.6°) or (4.47, 2.03), (d) (2.83, 225°) or (2.83, 5π/4), or (2.83, –135°) or (2.83, –3π/4), (e) (2, 0°) or (2, 0), (f) (2, 90°) or (2, π/2), (g) (4, 180°) or (4, π), (h) (4, 270°) or (4, 3π/2)
3 (a) (2.8, 2.8), (b) (1.5, 2.6), (c) (–1.7, 1.0), (d) (1.0, –2.8), (e) (1.4, 1.4), (f) (3.5, 2.0), (g) (–3.0, 0)
4 (a) As figure 14.6 with radius 3, (b) as figure 14.7 with angle π/6 rad, (c) as figure 14.8 with all radii doubled, (d) see figure A.13.

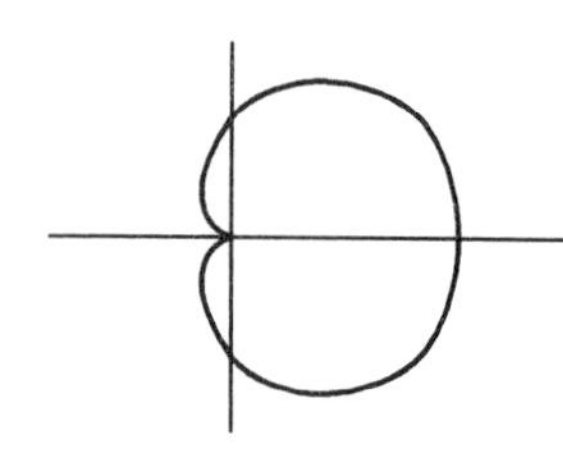

Figure A.13 *Revision problem 4(d)*

Problems

1 (a) (3, 90°) or (3, π/2), (b) (2, 180°) or (2, π), (c) (2.83, 135°) or (2.83, 3π/4), (d) (5, 53.1°) or (5, 0.93), (e) (5, 126.9°) or (5, 2.22), (f) (4.47, 333.4°) or (4.47, 5.82), or (4.47, –26.6°) or (4.47, –0.46), (g) (3, 0°) or (3, 0)
2 (a) (1.3, 1.5), (b) (–2.8, 1.0), (c) (–0.5, –3.0), (d) (3.8, –1.4), (e) (1.0, 1.7), (f) (0.5, 0.9), (g) (–3.0, 0), (h) (0, 4.0)
3 (a) As in figure 14.6 with radius 4, (b) as in figure 14.7 with angle π/4, (c) see figure A.14(a), (d) see figure A.14(b).
4 See figure A.15.
5 As in the problem.

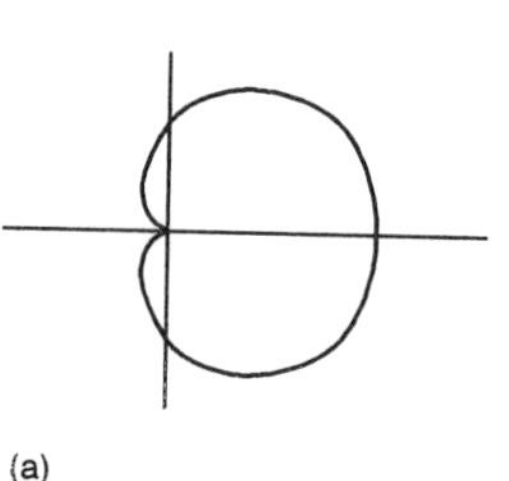

(a)

(b)

Figure A.14 *Chapter 14 Problem 3*

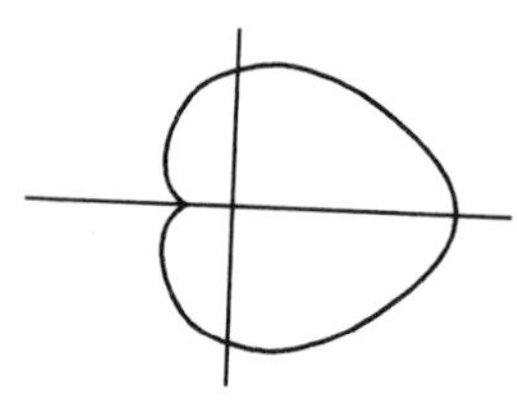

Figure A.15 *Chapter 14 Problem 4*

Chapter 15

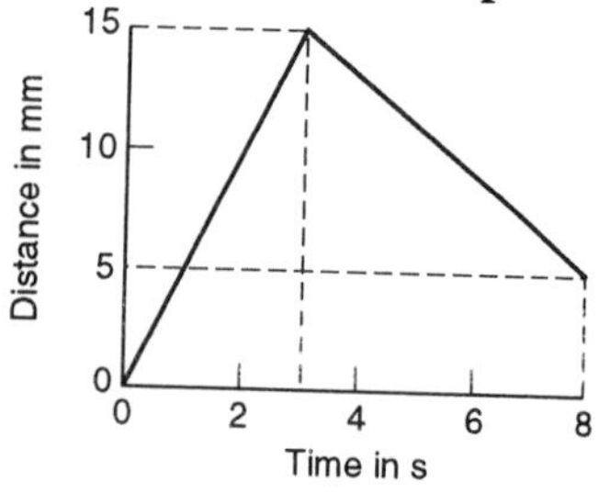

Figure A.16 Chapter 15 Revision problem 5

Revision

1 A, C changing velocity, B constant velocity, D at rest
2 (a) 2 m/s, (b) 0, (c) 1.5 m/s, (d) 1.1 m/s
3 (a) 19.6 m/s, (b) 29.4 m/s, (c) 39.2 m/s
4 (a) 16.8 m/s, (b) –8.3 m/s, (c) –15.1 m/s, (d) –5.6 m/s
5 See figure A.16.
6 See figure A.17. (a) 0.9 m/s^2, (b) –1.8 m/s^2, (c) 3150 m
7 (a) 1 m/s^2, (b) 60 m
8 See figure A.18. 12 m/s^2, 98.3 m
9 See figure A.19

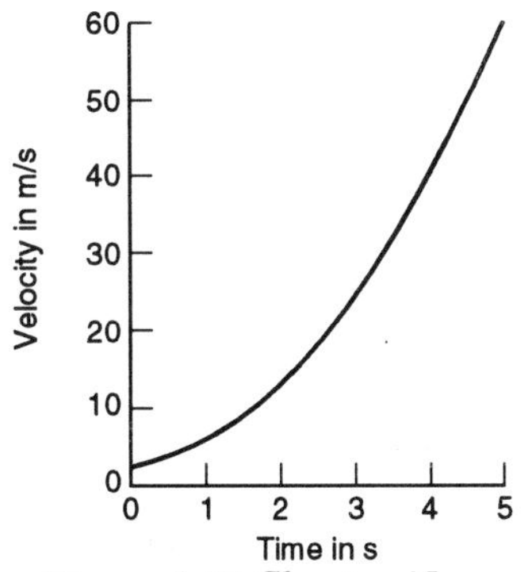

Figure A.18 Chapter 15 Revision problem 9

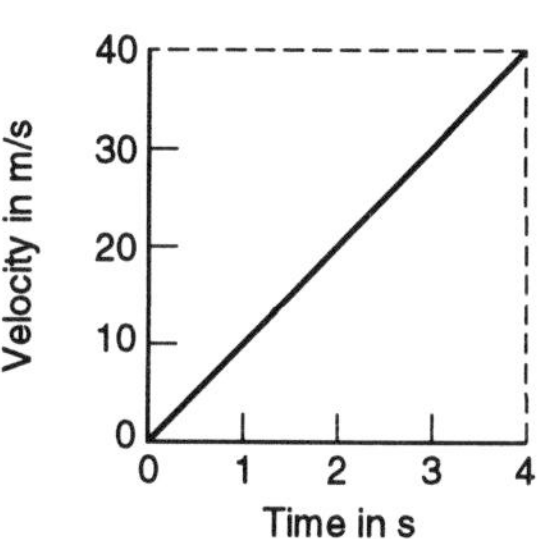

Figure A.19 *Chapter 15 Revision problem 9*

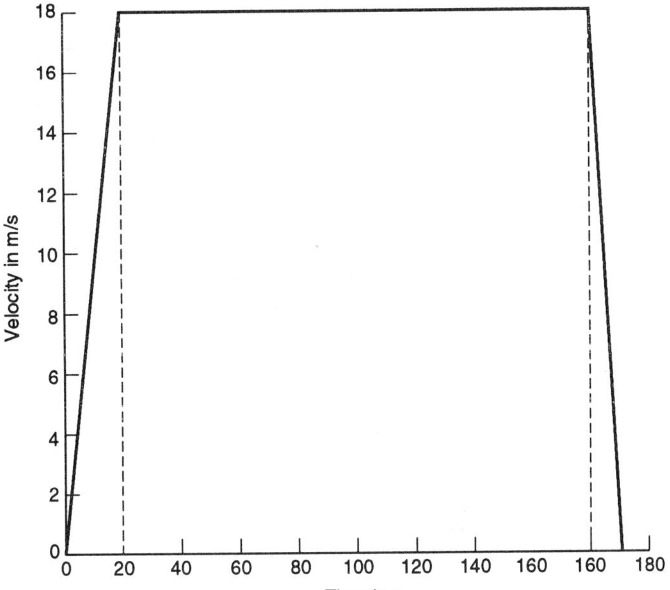

Figure A.17 Chapter 15 problem 6

Problems

1 (a) 34 mm/s, (b) 50 mm/s, (c) 66 mm/s
2 (a) 8 m/s, (b) 11 m/s, (c) 16 m/s
3 (a) 16 mm/s, (b) 24 mm/s, (c) 32 mm/s
4 (a) 4 mm/s, (b) 6 mm/s, (c) 32 mm/s
5 (a) 18 mm/s, (b) 26 mm/s, (c) 8 mm/s
6 -1.0 m/s^2, 11 m
7 (a) 10 m/s^2, (b) 10 m/s^2, (c) 11.25 m
8 (a) 0.69 m/s^2, (b) -1.39 m/s^2, (c) 903 m
9 1.2 m/s^2, 12.6 m
10 54.8 s
11 1.25 m/s^2, 256 m
12 -4 m/s^2, 54 m
13 (a) Distance–time graph non-zero gradient straight line, velocity–time graph straight line with zero gradient, (b) distance–time graph curved, velocity–time graph non-zero gradient straight line, (c) as (a), (d) as (b).

Chapter 16

Revision

1 (a) 1 V, (b) 0, (c) 2 A, (d) 1 A, (e) 0.9 V, (f) 1 V
2 As given in the problem
3 14 mA
4 8.2 V
5 As given in the problem
6 7.5 mA, 19 mA, 2.53

Problems

1 (a) 1.13 A, (b) 0

2 (a) 5 V, (b) 0
3 As given in the problem
4 As given in the problem
5 2.83 V
6 4 A
7 (a) 6.50 V, (b) 7.14 V, (c) 1.10
8 (a) 15.0 mA, (b) 17.1 mA, (c) 1.14
9 As given in the problem
10 As given in the problem
11 5 V, 7.07 V, 1.4
12 1 V
13 1.4 V
14 7.1 mA
15 141.4 mA
16 10 V

Chapter 17 *Revision*

1 (a) $6x$, 0, 6, (b) $2x + 1$, 1, 5, (c) 3, 3, 3, (d) $-1/2x^2$, infinity, $-1/2$, (e) 0, 0, 0, (f) 3, 3, 3, (g) 1/4, 1/4, 1/4
2 As given in the problem
3 (a) $6x^2$, (b) $6x^5$, (c) $-4x$, (d) $\frac{3}{2}x^{1/2}$, (e) $\frac{1}{2}x^{-3/4}$, (f) $30x^9$, (g) $-\frac{3}{5}x^{-6/5}$, (h) $0.4x^3$
4 10
5 24
6 $-1/2$
7 62.8 mm^2/mm
8 (a) $2x$, (b) $8x^3 + 6x$, (c) $12x^2 - 4x + 3$, (d) $12x^2 - 6x + 4$, (e) $10x^4 + 9x^2 - 2$, (f) $20x^4$, (g) $-2x^{-2} - 10x^{-3}$, (h) $4x + 3x^{-2}$
9 (a) $\cos x$, (b) $5 \cos 5x$, (c) $-4 \sin 4x$, (d) $-3 \cos 3x$, (e) $6 \cos 3x$, (f) $-40 \sin 4x$
10 $v = -50 \sin 50t$
11 $i = 200 \cos 100t$
12 (a) $2\ e^{2x}$, (b) $-3\ e^{-3x}$, (c) $\frac{1}{2}\ e^{x/2}$, (d) $6\ e^{3x}$, (e) $2\ e^{2x}$, (f) $1 + 6\ e^{2x}$, (g) $2\ e^x + 6\ e^{2x}$
13 $-\lambda N_0\ e^{-\lambda t} = -\lambda N$
14 $1/x$
15 (a) 2, (b) 4, (c) $6x + 4$, (d) $-\sin x$, (e) $4\ e^{2x}$
16 (a) $-\dfrac{CV}{RC}\ e^{-t/RC}$, (b) $\dfrac{CV}{R^2C^2}\ e^{-t/RC}$

Problems

1 (a) $-4/x^2$, (b) $-2/x^3$
2 (a) $12x^5$, (b) $-10/x$, (c) $\frac{7}{3}x^{4/3}$, (d) $0.6x^2$, (e) $x^{-1/2}$, (f) $-3x^{-5/2}$, (g) $8x + 1$, (h) $6x^2 + 6x$, (i) $9x^2 - 10x + 3$, (j) $-15x^{-4}$, (k) $6x^2 - 2x^{-3}$, (l) $-10x^{-3} + 2x^{-2}$, (m) $4 \cos 4x$, (n) $-2 \sin 2x$, (o) e^x, (p) $2\ e^{2x}$, (q) $-\frac{2}{3}\ e^{-2x/3}$, (r) $1/x$, (s) $\frac{3}{2}x^{-1/2}$, (t) $\cos x - 2 \sin 2x$
3 6
4 11

5 at
6 (a) 3, (b) –2.12, (c) 0
7 218.4
8 –0.30
9 20 mm^2/mm
10 $I\omega$
11 CV
12 $12 - 10t$
13 $-p_0 c\, e^{-ch}$
14 434.8 A/s
15 –1091 V/s
16 4.8 mV/°C
17 (3, –40), (–1, 12)
18 –112 m^3/min
19 (a) 6, (b) $4\, e^{2x}$, (c) $-4 \cos 2x$, (d) 6

Chapter 18

Revision

1 (a) $x = 3$, minimum, (b) $x = -1$ maximum, $x = 2$ minimum, (c) $x = 0$ minimum, $x = 2$ maximum, (d) $x = -3$ maximum, $x = 2$ minimum, (e) –3/2 minimum, (f) 1/4 maximum, (g) $\pi/2$ maximum, $3\pi/2$ minimum, $5\pi/2$ maximum, etc.
2 $x = \dfrac{a-b}{2}$
3 48 m^2
4 $L/2$
5 20, 20
6 4 s
7 $r = 0.34$ m, $h = 2.15$ m
8 15 cm, 15 cm
9 2.89 cm × 2.89 cm × 5.77 cm
10 $I = \dfrac{E}{2r}$
11 33.4 V
12 $-\dfrac{a}{2b}$

Problems

1 (a) $x = 1$, minimum, (b) $x = -3$ minimum, (c) $x = 0$, minimum, (d) $x = 1/\sqrt{3}$ minimum, $x = -1/\sqrt{3}$ maximum, (e) $x = 0$, maximum, $x = -2/3$ minimum, (f) $x = 3$ minimum, $x = -2$ maximim, (g) $x = -2$ minimum, (h) $x = 2$ minimum, $x = -2$ maximum, (i) $x = 0$ inflexion, (j) $x = 1$ minimum, (k) $x = 1$ minimum, (l) $x = 1$ minimum, $x = -1$ maximum, (m) $x = 1/2$ minimum, $x = -2$ maximum, (n) $x = -2$ maximum, $x = 4$ minimum
2 40
3 6, 12
4 $L = 100$ m
5 $t = 3/2$ s
6 39.2 mm
7 $r = 0.86$ m, $h = 0.86$ m

8 50 m
9 As given in the problem
10 500 000

Chapter 19

Revision

1 (a) $2x^2 + C$, (b) $2x^3 + C$, (c) $\frac{1}{2}x^2 + 2x + C$, (d) $-\cos x + C$, (e) $e^x + C$
2 (a) $\frac{1}{4}x^4 + C$, (b) $\frac{1}{2}x^4 + C$, (c) $-\frac{1}{6}x^{-6} + C$, (d) $2x^{1/2} + C$, (e) $\frac{3}{4}x^{4/3} + C$, (f) $-\frac{1}{3}\cos 3x + C$, (g) $\frac{1}{2}\sin 2x + C$, (h) $\frac{1}{4}e^{4x} + C$, (i) $-\frac{1}{5}e^{-5x} + C$
3 (a) $\frac{5}{2}x^2 + 3x + C$, (b) $x^4 + x^3 + 2x + C$, (c) $-\cos x + \frac{1}{2}\sin 2x + C$, (d) $2x - \frac{1}{3}e^{-3x} + C$, (e) $\ln x + \frac{3}{2}x^2 + C$, (f) $\frac{3}{2}x^2 - \frac{1}{4}\cos 4x + C$
4 (a) $y = x^2 + 3x - 7$, (b) $y = 3x^3 + x^2 + 2x + 1$, (c) $y = 2x^2 - 9x + 5$, (d) $y = 2x + \frac{1}{3}x^{-1} + \frac{2}{3}$
5 $x = 10t + t^2 + 2$
6 $x = t^2 + 5t$
7 (a) 20 sq units, (b) 7 sq units, (c) 12 sq units, (d) 8 sq units, (e) 1 sq unit, (f) 0.71 sq units, (g) 0.86 sq units, (h) 2.25 sq units, (i) 21.3 sq units, (j) 7.5 sq units
8 8 force × distance units
9 109.5 force × distance units

Problems

1 (a) $\frac{1}{6}x^6 + C$, (b) $-\frac{1}{2}x^2 + C$, (c) $\frac{2}{3}x^{3/2} + C$, (d) $\frac{1}{2}x^{1/2} + C$, (e) $x^2 + C$, (f) $-\frac{1}{3}e^{-3x} + C$, (g) $2\,e^{x/2} + C$, (h) $\sin x + C$, (i) $-\frac{1}{4}\cos 4x + C$, (j) $-\frac{1}{2}\cos 2x + C$, (k) $\frac{1}{2}\sin 2x + C$, (l) $x^2 + \frac{1}{3}x^3 + C$, (m) $\frac{1}{4}x^4 + x^2 + 4x + C$, (n) $3x - \frac{1}{2}e^{-2x} + C$, (o) $-\frac{1}{2}\cos 2x + \frac{1}{3}\sin 3x + C$
2 (a) $y = x^3 + 2$, (b) $y = 2x^2 - 2x + 3$, (c) $y = 2x^3 + 5x^2 - 5x - 1$, (d) $y = 2x - 1$
3 $q = -\frac{3}{100}\cos 100t$ mC
4 $x = 5t^2 + t$
5 $\theta = \frac{1}{2}cl^2$
6 (a) 64/3, (b) 1, (c) 8, (d) 0.366, (e) 1.099, (f) −2,25, (g) 50/3
7 (a) 52/6 sq units, (b)1/4 sq units, (c) 8 sq units, (d) 66 sq units, (e) 12 sq units, (f) 32 sq units, (g) 0.71 sq units, (h) 52 sq units, (i) 4 square units, (j) 25.5 square units

Chapter 20

Revision

1 (a) $v = 10 - 2t$, (b) $a = -2$, (c) $v = 6$ m/s, $a = -2$ m/s^2
2 (a) $v = \frac{1}{2}t^{-1/2}$, (b) $a = -\frac{1}{4}t^{-3/2}$
3 2 m/s upwards
4 12 m/s, −4 m/s^2
5 $v = \frac{1}{2}\cos 2t - \frac{1}{3}\sin 3t$, $a = -\frac{1}{4}\sin 2t - \frac{1}{9}\cos 3t$
6 (a) $v = 6t$, (b) $a = 6$, (c) $v = 6$ m/s, $a = 6$ m/s^2
7 27 m
8 36 m
9 38 m
10 −29/3 m
11 $v = t^2 + t + 4$

12 (a) $v = 3t^2 + 4$, (b) $x = t^2 + 4t + 10$

Problems

1 $v = 4t - 3$, $a = 4$
2 $v = -5 \sin 2t$, $a = -2.5 \cos 2t$
3 $v = 2\, e^{-2t}$, $a = -4\, e^{-2t}$
4 $v = 9t^2 - 1$, $a = 18t$
5 $v = 6 - 12t$, $a = -12$
6 2 m/s, -4 m/s^2
7 4 m/s, -8 m/s^2
8 10 s
9 2 m/s^2
10 2 s
11 $x = 36t - 2t^2 + 20$
12 180 m
13 3 m
14 1.96 m
15 32 s
16 $v = \frac{4}{3}t^3 + \frac{11}{3}$
17 $v = 3t^2 + 2t + 10$
18 $v = 4t^{3/2} + 32$

Chapter 21

Revision

1 $i = 20t$ A
2 $i = -12\, e^{-3t}$ A
3 $i = 2 - 5\, e^{-5t}$ A
4 $i = -10 \sin 5t$ A
5 3.0 C
6 3 C
7 5 mA, constant
8 $0.04 \cos 10t$ A
9 $-0.04 \sin 10t$ A
10 See figure A.20
11 $v = 10$ V
12 $v = 20\, e^{-100t}$ V
13 $v = -0.2\, e^{-2t}$ V
14 $i = 3t^3$

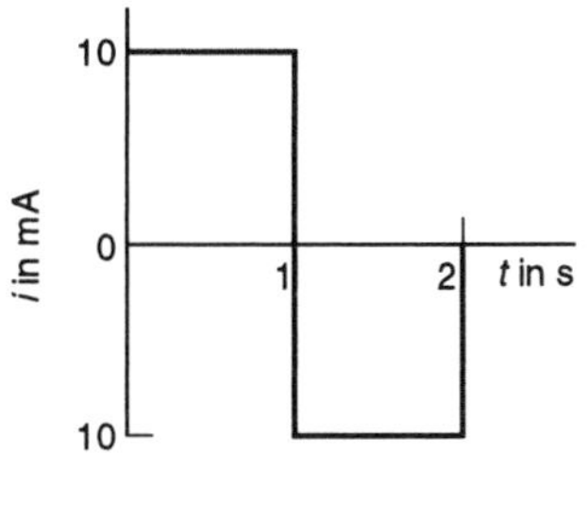

Figure A.20 *Chapter 21 Revision problem 10*

Problems

1 $i = 4t$ A
2 $i = -10\, e^{-2t}$ A
3 $i = 2 - 10\, e^{-2t}$ A
4 $i = 10 \cos 2t$ A
5 0.024 C
6 0.4 C
7 (a) 0, (b) Ca, (c) $-CV\omega \sin \omega t$
8 $i = 0.2 \cos 100t$ A
9 $i = 16t$ mA
10 $v = 0.02$ V

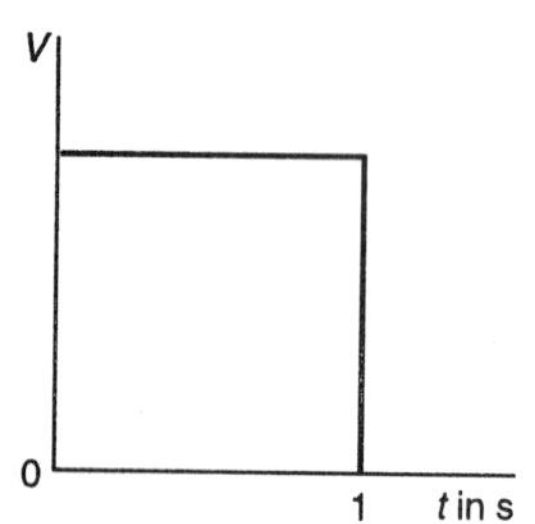

Figure A.21 *Chapter 21 Problem 13*

11 $v = -0.3\ e^{-3t}$ V
12 $v = -16\ e^{-20t} + 16\ e^{-40t}$ V
13 See figure A.21
14 $i = 2t^2$

Chapter 22 *Revision*

1 $\frac{1}{3}kx^3$
2 30 J
3 18 J
4 $-\frac{3}{2}k\left(V_2^{-2/3} - V_1^{-2/3}\right)$

Problems

1 $\frac{2}{3}kx^{3/2}$
2 10 J
3 h^2
4 44 J
5 $-\frac{k}{r}$
6 $-0.4k\left(V_2^{-0.4} - V_1^{-0.4}\right)$

Index